INTRODUCTION TO ELECTRONICS

Third Edition

Delmar Publishers' Online Services

To access Delmar on the World Wide Web, point your browser to:

http://www.delmar.com/delmar.html

To access through Gopher: gopher://gopher.delmar.com

(Delmar Online is part of "thomson.com", an Internet site with information on more than 30 publishers of the International Thomson Publishing organization.)

For more information on our products and services:

e-mail: info@delmar.com

Or call 800-347-7707

INTRODUCTION TO
ELECTRONICS

A Practical Approach
Third Edition

Earl D. Gates

Delmar Publishers

an International Thomson Publishing company

Albany • Bonn • Boston • Cincinnati • Detroit • London • Madrid
Melbourne • Mexico City • New York • Pacific Grove • Paris • San Francisco
Singapore • Tokyo • Toronto • Washington

NOTICE TO THE READER

Cover Design: Susan Mathews, Stillwater Studio
Cover Image by Kari Kivisalo

Delmar Staff
Publisher: Robert D. Lynch
Acquisitions Editor: Paul Shepardson
Developmental Editor: Michelle Ruelos Cannistraci
Production Manager: Larry Main
Art & Design Coordinator: Mary Beth Vought

COPYRIGHT © 1998
By Delmar Publishers
a division of International Thomson Publishing Inc.
The ITP logo is a trademark under license

Printed in the United States of America

For more information contact:

Delmar Publishers
3 Columbia Circle, Box 15015
Albany, New York 12212–5015

International Thomson Publishing Europe
Berkshire House 168-173
High Holborn
London WC1V 7AA   England

Thomas Nelson Australia
102 Dodds Street
South Melbourne, 3205
Victoria, Australia

Nelson Canada
1120 Birchmount Road
Scarborough, Ontario
Canada M1K 5G4

International Thomson Editores
Campos Eliseos 385, Piso 7
Col Polanco
11560 Mexico D F Mexico

International Thomson Publishing GmbH
Königswinterer Strasse 418
53227 Bonn   Germany

International Thomson Publishing Asia
221 Henderson Road
#05–10 Henderson Building
Singapore 0315

International Thomson Publishing-Japan
Hirakawacho Kyowa Building, 3F
2-2-1 Hirakawacho
Chiyoda-ku, Tokyo 102   Japan

    4 5 6 7 8 9 10 XXX 10 00 99 98

Library of Congress Cataloging-in-Publication Data

Gates, Earl D., 1945–
    Introduction to electronics : a practical approach / Earl D.
Gates. –– 3rd ed.
        p.   cm.
    Includes index.
    ISBN 0–8273–6789–9
    1. Electronics.   I. Title.
TK7816.G36 1997
621.381 –– dc21   96–47948
                                 CIP

CONTENTS

PREFACE

Several years ago I completed a survey with approximately twenty area electronics industries. From this study, I determined that industry wants a student graduating with a background in electronics to be able to:

- troubleshoot
- make various measurements with different pieces of test equipment, especially the oscilloscope
- be able to solder
- know where to find information
- know the references used in electronics

I found industry values the student's ability to *do* more than it values their ability to *know*. In short, I found that less time should be spent teaching theory and more time should be spent instructing hands-on applications.

The third edition of *Introduction to Electronics* continues to give students the basic background that more closely relates to the needs of industry. It provides the hands-on instruction required by industry along with the required theory. This text has been carefully developed and designed to make learning easier and teaching more efficient. It is intended to meet both the needs of a one-year program in electronics for high school, junior college, vocational and adult centers, and the needs for college. It is intended to be taught in a more exploratory fashion than for specific skill development. A three-semester sequence could emerge if a student focuses on DC and AC circuits the first semester, semiconductors and linear circuits the second semester, and digital circuits the third semester.

The following list provides some of the salient features of this text:

- Chapters are kept brief and focused.
- Objectives clearly state the learning goals at the beginning of each chapter.
- Illustrations are used generously to amplify the concepts learned.
- Review questions appear throughout each chapter so students may check their comprehension.
- The mathematical inclusions are written using only basic formulas.
- Frequent examples show math and formulas in use.
- Summaries at the end of each chapter review important concepts.
- Self-tests complete the learning tools for each chapter.
- The two-color design calls attention to the important features of this text.

Through a detailed reviewing process, the contents and organization of this text have been tailored to fit the current needs of students and teachers:

- A new and expanded introduction includes career exploration, using a calculator, and safety precautions.
- State-of-the-art graphic calculators are used for explanations.
- Numerous examples of real-life applications for the chapter materials are integrated.

This text is structured to provide a logical progression of material. However, because each chapter is a self-contained unit, the sequence may be varied to suit an appropriate teaching style.

I had a math teacher check the accuracy of all examples and self-test answers, and these examples were prepared with the assistance of this teacher. This approach creates examples that will help students correlate the math learned in mathematics class with the mathematics used in electronics.

In the lab, students transfer theory learned in class to hands-on practical applications. Therefore, a lab manual was developed to provide for this hands-on experience demanded by industry. Ambitious projects reinforce students' learning and help them see theory become practice. Many projects are available on the market that will provide the required reinforcement.

This textbook and the lab manual help students develop an interest in the field of electronics. A curriculum guide is included in the Instructor's Guide to serves as a foundation for a program in electronics.

I would like to thank Cheryl Scholand, a mathematics teacher, and Rolf Tiedemann, an electronics/technology education teacher, both of Greece Central School, whose help and support made this revision possible. I would also like to recognize the following people from the industry who continue to provide me with needed support: Gerald Buss, President of EIC Electronics, and Thomas Fegadel, Owner of Glenwood Sales.

Thanks are also due to the numerous teachers who use this text and have brought discrepancies to my attention and have identified areas to include or expand upon.

I would also like to thank the following reviewers for their valuable suggestions:

Tom S. Best
Glasgow High School
Glasgow, MO

Ronald J. Fronckowiak
Orleans Educational Center
Corfu, NY

Larry Goldberg
Rancho Santiago Community College
Santa Ana, CA

John Hammer
Rhinelander High School
Rhinelander, WI

Paul Jansson
Boise State University
School of Applied Technology
Boise, ID

Raymond Klein
Indiana Vocational Technical College
Gary, IN

Bruce Koller
Diablo Valley College
Pleasant Hill, CA

David Mixer
Leander High School
Leander, TX

Richard Rime
Dawson County High School
Glendive, MT

Bob Snider
Austin, TX

James Snively
Altoona Area Vo-Tech
Altoona, PA

Richard Steinmeier
Electronics Institute
Middletown, PA

Finally, I would like to thank my wife, Shirley; my daughters, Kimberly and Susan; and my son, Timothy, who have supported me in the development of this text.

Earl D. Gates
Rochester, NY
1997

CAREERS IN ELECTRONICS

Many exciting career opportunities exist in the electrical/electronics field. A sample of these available opportunities are provided in the following information. Check for other career opportunities at the career information center in your school or community.

Automation Mechanic

An automation mechanic maintains controllers, assembly equipment, copying machines, robots, and other automated or computerized devices. A person with this job installs, repairs, and services machinery with electrical, mechanical, hydraulic, or pneumatic components. Precision measuring instruments, test equipment, and handtools are used. A knowledge of electronics and the ability to read wiring diagrams and schematics are required.

Becoming an automation mechanic requires formal training which is offered by the military, junior/community colleges, vocational-technical schools, and in-house apprenticeship programs. Although most training is provided through formal classroom instruction, some of the training may only be obtained through on-the-job training.

Automation mechanic is one of the fastest growing vocations in the industry. This rapid growth is expected to continue through the year 2000.

Computer Technician

A computer technician installs, maintains, and repairs computer equipment and systems. Initially, the computer technician is responsible for laying cables and making equipment connections. This person must thoroughly test the new system(s), resolving all problems before the customer uses the equipment. At regular intervals, the computer technician maintains the equipment to ensure that everything is operating efficiently. A knowledge of basic and specialized test equipment and handtools is necessary.

Computer technicians spend much of their time working with people—listening to complaints, answering questions, and sometimes offering advice on both equipment system purchases and ways to keep equipment operating efficiently. Experienced computer technicians often train new technicians and sometimes have limited supervisory roles before moving into a supervisory or service managerial position.

A computer technician is required to have one or two years of training in basic electronics or electrical engineering from a junior college, college or vocational training center, or military installation. The computer technician must be able to keep up with all the new hardware and software.

Projections indicate that employment for computer technicians will be high through the year 2000. The nation's economy is expanding, so the need for computer equipment will increase; therefore, more computer technicians will be required to install and maintain equipment. Many job openings for computer technicians may develop from the need to replace technicians who leave the labor force, transfer to other occupations or fields, or move into management.

Electrical Engineer

Electrical engineers make up the largest branch of engineering. An electrical engineer designs new products, writes performance specifications, and develops maintenance requirements. Electrical engineers also test equipment and solve operating problems within a system, and predict how much time a project will require. Then, based on the time estimate, the electrical engineer determines how much the project will cost.

The electrical engineering field is divided into two specialty groups: electrical engineering and electronic engineering. An electrical engineer works in one or more areas of power-generating equipment, power-transmitting equipment, electric motors, machinery control, and lighting and wiring installation. An electronics engineer works with electronic equipment associated with radar, computers, communications, and consumer goods.

The number of engineers in demand is expected to increase through the year 2000. This projected growth is attributed to an increase in demand for computers, communication equipment, and military equipment. Additional jobs are being created through research and development of new types of automation and industrial robots. Despite this rapid growth, a majority of openings will result from a need to replace electrical and electronics engineers who leave the labor force, transfer to other occupations or fields, or move into management.

Electrician

An electrician may specialize in construction, maintenance, or both. Electricians assemble, install, and maintain heating, lighting, power, air-conditioning, or refrigeration components. The work of an electrician is active and sometimes strenuous. An electrician risks injury from electrical shock, falls, and cuts from sharp objects. To decrease the risk of these job-related hazards, an electrician is taught to use protective equipment and clothing to prevent shocks and other injuries. An electrician must adhere to the *National Electrical Code (NEC)®* specifications and procedures, as well as to the requirements of state, county, and municipal electric codes.

A large proportion of electricians are trained through apprenticeship programs. These programs are comprehensive, and people who complete them are qualified for both maintenance and construction work. Most localities require that an electrician be licensed. To obtain the license, electricians must pass an examination that tests their knowledge of electrical theory, the *National Electrical Code®*, and local electrical and building codes. After electricians are licensed, it is their responsibility to keep abreast of changes in the *National Electrical Code®*, with new materials, and with methods of installation.

Employment for an electrician is expected to increase through the year 2000. As population increases and the economy grows, more electricians will be needed to maintain the electrical systems used in industry and in homes.

Electronics Technician

Electronics technicians develop, manufacture, and service electronic equipment and they use sophisticated measuring and diagnostic equipment to test, adjust, and repair electronic equipment. This equipment includes radio, radar, sonar, television, and computers, as well as industrial and medical measuring and controlling devices.

One of the largest areas of employment for electronics technicians is in research and development. Technicians work with engineers to set up experiments and equipment, and calculate the results. They also assist engineers by making prototypes of newly developed equipment, as well as by performing routine design work. Some elec-

National Electrical Code and *(NEC)®* are registered trademarks of the National Fire Protection Association, Inc., Quincy, MA.

tronics technicians work as sales or field representatives to give advice on installation and maintenance of complex equipment. Most electronics technicians work in laboratories, electronics shops, or industrial plants. Ninety percent of electronics technicians work in private industry.

Becoming an electronics technician requires formal training, which is offered by the military, junior/community colleges, vocational-technical schools, or in-house apprenticeship programs.

Employment of electronics technicians is expected to increase through the year 2000 due to an increased demand for computers, communication equipment, military electronics, and electronic consumer goods. Increased product demand will provide job opportunities, but the need to replace technicians who leave the labor force, transfer to other occupatons or fields, or move into management may also increase.

USING A CALCULATOR

Due to a decrease in cost, the hand-held electronic calculator has become very popular. Many students have rejoiced that all of their mathematical work is now mastered. In just a few keystrokes, a calculator will give the correct answer. However, students fail to realize that a calculator is just a tool to perform calculations very quickly, but with no guarantees for a correct answer. A calculator gives the correct answer only when the correct numbers are entered, in the correct order, and with the correct function keys used at the appropriate time.

If operators do not understand principles of the mathematical process, they will not be able to properly enter data into a calculator, nor will they be able to correctly interpret the results. Mathematical skills still count. Even when all data is entered correctly, the answer may be incorrect due to battery failure, and so forth.

Selecting a calculator appropriate for electronics is an important decision. The marketplace is flooded with many makes and models. Which is the right one? What are the functions that will prove to be the most useful? For this course, choose one that has the following functions: $+, -, \times, \div, 1/x, x^2$, and $\sqrt{}$. A memory function is optional. Scientific and programmable calculators have become popular. Although they are not needed for this course, they typically include formulas and functions used in trigonometry and statistics. If you decide to purchase one, study the manual carefully so you may use the calculator to its fullest extent. All calculators generally come with a manual, which should be kept handy.

The following examples show how a calculator is used to solve various types of problems in electronics. Turn on your calculator. Examine the keyboard. Let's do some calculating.

Addition

Example 1   Add: 39,857 + 19,733

Solution

Enter	Display
39857	39857
+	39857
19733	19733
=	59590

Subtraction

Example 2   Subtract: 30,102 − 15,249

Solution

Enter	Display
30102	30102
−	30102
15249	15249
=	14853

Multiplication

Example 3   Multiply: 33,545 × 981

Solution

Enter	Display
33545	33545
×	33545
981	981
=	32907645

Division

Example 4   Divide: 36,980 by 43 or $\dfrac{36,980}{43}$ or $\dfrac{43}{36,980}$

Solution

Enter	Display
36980	36980
÷	36980
43	43
=	860

Square Root

Example 5   Find the square root of 35,721

Solution

Enter	Display
35721	35721
√	189

Total Resistance (Parallel Circuit)

The total resistance of a parallel circuit may be calculated by first computing the reciprocal of each branch and then taking the reciprocal of the branch total.

Parallel circuits are made up of resistors that are sold in resistance values of ohms. Calculating parallel circuit total resistance involves the use of re-ciprocals (1/R) as shown in the parallel circuit formula:

$$\frac{1}{R_T} = \frac{1}{R_1} + \frac{1}{R_2} + \frac{1}{R_3} \cdots + \frac{1}{R_n}$$

A calculator gives the reciprocal of a number by simply pressing the 1/X key. If the calculator does not have a 1/X key, then each reciprocal value will be found separately by dividing 1 by the resistance value.

Example 6   Calculate the total equivalent resistance of the parallel circuit shown

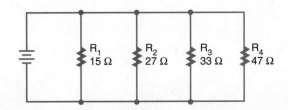

Solution

	Enter	Display
Reciprocal of R_1	15	15
	1/X	0.0666667
Reciprocal of R_2	27	27
	1/X	0.037037
Reciprocal of R_3	33	33
	1/X	0.030303
Reciprocal of R_4	47	47
	1/X	0.0212766
	Enter	Display
Totals of reciprocals	0.0666667	0.0666667
	+	
	0.037037	0.1037037
	+	
	0.030303	0.1340067
	+	
	0.0212766	0.1552833
	Enter	Display
Reciprocal of totals	0.1552833	0.1552833
	1/X 6.4398425	Round answer to 6.44 Ω

Example 7   Using a Calculator with memory function

Solution

	Enter	Display
Reciprocal of R_1	15	15
	1/X	0.0666667
	M+	0.0666667 M
	C	0
Reciprocal of R_2	27	27
	1/X	0.037037
	M+	0.037037 M
	C 0	
Reciprocal of R_3	33	33
	1/X	0.030303
	M+	0.030303 M
	C	0
Reciprocal of R_4	47	47
	1/X	0.0212766
	M+	0.0212766 M
	C	0
Totals of reciprocals	RM	0.0.155283329
Reciprocal of totals	1/X	6.439841299
		Round answer to 6.44 Ω

Rounding

Note: Rounding is not a calculator function and must be done mentally. The number of significant digits can be reduced by *rounding off.* This means dropping the least significant digits until the desired number of digits remain. The new least significant digit may be changed using the following rules:

If the highest significant digit dropped is

■ less than 5, the new significant digit is not changed.

■ greater than 5, the new significant digit is increased by one.

■ 5, the new significant digit is not changed if it is even.

■ 5, the new significant digit is increased by one if it is odd.

Example   Round 352.580

Round to the nearest tenth	352.6
Round to the nearest whole number	352
Round to the nearest hundred	400

These rules result in a rounding off technique that on the average gives the most consistent reliability.

SAFETY PRECAUTIONS

The following safety precautions are not intended as a replacement for information given in class or lab manuals. If at any time you question what steps or procedures to follow, consult your teacher.

General Safety Precautions

Because of the possibility of personal injury, danger of fire, and possible damage to equipment and materials, all work on electrical and electronic circuits should be conducted following these basic safety precedures.

1. *Remove power from the circuit or equipment prior to working on it.* Never override interlock safety devices. Never assume the circuit is off; check it with a voltmeter.
2. *Remove and replace fuses only after the power to the circuit has been deenergized.*
3. *Make sure all equipment is properly grounded.*
4. *Use extreme caution when removing or installing batteries containing acid.*
5. *Use cleaning fluids only in well-ventilated spaces.*
6. *Dispose of cleaning rags and other flammable materials in tightly closed metal containers.*
7. *In case of an electrical fire, deenergize the circuit and report it immediately to the appropriate authority.*

High Voltage Safety Precautions

As people become familiar with working on circuits, it is human nature to become careless with routine procedures. Many pieces of electrical equipment use voltages that are dangerous and can be fatal if contacted. The following precautions should be followed at all times when working on or near high voltage circuits:

1. *Consider the result of each act.* There is absolutely no reason for individuals to take chances that will endanger their life or the lives of others.
2. *Keep away from live circuits.* Do not work on or make adjustments with high voltage on.
3. *Do not work alone.* Always work in the presence of another person capable of providing assistance and first aid in case of an emergency. People who are considering a career working in the electricity and electronics field should become CPR certified.
4. *Do not tamper with interlocks.*
5. *Do not ground yourself.* Make sure you are not grounded when making adjustments or using measuring instruments. Use only one hand when connecting equipment to a circuit. Make it a practice to put one hand in your rear pocket.
6. *Never energize equipment in the presence of water leakage.*

Personal Safety Precautions

Take time to be safe when working on electrical and electronic circuits. Do not work on any circuits or equipment unless the power is secured.

1. *Work only in clean dry areas.* Avoid working in damp or wet locations because the resistance of the skin will be lower; this increases the chance of electrical shock.
2. *Do not wear loose or flapping clothing.* Not only may it get caught, but it might also serve as a path for the conduction of electricity.
3. *Wear only nonconductive shoes.* This will reduce the chance of electrical shock.
4. *Remove all rings, wristwatches, bracelets, ID chains and tags, and similar metal items.* Avoid clothing that contain exposed metal zippers, buttons, or other types of metal fasteners. The metal can act as a conductor, heat up and cause a bad burn.
5. *Do not use bare hands to remove hot parts.*
6. *Use a shorting stick to remove high voltage charges on capacitors.* Capacitors can hold a charge for long periods of time and are frequently overlooked.
7. *Make certain that the equipment being used is properly grounded.* Ground all test equipment to the circuit and/or equipment under test.
8. *Remove power to a circuit prior to connecting alligator clips.* Handling uninsulated alligator clips could cause potential shock hazards.
9. *When measuring voltages over 300 volts, do not hold the test prods.* This eliminates the possibility of shock from leakage on the probes.

Safety is everyone's responsibility. It is the job of everybody in and out of class to exercise proper precautions to insure that no one will be injured and no equipment will be damaged. **Every class in which you work should emphasize and practice safety.**

INTRODUCTION TO ELECTRONICS

Third Edition

Section 1

1

DC Circuits

1

FUNDAMENTALS OF ELECTRICITY

Objectives

After completing this chapter, the student will be able to:

- Define *atom, matter, element,* and *molecule.*
- List the parts of an atom.
- Define the *valence shell* of an atom.
- Identify the unit for measuring current.
- Draw the symbol used to represent current flow in a circuit.
- Describe the difference between conductors, insulators, and semiconductors.
- Define *difference of potential, electromotive force,* and *voltage.*
- Draw the symbol used to represent voltage.
- Identify the unit used to measure voltage.
- Define *resistance.*
- Identify characteristics of resistance in a circuit.
- Identify the unit for measuring resistance.
- Draw the symbol used to represent resistance in a circuit.

Everything, whether natural or man-made, can be broken down into either an element or a compound. However, the smallest part of each of these is the atom.

The atom is made up of protons, neutrons, and electrons. The protons and neutrons group together to form the center of the atom called the nucleus. The electrons orbit the nucleus in shells located at various distances from the nucleus.

When appropriate external force is applied to electrons in the outermost shell, they are knocked loose and become free electrons. The movement of free electrons is called current. The external force needed to create this current is called voltage. As it travels along its path, the current encounters some opposition, called resistance.

This chapter looks at how current, voltage, and resistance collectively form the fundamentals of electricity.

1–1 Matter, Elements, and Compounds

Matter is anything that occupies space and has weight. It may be found in any one of three states: solid, liquid, or gas. Examples of matter include the air we breathe, the water we drink, the clothing we wear, and ourselves. *Matter* may be either an element or a compound.

An **element** is the basic building block of nature. It is a substance that cannot be reduced to a simpler substance by chemical means. There are now over 100 known elements (Appendix 2). Examples of elements are gold, silver, copper, and oxygen.

The chemical combination of two or more elements is called a **compound** (Figure 1–1). A *compound* can be separated by chemical but not by physical means. Examples of compounds are water, which consist of hydrogen and oxygen, and salt, which consists of sodium and chlorine. The smallest part of the compound that still retains the properties of the compound is called a **molecule.** A *molecule* is the chemical combination of two or more atoms. An **atom** is the smallest particle of an element that retains the characteristic of the element.

The physical combination of elements and compounds is called a **mixture.** Examples of *mixtures* include air, which is made up of oxygen, nitrogen, carbon dioxide, and other gases, and salt water, which consists of salt and water.

1–1 Questions

1. In what forms can matter be found?
2. What is a substance called that cannot be reduced to a simpler substance by chemical means?
3. What is the smallest possible particle that retains the characteristic of a compound?
4. What is the smallest possible particle that retains the characteristic of an element?

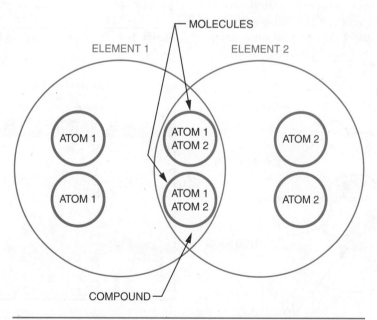

Figure 1–1 The chemical combination of two or more elements is called a compound. A molecule is the chemical combination of two or more atoms. Examples are water (H_2O) and salt (NaCl).

1–2 A Closer Look at Atoms

As previously stated, an *atom* is the smallest particle of an *element*. Atoms of different elements differ from each other. If there are over 100 known elements, then there are over 100 known atoms.

Every atom has a **nucleus.** The *nucleus* is located at the center of the atom. It contains positively charged particles called **protons** and uncharged particles called **neutrons.** Negatively charged particles called **electrons** orbit around the nucleus (Figure 1–2).

The number of *protons* in the nucleus of the atom is called the element's **atomic number.** Atomic numbers distinguish one element from another.

Each element also has an atomic weight. The **atomic weight** is the mass of the atom. It is determined by the total number of protons and neutrons in the nucleus. Electrons do not contribute to the total mass of the atom; an electron's mass is only 1/1845 that of a proton and is not significant enough to consider.

The electrons orbit in concentric circles about the nucleus. Each orbit is called a **shell.** These *shells* are filled in sequence; K is filled first, then L, M, N, and so on (Figure 1–3). The maximum number of electrons that each shell can accommodate is shown in Figure 1–4.

The outer shell is called the **valence shell** and the number of electrons it contains is the **valence.** The farther the *valence shell* is from the nucleus, the less attraction the nucleus has on each valence electron. Thus the potential for the atom to gain or

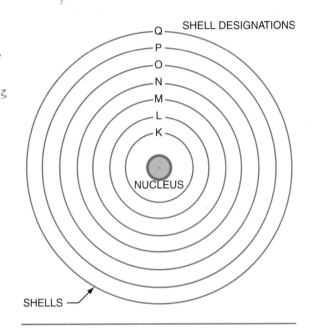

Figure 1–3 The electrons are held in shells around the nucleus.

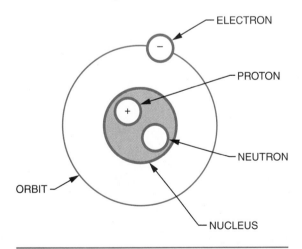

Figure 1–2 Parts of an atom.

SHELL DESIGNATION	TOTAL NUMBER OF ELECTRONS
K	2
L	8
M	18
N	32
O	18
P	12
Q	2

Figure 1–4 The number of electrons each shell can accommodate.

lose electrons increases if the valence shell is not full and is located far enough away from the nucleus. Conductivity of an atom depends on its valence band. The greater the number of electrons in the valence shell, the less it conducts. For example, an atom having seven electrons in the valence shell is less conductive than an atom having three electrons in the valence shell.

Electrons in the valence shell can gain energy. If these electrons gain enough energy from an external force, they can leave the atom and become free electrons, moving randomly from atom to atom. Materials that contain a large number of free electrons are called **conductors.** Figure 1–5 compares the conductivity of various metals used as *conductors*. On the chart, silver, copper, and gold have a valence of one (Figure 1–6). However, silver is the best conductor because its free electron is more loosely bonded.

Insulators, the opposite of conductors, prevent the flow of electricity. *Insulators* are stabilized by absorbing valence electrons from other atoms to fill their valence shells, thus eliminating free electrons. Materials classified as insulators are compared in Figure 1–7. Mica is the best insulator because it has the fewest free electrons in its valence shell. A perfect insulator will have atoms with full valence shell. This means it cannot gain electrons.

Halfway between conductors and insulators are **semiconductors.** *Semiconductors* are neither

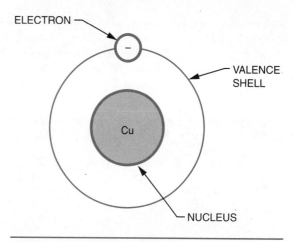

Figure 1–6 Copper has a valence of 1.

good conductors nor good insulators but are important because they can be altered to function as conductors or insulators. Silicon and germanium are two semiconductor materials.

An atom that has the same number of electrons and protons is said to be electrically balanced. A balanced atom that receives one or more electrons is no longer balanced. It is said to be negatively charged

MATERIAL	CONDUCTANCE
Silver	High
Copper	
Gold	
Aluminum	
Tungsten	
Iron	
Nichrome	Low

Figure 1–5 Conductivity of various metals used as conductors.

MATERIAL	INSULATION PROPERTIES
Mica	High
Glass	
Teflon	
Paper (Paraffin)	
Rubber	
Bakelite	
Oils	
Procelain	
Air	Low

Figure 1–7 Insulation properties of various materials used as insulators.

and is called a **negative ion.** A balanced atom that loses one or more electrons is said to be positively charged and is called a **positive ion.** The process of gaining or losing electrons is called **ionization.** *Ionization* is significant in current flow.

1–2 Questions

1. What atomic particle has a positive charge and a large mass?
2. What atomic particle has no charge at all?
3. What atomic particle has a negative charge and a small mass?
4. What does the number of electrons in the outermost shell determine?
5. What is the term for describing the gaining or losing of electrons?

1–3 Current

Given an appropriate external force, the movement of electrons is from negatively charged atoms to positively charged atoms. This flow of electrons is called **current** (I). The symbol I is used to represent current. The amount of *current* is the sum of the charges of the moving electrons past a given point.

An electron has a very small charge, so the charge of 6.24×10^{18} electrons is added together and called a **coulomb** (C). When one *coulomb* of charge moves past a single point in one second it is called an **ampere** (A). The *ampere* is named for a French physicist named André Marie Ampère (1775-1836). Current is measured in amperes.

1–3 Questions

1. What action causes current in an electric circuit?
2. What action results in an ampere of current?
3. What symbol is used to represent current?
4. What symbol is used to represent the unit ampere?

1–4 Voltage

When there is an excess of electrons (negative charge) at one end of a conductor and a deficiency of electrons (positive charge) at the opposite end, a current flows between the two ends. A current flows through the conductor as long as this condition persists. The source that creates this excess of electrons at one end and the deficiency at the other end represents the **potential.** The *potential* is the ability of the source to perform electrical work.

The actual work accomplished in a circuit is a result of the **difference of potential** available at the two ends of a conductor. It is this *difference of potential* that causes electrons to move or flow in a circuit (Figure 1–8). The difference of potential is referred to as **electromotive force** (emf) or **voltage.** *Voltage* is the force that moves the electrons in the circuit. Think of voltage as the pressure or pump that moves the electrons.

The symbol **E** is used in electronics to represent voltage. The unit for measuring voltage is the **volt** (V), named for Count Alessandro Volta (1745–1827), inventor of the first cell to produce electricity.

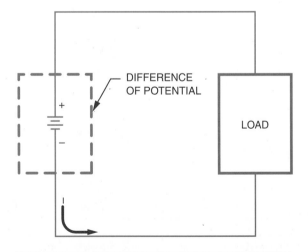

Figure 1–8 Electrons flow in a circuit because of the difference of potential.

1—4 Questions

1. What is the device called that supplies a voltage?
2. What is the term that represents the potential between the two ends of a conductor?
3. What symbol is used to represent voltage?
4. What symbol is used to represent the unit volt?

1—5 Resistance

As the free electrons move through the circuit, they encounter atoms that do not readily give up electrons. This opposition to the flow of electrons (the current) is called **resistance** (R).

Every material offers some resistance or opposition to current flow. The degree of resistance of a material depends on its size, shape, and temperature.

Materials with a low resistance are called **conductors.** *Conductors* have many free electrons and offer little resistance to current flow. As previously mentioned, silver, copper, gold, and aluminum are examples of good conductors.

Materials with a high resistance are called **insulators.** *Insulators* have few free electrons and offer a high resistance to current flow. As previously mentioned, glass, rubber, and plastic are examples of good insulators.

Resistance is measured in **ohms,** a unit named for the German physicist George Simon Ohm (1787–1854). The symbol for the ohm is the Greek letter omega (Ω).

1—5 Questions

1. What is the term used to describe opposition to current flow?
2. What is the main difference between conductors and insulators?
3. What is the symbol used to represent resistance?
4. What is the symbol used to represent the unit of resistance?

SUMMARY

■ Matter is anything that occupies space.
■ Matter can be an element or compound.
■ An element is the basic building block of nature.
■ A compound is a combination of two or more elements.
■ A molecule is the smallest unit of a compound that retains the properties of the compound.
■ An atom is the smallest unit of matter that retains the structure of the element.
■ An atom consists of a nucleus, which contains protons and neutrons. It also has one or more electrons that orbit around the nucleus.
■ Protons have a positive charge, electrons have a negative charge, and neutrons have no charge.
■ The atomic number of an element is the number of protons in the nucleus.
■ The atomic weight of an atom is the sum of protons and neutrons.
■ The orbits of the electrons are called shells.
■ The outer shell of an atom is called the valence shell.
■ The number of electrons in the valence shell is called the valence.
■ An atom that has the same number of protons as electrons is electrically balanced.
■ The process by which atoms gain or lose electrons is called ionization.
■ The flow of electrons is called current.
■ Current is represented by the symbol I.
■ The charge of 6,240,000,000,000,000,000 (or 6.24×10^{18}) electrons is called a coulomb.
■ An ampere of current is measured when one coulomb of charge moves past a given point in one second.
■ Ampere is represented by the symbol A.
■ Current is measured in amperes.
■ An electric current flows through a conductor when there is an excess of electrons at one end and a deficiency at the other end.
■ A source that supplies excess electrons represents a potential or electromotive force.

- The potential or electromotive force is referred to as voltage.
- Voltage is the force that moves electrons in a circuit.
- The symbol E is used to represent voltage.
- A volt (V) is the unit for measuring voltage.
- Resistance is the opposition to current flow.
- Resistance is represented by the symbol R.
- All materials offer some resistance to current flow.

- The resistance of a material is dependent on the material's size, shape, and temperature.
- Conductors are materials with low resistance.
- Insulators are materials with high resistance.
- Resistance is measured in ohms.
- The Greek letter omega (Ω) is used to represent ohms.

Chapter 1 Self-Test

1. What criteria determines whether an atom is a good conductor?

2. What determines whether a material is a conductor, semiconductor, or insulator?

3. Why is it essential to understand the relationship between conductors, semiconductors, and insulators?

4. Explain the difference between current, voltage, and resistance.

5. Describe how the resistance of a material is determined.

Chapter

2

CURRENT

Objectives

After completing this chapter, the student will be able to:

- State the two laws of electrostatic charges.
- Define *coulomb*.
- Identify the unit used to measure current flow.
- Define the relationship of amperes, coulombs, and time through a formula.
- Describe how current flows in a circuit.
- Describe how electrons travel in a conductor.
- Define and use *scientific notation*.
- Identify commonly used prefixes for powers of ten.

The atom has been defined as the smallest particle of an element. It is composed of electrons, protons, and neutrons.

Electrons breaking away from atoms and flowing through a conductor produce an electric current.

This chapter examines how electrons break free from atoms to produce a current flow and how to use scientific notation. Scientific notation expresses very large and small numbers in a form of mathematical shorthand.

2–1 Electrical Charge

Two electrons together or two protons together represent "like" charges. Like charges resist being brought together and instead move away from each other. This movement is called *repelling*. This is the first law of electrostatic charges: like charges repel each other (Figure 2–1). According to the second law of electrostatic charges, unlike charges attract each other.

The negative electrons are drawn toward the positive protons in the nucleus of an atom. This attractive force is balanced by the centrifugal force caused by the electron's rotation about the nucleus. As a result the electrons remain in orbit and are not drawn into the nucleus.

The amount of attracting or repelling force that acts between two electrically charged bodies depends on two factors: their charge and the distance between them.

Single electrons have a charge too small for practical use. The unit adopted for measuring charges is the **coulomb** (C), named for Charles Coulomb. The electrical charge (Q) carried by 6,240,000,000,000,000,000 electrons (six quintillion, two hundred eighty quadrillion or 6.24×10^{18}) represents one *coulomb*.

$$1C = 6.24 \times 10^{18} \text{ electrons}$$

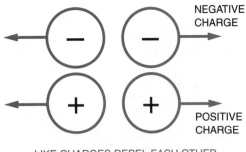

NEGATIVE CHARGE

POSITIVE CHARGE

LIKE CHARGES REPEL EACH OTHER

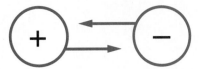

UNLIKE CHARGES ATTRACT EACH OTHER

Figure 2–1 Basic laws of electrostatic charges.

Electrical charges are created by the displacement of electrons. When there is an excess of electrons at one point and a deficiency of electrons at another point, a difference of potential exists between the two points. When a difference of potential exists between two charged bodies connected by a conductor, electrons will flow along the conductor. This flow of electrons is called current.

2–1 Questions

1. What are the two laws of electrostatic charges?
2. What does an electrical charge represent?
3. Define *coulomb*.

2–2 Current Flow

An electric **current** consists of the drift of electrons from an area of negative charge to an area of positive charge. The measure of current flow is the **ampere** (A). An ampere represents the amount of current in a conductor when one coulomb of charge moves past a point in one second. The relationship between amperes and coulombs per second can be expressed as:

$$I = \frac{Q}{t}$$

where: I = current measured in amperes
    Q = quantity of electrical charge in coulombs
    t = time in seconds

EXAMPLE What is the current in amperes if 9 coulombs of charge flow past a point in an electric circuit in 3 seconds?

Given:	*Solution:*
I = ?	$I = \dfrac{Q}{t}$
Q = 9 coulombs	$I = \dfrac{9}{3}$
t = 3 seconds	I = 3 amperes

EXAMPLE A circuit has a current of 5 amperes. How long does it take for one coulomb to pass a given point in the circuit?

Given:

$$I = 5 \text{ amperes}$$
$$Q = 1 \text{ coulomb}$$
$$t = ?$$

Solution:

$$I = \frac{Q}{t}$$
$$5 = \frac{1}{t}$$
$$\frac{5}{1} \times \frac{1}{t} \text{ (cross multiply)}$$
$$(1)(1) = (5)(t)$$
$$1 = 5t$$

$$\frac{1}{5} = \frac{5t}{5} \text{(divide both sides by 5)}$$
$$\frac{1}{5} = t$$
$$0.2 \text{ seconds} = t$$

Electrons, with their negative charge, represent the charge carrier in an electric circuit. Therefore, electric current is the flow of negative charges. Scientists and engineers once thought that current flowed in a direction opposite to electron flow. Later work revealed that the movement of an electron from one atom to the next created the appearance of a positive charge, called a **hole,** moving in the opposite direction (Figure 2–2, 2–3).

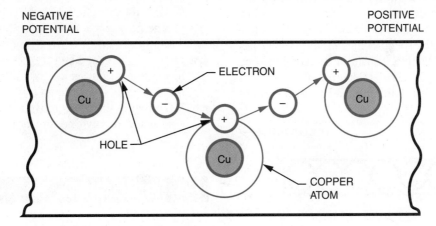

Figure 2–2 As electrons move from one atom to another, they create the appearance of a positive charge, called a *hole.*

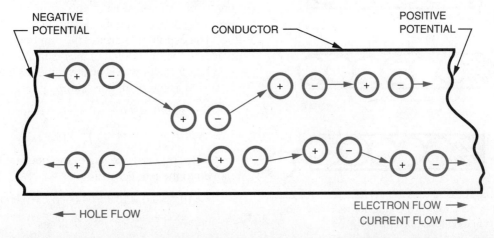

Figure 2–3 Electron movement occurs in the opposite direction to hole movement.

Electron movement and current were found to be the same.

If electrons are added to one end of a conductor, a provision is made to take electrons from the other end, an electric current flows through the conductor. As free electrons move slowly through the conductor, they collide with atoms, knocking other electrons free. These new free electrons travel toward the positive end of the conductor and collide with other atoms. The electrons drift from the negative to the positive end of the conductor because like charges repel. In addition, the positive end of the conductor, which represents a deficiency in electrons, attracts the free electrons because unlike charges attract.

The drift of electrons is slow (approximately an eighth of an inch per second), but individual electrons ricochet off atoms, knocking other electrons loose, at the speed of light (186,000 miles per second). For example, visualize a long hollow tube filled with Ping-Pong balls (Figure 2–4). As a ball is added to one end of the tube, a ball is forced out the other end of the tube. Although an individual ball takes time to travel down the tube, the speed of its impact can be far greater.

The device that supplies electrons from one end of a conductor (the negative terminal) and removes them from the other end of the conductor (the positive terminal) is called the *voltage source*. It can be thought of as a kind of pump (Figure 2–5).

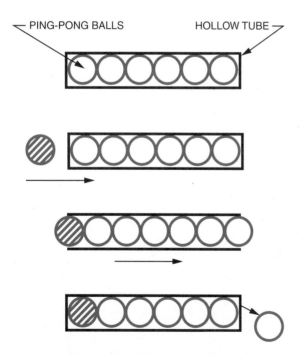

Figure 2–4 Electrons in a conductor react like Ping-Pong balls in a hollow tube.

2–2 Questions

1. Define *electric current*.
2. What is the unit for measuring flow?
3. What is the relationship of current, coulombs, and time?
4. What is the current if 15 coulombs of charge flow past a point in a circuit in 5 seconds?
5. How long does it take for 3 coulombs to move past a point in a circuit if the circuit has 3 amperes of current flow?
6. What makes electrons move through a conductor in only one direction?

2–3 Scientific Notation

In electronics, it is common to encounter very small and very large numbers. **Scientific notation** is a means of using single-digit numbers plus powers of ten to express large and small numbers. For example, 300 in scientific notation is 3×10^2.

The exponent indicates the number of decimal places to the right or left of the decimal point in the number. If the power is positive, the decimal point is moved to the right. For example:

$$3 \times 10^3 = 3.0 \times 10^3 = 3.000 = 3000$$
$$\text{3 places}$$

If the power is negative, the decimal point is moved to the left. For example:

$$3 \times 10^{-6} = 3.0 \times 10^{-6} = 0.000003. = 0.000003$$
$$\text{6 places}$$

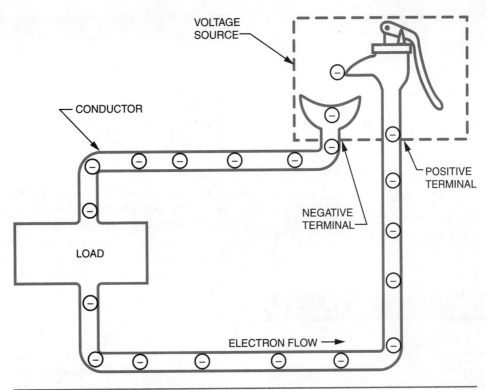

VOLTAGE
SOURCE

CONDUCTOR

POSITIVE
TERMINAL

NEGATIVE
TERMINAL

LOAD

ELECTRON FLOW →

Figure 2–5 A voltage source can be considered a pump that supplies electrons to the load and recycles the excess electrons.

Figure 2–6 lists some commonly used powers of ten, both positive and negative, and the prefixes and symbols associated with them. For example, an ampere (A) is a large unit of current that is not often found in low-power electronic circuits. More frequently used units are the **milliampere** (mA) and the **microampere** (µA). A milliampere is equal to one-thousandth (1/1000) of an ampere or 0.001 A. In other words, it takes 1000 milliamperes to equal one ampere. A microampere is equal to one-millionth (1/1,000,000) of an ampere or 0.000001A; it takes 1,000,000 microamperes to equal one ampere.

EXAMPLE How many milliamperes are there in 2 amperes?

Solution:

$$\frac{1000 \text{ mA}}{1 \text{ A}} = \frac{X \text{ mA}}{2 \text{ A}} (1000 \text{ mA} = 1 \text{ A})$$

$$\frac{1000}{1} = \frac{X}{2}$$

$$(1)(X) = (1000)(2)$$

$$X = 2000 \text{ mA}$$

PREFIX	SYMBOL	VALUE	DECIMAL VALUE
Giga-	G	10^9	1,000,000,000
Mega-	M	10^6	1,000,000
Kilo-	k	10^3	1,000
Milli-	m	10^{-3}	0.001
Micro-	n	10^{-6}	0.000001
Nano-	µ	10^{-9}	0.000000001
Pico-	p	10^{-12}	0.000000000001

Figure 2–6 Prefixes commonly used in electronics.

EXAMPLE How many amperes are there in 50 micro-amperes?

Solution:

$$\frac{1,000,000\ \mu A}{1\ A} = \frac{50\ \mu A}{X\ A}$$

$$\frac{1,000,000}{1} = \frac{50}{X}$$

$$(1)(50) = (1,000,000)(X)$$

$$\frac{50}{1,000,000} = X$$

$$0.00005 = X$$

$$0.00005\ A = X$$

2—3 Questions

1. Define *scientific notation*.

2. In scientific notation:
   a. What does a positive exponent mean?
   b. What does a negative exponent mean?

3. Convert the following numbers to scientific notation:
   a. 500
   b. 3768
   c. 0.0056
   d. 0.105
   e. 356.78

4. Define the following prefixes:
   a. Milli-
   b. Micro-

5. Perform the following conversions:
   a. 1.5 A = _____ mA
   b. 1.5 A = _____ μA
   c. 150 mA = _____ A
   d. 750 μA = _____ A

SUMMARY

■ Laws of electrostatic charges: like charges repel, unlike charges attract.

■ Electrical charge (Q) is measured in coulombs (C).

■ One coulomb is equal to 6.24×10^{18} electrons.

■ An electric current is the slow drift of electrons from an area of negative charge to an area of positive charge.

■ Current flow is measured in amperes.

■ One ampere (A) is the amount of current that flows in a conductor when one coulomb of charge moves past a point in one second.

■ The relationship between current, electrical charge, and time is represented by the formula:

$$I = \frac{Q}{t}$$

■ Electrons (negative charge) represent the charge carrier in an electrical circuit.

■ Hole movement (positive charge) occurs in the opposite direction to electron movement.

■ Current flow in a circuit is from negative to positive.

■ Electrons travel very slowly through a conductor, but individual electrons move at the speed of light.

■ Scientific notation expresses a very large or small number as a numeral from 1 to 9 to a power of ten.

■ If the power-of-ten exponent is positive, the decimal point is moved to the right.

■ If the power-of-ten exponent is negative, the decimal point is moved to the left.

■ The prefix *milli-* means one-thousandth.

■ The prefix *micro-* means one-millionth.

Chapter 2 Self-Test

1. How much current is in a circuit if it takes 5 seconds for 7 coulombs to flow past a given point?
2. Describe how electrons flow in a circuit with reference to the potential in the circuit.
3. Convert the following numbers to scientific notation:
   a. 235
   b. 0.002376
   c. 56323.786
4. What do the following prefixes represent?
   a. Milli-
   b. Micro-

3

VOLTAGE

Objectives

After completing this chapter, the student will be able to:

- Identify the six most common voltage sources.
- Describe six different methods of producing electricity.
- Define a cell and a battery.
- Describe the difference between primary and secondary cells.
- Describe how cells and batteries are rated.
- Identify ways to connect cells or batteries to increase current or voltage output or both.
- Define *voltage rise* and *voltage drop*.
- Identify the two types of grounds associated with electrical circuits.

In a piece of copper wire, the electrons are in random motion with no direction. To produce a current flow, the electrons must all move in the same direction. To produce motion in a given direction, energy must be imparted to the electrons in the copper wire. This energy comes from a source connected to the wire.

The force that causes the electrons to move in a common direction is referred to as *difference of potential,* or *voltage.* This chapter examines how voltage is produced.

3—1 Voltage Sources

A current is produced when an electron is forced from its orbit around an atom. Any form of energy that dislodges electrons from atoms can be used to produce current. It is important to note that energy is not created; rather, there is simply a transfer of energy from one form to another. The source supplying the voltage is not simply a source of electrical energy. Instead, it is the means of converting some other form of energy into electrical energy. The six most common voltage sources are friction, magnetism, chemicals, light, heat, and pressure.

Friction is the oldest known method of producing electricity. A glass rod can become charged when rubbed with a piece of fur or silk. This is similar to the charge you can generate by scuffing your feet across a carpet in a dry room. A **Van de Graaf generator** is a device that operates using the same principles as the glass rod and is capable of producing millions of volts (Figure 3–1).

Magnetism is the most common method of producing electrical energy today. If a wire is passed through a magnetic field, voltage is produced, as long as there is motion between the magnetic field and the conductor. A device based on this principle is called a **generator** (Figure 3–2). A generator can produce either direct current or alternating current, depending on how it is wired. When electrons flow in only one direction, the current is called **direct current (DC).** When electrons flow in one direction then in the opposite direction, the current is called **alternating current (AC).** A generator may be powered by steam from nuclear power or coal, water, wind, or gasoline or diesel engines. The schematic symbol for an AC generator is shown in Figure 3–3.

The second most common method of producing electrical energy today is by the use of a chemical **cell.** The cell consists of two dissimilar metals, copper and zinc, immersed in a salt, acid, or alkaline solution. The metals, copper and zinc, are the electrodes. The electrodes establish contact with the electrolyte (the salt, acid, or alkaline solution) and the circuit. The electrolyte pulls the free electrons from the copper electrode, leaving it with a positive charge. The zinc electrode attracts free

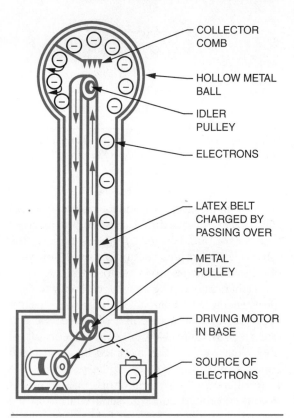

COLLECTOR COMB

HOLLOW METAL BALL

IDLER PULLEY

ELECTRONS

LATEX BELT CHARGED BY PASSING OVER

METAL PULLEY

DRIVING MOTOR IN BASE

SOURCE OF ELECTRONS

Figure 3–1 A Van de Graaf generator is capable of producing millions of volts. *(From Loper,* Direct Current Fundamentals *by Delmar Publishers.)*

Figure 3–2 A generator uses magnetism to produce electricity. *(Courtesy of Westinghouse Electric Corporation.)*

Figure 3–3 Schematic symbol for an alternating current generator.

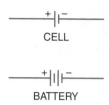

CELL

BATTERY

Figure 3–4 Schematic symbol for a cell and a battery. The combination of two or more cells forms a battery.

electrons in the electrolyte and thus acquires a negative charge. Several of these cells can be connected together to form a battery. Figure 3–4 shows the schematic symbol for a cell and battery. Many types of cells and batteries are in use today (Figure 3–5).

Light energy can be converted directly to electrical energy by light striking a photosensitive (light-sensitive) substance in a **photovoltaic cell** (solar cell) (Figure 3–6). A solar cell consists of photosensitive materials mounted between metal contacts. When the surface of the photosensitive material is exposed to light, it dislodges electrons from their orbits around the surface atoms of the material. This occurs because light has energy. A single cell can produce a small voltage. Figure 3–7 shows the schematic symbol for a solar cell. Many cells must be linked together to produce a usable voltage and current. Solar cells are used primarily in satellites and cameras. The high cost of con-

Figure 3–5 Some of the more common chemical batteries and cells in use today. (*Courtesy of the Union Carbide Corporation, Battery Division.*)

Figure 3–6 A photovoltaic cell can convert sunlight directly into electricity. (*Courtesy of Bell Laboratories.*)

Figure 3–7 Schematic symbol of a photovoltaic cell (solar cell).

struction has limited their general application. However, the price of solar cells is declining.

Heat can be converted directly to electricity with a device called a **thermocouple** (Figure 3–8). The schematic symbol for a thermocouple is shown in Figure 3–9. A thermocouple consists of two dissimilar metal wires, twisted together. One wire is copper and the other wire is zinc or iron. When heat is applied to the twisted connection, the copper wire readily gives up free electrons, which are transferred to the other wire. Thus the copper wire develops a positive charge and the other wire develops a negative charge, and a small voltage occurs. The voltage is directly proportional to the amount of heat applied. One application of the thermocouple is as a thermometer. Also called a pyrometer, these devices are often used in high temperature kilns and foundries.

When pressure is applied to certain crystalline materials such as quartz, tourmaline, Rochelle salts, or barium titanate, a small voltage is produced. This

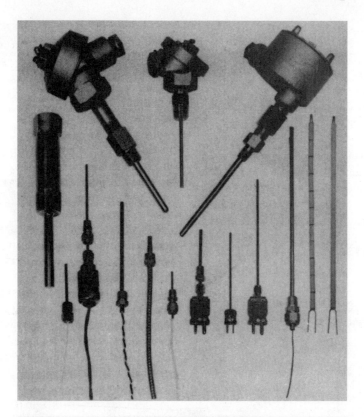

Figure 3–8 Thermocouples convert heat energy directly into electrical energy. *(Courtesy of Hy-Cal Engineering.)*

Figure 3–9 Schematic symbol for a thermocouple.

is referred to as the **piezoelectric effect.** Initially, negative and positive charges are distributed randomly throughout a piece of crystalline material and no overall charge can be measured. However, when pressure is applied, electrons leave one side of the material and accumulate on the other side. A charge is produced as long as the pressure remains. When the pressure is removed, the charge is again distributed, so no overall charge exists. The voltage produced is small and must be amplified to be useful. Uses of the piezoelectric effect include crystal microphones, phonograph pickups (crystal cartridges), and precision oscillators (Figures 3–10, 3–11).

Figure 3–10 Crystal microphone. *(Courtesy of CTI.)*

Figure 3–11 Schematic diagram of piezoelectric crystal.

Note that while a voltage can be produced by these means, the reverse is also true; that is, a voltage can be used to produce magnetism, chemicals, light, heat, and pressure. Magnetism is evident in motors, speakers, solenoids, and relays. Chemical activities can be produced through electrolysis and electroplating. Light is produced with light bulbs and other optoelectric devices. Heat is produced with heating elements in stoves, irons, and soldering irons. And voltage can be applied to bend or twist a crystal.

3—1 Questions

1. What are the six most common voltage sources?

2. What is the most common method for producing a voltage?
3. What is the second most common method for producing a voltage?
4. Why are solar cells not used more for producing a voltage?

3—2 Cells and Batteries

As mentioned in the previous unit, a cell contains a positive and a negative electrode separated by an electrolytic solution. A **battery** is a combination of two or more cells.

There are two basic types of cells. Cells that cannot be recharged are called **primary cells.** Cells that can be recharged are called **secondary cells.**

An example of a primary cell is a Leclanche cell, also called a dry cell (Figure 3–12). This type of cell is not actually dry. It contains a moist paste

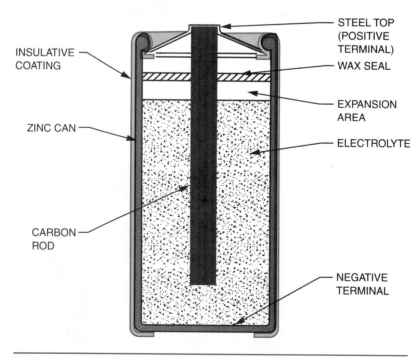

INSULATIVE COATING

ZINC CAN

CARBON ROD

STEEL TOP (POSITIVE TERMINAL)

WAX SEAL

EXPANSION AREA

ELECTROLYTE

NEGATIVE TERMINAL

Figure 3–12 Diagram of a dry cell (shown cut away).

as the electrolyte. A seal prevents the paste from leaking out when the cell is turned sideways or upside down. The electrolyte in a dry cell is a solution of ammonium chloride and manganese dioxide. The electrolyte dissolves the zinc electrode (the case of the cell) leaving an excess of electrons with the zinc. As the current is removed from the cell, the zinc, ammonium chloride, and manganese dioxide produce manganese dioxide, water, ammonia, and zinc chloride. The carbon rod (center electrode) gives up the extra electrons that accumulate on the zinc electrode. This type of cell, produces as much as 1.75 to 1.8 volts when new. A typical Leclanche cell has an energy density of approximately 30 watts-hours per pound. As the cell is used, the chemical action decreases, and eventually the resulting current ceases. If the cell is not used, the electrolytic paste eventually dries out. The cell has a shelf life of about two years. The output voltage of this type of cell is determined entirely by the materials used for the electrolyte and the electrodes. The AA cell, C cell, D cell, and No. 6 dry cell (Figure 3–13) are all constructed of the same materials and therefore produce the same voltage. It should be noted that although the Laclanche cell is frequently referred to as a carbon-zinc (or zinc-carbon) cell, the carbon does not take any part in the chemical reaction that produces electricity.

The alkaline cell is named because of the highly caustic base, potassium hydroxide (KOH) used as the electrolyte. The design of an alkaline cell on the outside is very similar to that of a carbon-zinc cell. However, the inside of the alkaline cell is significantly different (Figure 3–14). Alkaline cells have an open-circuit rating of approximately 1.52 volts, and an energy density of about 45 watt-hours per pound. The alkaline cell performs much better over temperature extremes than carbon-zinc cells. Alkaline cells perform best where moderate-to-high currents are drawn over extended periods of time.

Lithium cells (Figure 3–15) have overcome the inherent properties associated with lithium. Lithium is extremely reactive with water. Lithium cell formation uses lithium, manganese dioxide (MnO_2) and a lithium perchlorate ($LiClO_4$) in an organic solvent (water cannot be used). The output of a lithium cell is approximately 3 volts. Lithium cells are very efficient, with energy densities of

Figure 3–13 Some common examples of dry cells. (*Courtesy of RAYOVAC Corporation.*)

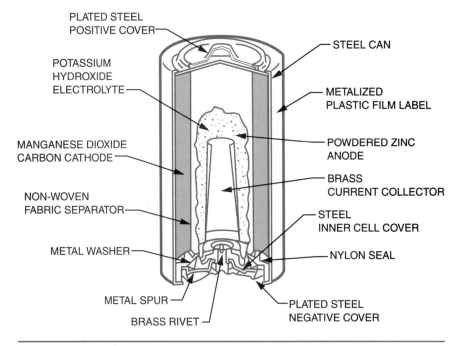

PLATED STEEL
POSITIVE COVER

POTASSIUM
HYDROXIDE
ELECTROLYTE

MANGANESE DIOXIDE
CARBON CATHODE

NON-WOVEN
FABRIC SEPARATOR

METAL WASHER

METAL SPUR

BRASS RIVET

STEEL CAN

METALIZED
PLASTIC FILM LABEL

POWDERED ZINC
ANODE

BRASS
CURRENT COLLECTOR

STEEL
INNER CELL COVER

NYLON SEAL

PLATED STEEL
NEGATIVE COVER

Figure 3–14 Alkaline cells are constructed inside out. The cathode surrounds the anode.

about 90 watt-hours per pound. The greatest benefit of lithium cells is their extremely long shelf life of five to ten years.

A secondary cell is a cell that can be recharged by applying a reverse voltage. An example is the lead-acid battery used in automobiles (Figure 3–16). It is made of six 2-volt secondary cells connected in series. Each cell has a positive electrode of lead peroxide (PbO_2) and a negative electrode of spongy lead (Pb). The electrodes are separated by plastic or rubber and immersed in an electrolytic solution of sulfuric acid (H_2SO_4) and distilled water (H_2O). As the cell is discharged, the sulfuric acid combines with the lead sulfate, and the electrolyte converts to water. Recharging the cell involves applying a source of DC voltage greater than that produced by the cell. As the current flows through the cell, it changes the electrode back to lead peroxide and spongy lead and converts the electrolyte back to sulfuric acid and water. This type of cell is also referred to as a wet cell.

Another type of secondary cell is the nickel-cadmium (Ni-Cad) cell (Figure 3–17). This is a dry cell that can be recharged many times and can hold its charge for long periods of time. It consists of a positive and negative electrode, a separator, electrolyte, and package. The electrodes consist of a deposit of powdered nickel on a nickel wire screen, which is coated with a nickel salt solution for the positive electrode and a cadmium salt solution for a negative electrode. The separator is made of an absorbent insulating material. The electrolyte is potassium hydroxide. A steel can forms the package and is sealed tightly. A typical voltage from this type of cell is 1.2 volts.

The ability of a battery to deliver power continuously is expressed in **ampere-hours.** A battery rated at 100 ampere-hours can continuously supply any of the following: 100 amperes for 1 hour (100 × 1 = 100 ampere-hours), 10 amperes for 10 hours (10 × 10 = 100 ampere-hours), or 1 ampere for 100 hours (1 × 100 = 100 ampere-hours).

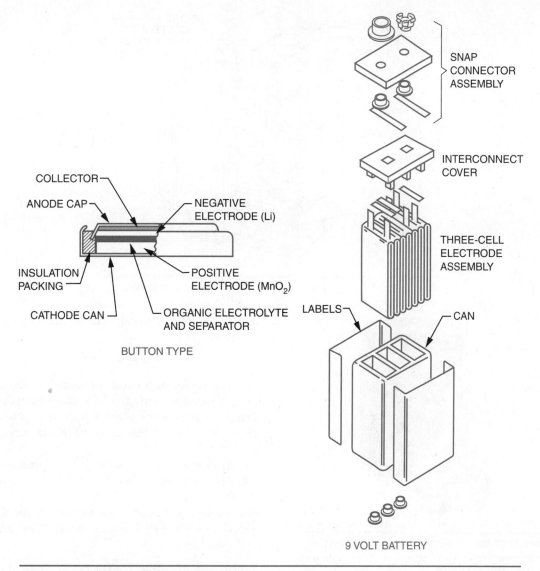

COLLECTOR

ANODE CAP

NEGATIVE ELECTRODE (Li)

INSULATION PACKING

POSITIVE ELECTRODE (MnO₂)

CATHODE CAN

ORGANIC ELECTROLYTE AND SEPARATOR

BUTTON TYPE

SNAP CONNECTOR ASSEMBLY

INTERCONNECT COVER

THREE-CELL ELECTRODE ASSEMBLY

LABELS

CAN

9 VOLT BATTERY

Figure 3–15 Lithium cells have extremely high energy density.

3–2 Questions

1. What are the components of a cell?
2. what are the two basic types of cells?
3. What is the major difference between the two types of cells?
4. List some examples of a primary cell.
5. List some examples of a secondary cell.

3–3 Connecting Cells and Batteries

Cells and batteries can be connected together to increase voltage and/or current. They can be connected in series, in parallel, or in series-parallel.

In series, cells or batteries can be connected in either series-aiding or series-opposing configurations. In a **series-aiding** configuration, the positive terminal of the first cell is connected to the negative

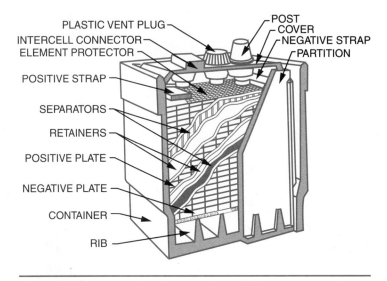

PLASTIC VENT PLUG
INTERCELL CONNECTOR
ELEMENT PROTECTOR
POSITIVE STRAP
SEPARATORS
RETAINERS
POSITIVE PLATE
NEGATIVE PLATE
CONTAINER
RIB

POST
COVER
NEGATIVE STRAP
PARTITION

Figure 3–16 An example of a secondary cell (shown cut away).

Figure 3–17 A nickel-cadmium (Ni-Ca) battery is another example of a secondary cell. *(Courtesy of RAYOVAC Corporation.)*

terminal of the second cell; the positive terminal of the second cell is connected to the negative terminal of the third cell and so on (Figure 3–18). In a series-aiding configuration the same current flows through all the cells or batteries. This can be expressed as:

$$I_T = I_1 = I_2 = I_3$$

The subscript numbers refer to the number of each individual cell or battery. The total voltage is the sum of the individual cell voltages and can be expressed as:

$$E_T = E_1 + E_2 + E_3$$

In a **series-opposing** configuration, the cells or batteries are connected with like terminals together, negative to negative or positive to positive. However, this configuration has little practical application.

In a **parallel** configuration, all the positive terminals are connected together and all the negative terminals are connected together (Figure 3–19). The total current available is the sum of the individual currents of each cell or battery. This can be expressed as:

$$I_T = I_1 + I_2 + I_3$$

The total voltage is the same as the voltage of each individual cell or battery. This can be expressed as:

$$E_T = E_1 = E_2 = E_3$$

If both a higher voltage and a higher current are desired, the cells or batteries can be connected in a series-parallel configuration. Remember, connecting cells or batteries in series increases the voltage, and connecting cells or batteries in parallel increases the current. Figure 3–20 shows four 3-volt

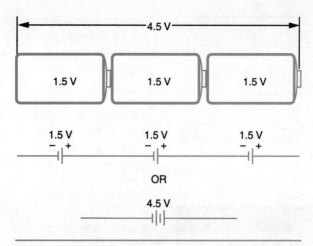

Figure 3–18 Cells or batteries can be connected in series to increase voltage.

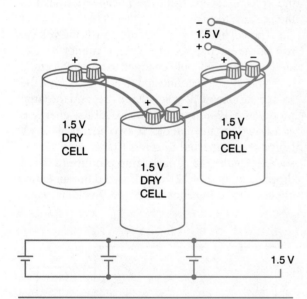

Figure 3–19 Cells or batteries can be connected in parallel to increase current flow.

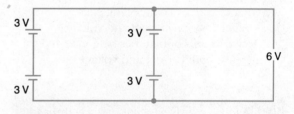

Figure 3–20 Cells and batteries can be connected in series-parallel to increase current and voltage outputs.

Figure 3–21 The voltage increases when cells are connected in series.

batteries connected in a series-parallel configuration. This configuration produces a total voltage of 6 volts with a current twice that of an individual battery. It is necessary to connect the two 3-volt batteries in series to get the 6 volts (Figure 3–21).

To increase the current, a second pair of 3-volt batteries is connected in series, and the resulting series-connected batteries are connected in parallel (Figure 3–22). The overall result is a **series-parallel** configuration.

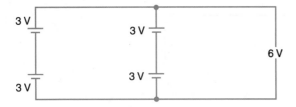

Figure 3–22 Connecting the series-connected cells in parallel increases the output current. The net result is a series-parallel configuration.

3–3 Questions

1. Draw three cells connected in a series-aiding configuration.
2. What effect does a series-aiding configuration have on current and voltage?
3. Draw three cells connected in parallel.
4. What effect does connecting cells in parallel have on current and voltage?
5. How can cells or batteries be connected to increase both current and voltage?

3–4 Voltage Rises and Voltage Drops

In electric and electronic circuits, there are two types of voltage: voltage rise and voltage drop.

Potential energy, or voltage, introduced into a circuit is called a **voltage rise** (Figure 3–23). The voltage is connected to the circuit so that the current flows from the negative terminal of the

voltage source and returns to the positive terminal of the voltage source. A 12-volt battery connected to a circuit gives a voltage rise of 12 volts to the circuit.

As the electrons flow through the circuit, they encounter a load, some resistance to the flow of electrons. As electrons flow through the load, they give up their energy. The energy given up is called a **voltage drop** (Figure 3–24). The energy is given up in most cases as heat. The energy that the electrons give up to a circuit is the energy given to them by the source.

To repeat, energy introduced into a circuit is called a voltage rise. Energy used up in a circuit by the load is called a voltage drop. A voltage drop occurs when there is a current flow in the circuit. Current moves through a circuit from the negative polarity to the positive polarity. It moves through the voltage source from the negative terminal to the positive terminal.

The voltage drop in a circuit equals the voltage rise of the circuit, because energy cannot be created or destroyed, only changed to another form. If a 12-volt source is connected to a 12-volt lamp, the source supplies a 12-volt voltage rise, and the lamp produces a 12-volt voltage drop. All the energy is consumed in the circuit. If two identical 6-volt lamps are connected in series to the same 12-volt source (Figure 3–25), each lamp produces a 6-volt drop, for a total of 12 volts. If two different lamps are connected in series, such as a 9-volt lamp and

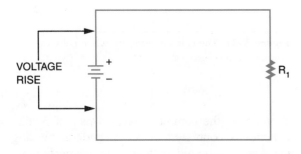

Figure 3–23 A potential applied to a circuit is called a voltage rise.

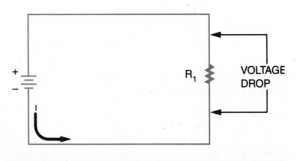

Figure 3–24 The energy used by the circuit in passing current through the load (resistance) is called a voltage drop. A voltage drop occurs when current flows in the circuit.

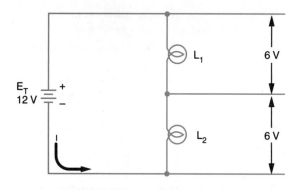

Figure 3–25 Two identical 6-volt lamps each produce a 6-volt drop when connected in series to a 12-volt source.

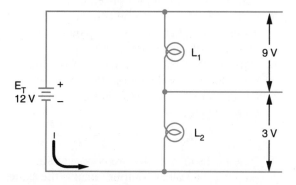

Figure 3–26 When two lamps of different voltage are connected in series to a 12-volt source, the voltage drop across each lamp will differ, based on the voltage requirement of the lamp.

a 3–volt lamp (Figure 3–26), the 9-volt lamp produces a voltage drop of 9 volts, and the 3–volt lamp produces a voltage drop of 3 volts. The sum of the voltage drops equals the voltage rise of 12 volts.

3–4 Questions

1. What is a voltage rise?
2. What is a voltage drop?
3. If two identical resistors are connected in series to a voltage source, what is the voltage drop across each of the resistors?

3–5 Ground as a Voltage Reference Level

Ground is a term used to identify zero potential. All other potentials are either positive or negative with respect to ground. There are two types of grounds: earth and electrical.

In the home, all electrical circuits and appliances are earth grounded. Consequently, no difference of potential exists between any two appliances or circuits. All circuits are tied to a common point in the circuit panel box (circuit-breaker or fuse box) (Figure 3–27). This common point (the neutral bus) is then connected by a heavy copper wire to a copper rod driven into the earth or fastened to the metal pipe that supplies water to the home. This protects the user from electrical shock in case of a faulty connection.

Electrical grounding is used in automobiles. Here the chassis of the automobile is used as a ground. This can be verified by seeing where the battery cables are attached to the automobile. Generally, the negative terminal is bolted directly to the frame of the automobile. This point or any other point on the frame of the automobile is considered

Figure 3–27 In a residential circuit panel box, all circuits are tied to a common point (the neutral bus). *(From Mullin,* Electrical Wiring—Residential *by Delmar Publishers.)*

to be ground. Ground serves as part of the complete circuit.

In electronics, electrical ground serves a different purpose: *Ground* is defined as the zero reference point against which all voltages are measured. Therefore, the voltage at any point in a circuit may be measured with reference to ground. The voltage measured may be positive or negative with respect to ground.

In larger pieces of electronic equipment, the chassis or metal frame serves as the ground point (reference point), as in the automobile. In smaller electronic devices plastic is used for the chassis, and all components are connected to a printed circuit board. In this case, ground is a copper pad on the circuit board that acts as a common point to the circuit.

3—5 Questions

1. What are the two types of grounds?
2. What is the purpose of earth grounding?
3. How is an electrical ground used in an automobile?
4. How is an electrical ground used in a piece of electronic equipment?
5. In electronics, what function does ground serve when taking voltage measurements?

SUMMARY

■ Current is produced when an electron is forced from its orbit.
■ Voltage provides the energy to dislodge electrons from their orbit.
■ A voltage source provides a means of converting some other form of energy into electrical energy.
■ Six common voltage sources are: friction, magnetism, chemicals, light, heat, and pressure.
■ Voltage can be used to produce magnetism, chemicals, light, heat, and pressure.

■ Magnetism is the most common method used to produce a voltage.
■ Chemical cells are the second most common means of producing a voltage.
■ A cell contains positive and negative electrodes separated by an electrolytic solution.
■ A battery is a combination of two or more cells.
■ Cells that cannot be recharged are called primary cells.
■ Cells that can be recharged are called secondary cells.
■ Dry cells are primary cells.
■ Lead-acid batteries and nickel-cadmium (Ni-Cad) cells are examples of secondary cells.
■ Cells and batteries can be connected in series, in parallel, or in series-parallel to increase voltage, current, or both.
■ When cells or batteries are connected in a series-aiding configuration, the output current remains the same, but the output voltage increases.

$$I_T = I_1 = I_2 = I_3 \qquad E_T = E_1 = E_2 = E_3$$

■ When cells or batteries are connected in parallel, the voltage output remains the same but the output current available increases.

$$I_T = I_1 + I_2 + I_3 \qquad E_T = E_1 = E_2 = E_3$$

■ A series-parallel combination increased both the output voltage and the output current.
■ Voltage applied to a circuit is referred to as a voltage rise.
■ The energy used by a circuit is referred to as a voltage drop.
■ The voltage drop in a circuit equals the voltage rise.
■ Two types of ground are earth and electrical.
■ Earth grounding is used to prevent electric shock by keeping all appliances and equipment at the same potential.
■ Electrical grounding provides a common reference point.

Chapter 3 Self-Test

1. Does current or voltage actually perform the work in a circuit?

2. List six forms of energy that can be used to produce electricity.

3. How are secondary cells rated?

4. Draw a series-parallel combination that will supply 9 volts at 1 ampere. Use 1½-volt cells rated at 250 milliamperes.

5. What is the voltage drop across three lamps of 3 volts, 3 volts, and 6 volts with 9 volts applied?

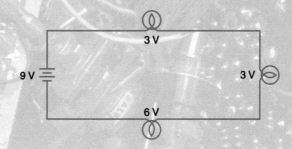

4

RESISTANCE

Objectives

After completing this chapter, the student will be able to:

- Define resistance and explain its affect in a circuit.
- Determine the tolerance range of a resistor.
- Identify carbon composition, wirewound, and film resistors.
- Identify potentiometers and rheostats.
- Describe how a variable resistor operates.
- Decode a resistor's value using the color code or alphanumerical code.
- Identify the three types of resistor circuits.
- Calculate total resistance in series, parallel, and series-parallel circuits.

Resistance is opposition to the flow of current. Some materials such as glass and rubber offer great opposition to current flow. Other materials such as silver and copper offer little opposition to current flow. This chapter examines the characteristics of resistance, types of resistance, and the effects of connecting resistors together by a conductor to form a circuit.

4–1 Resistance

As previously mentioned, every material offers some resistance or opposition to the flow of current. Some conductors such as copper, silver, and aluminum offer very little resistance to current flow. Insulators such as glass, wood, and paper offer high resistance to current flow.

The size and type of wires in an electric circuit are chosen to keep the electrical resistance as low as possible. This allows the current to flow easily through the conductor. In an electric circuit, the larger the diameter of the wire, the lower the electrical resistance to current flow.

Temperature also affects the resistance of an electrical conductor. In most conductors (copper, aluminum, and so on), resistance increases with temperature. Carbon is an exception because the resistance decreases as temperature increases. Certain alloys of metals (manganin and constantan) have resistance that does not change with temperature.

The relative resistance of several conductors of the same length and cross section is shown in Figure 4–1. Silver is used as a standard of 1 and the remaining metals are arranged in order of ascending resistance.

The resistance of an electric circuit is expressed by the symbol R. Manufactured circuit parts containing definite amounts of resistance are called resistors. Resistance (R) is measured in ohms. One ohm is the resistance of a circuit or circuit element that permits a steady current flow of one ampere (one coulomb per second) when one volt is applied to the circuit.

CONDUCTOR MATERIAL	RESISTIVITY
Silver	1.000
Copper	1.0625
Lead	1.3750
Gold	1.5000
Aluminum	1.6875
Iron	6.2500
Platinum	6.2500
Manganin	27.500
Carbon	2187.50

Figure 4–1 Resistance of several conductors of the same length and cross-section area.

4–1 Questions

1. What is the difference between conductors and insulators?
2. How does the diameter of a piece of wire affect its resistance?
3. What factors affect the resistance of a conductor?
4. What material makes the best conductor?

4–2 Conductance

The term in electricity that is opposite of resistance is *conductance (G)*. Conductance is the ability of a material to pass electrons. The unit of conductance is Mho, ohm spelled backwards. The symbol used to represent conductance is the inverted Greek letter omega (℧). Conductance is the reciprocal of resistance and is measured in Siemens (S). A reciprocal is obtained by dividing the number into one.

$$R = 1/G$$
$$G = 1/R$$

If the resistance of a material is known, dividing its value into one will give its conductance. Similarly, if the conductance is known, dividing its value into one will give its resistance.

4–2 Questions

1. Define the term conductance.
2. What is the significance of conductance in a circuit?
3. What symbol is used to represent conductance?
4. What is the unit of conductance?

4–3 Resistors

Resistance is a property of all electrical components. Sometimes the effect of resistance is undesirable, other times it is constructive. **Resistors** are components manufactured to possess a specific value of resistance to the flow of current. A resistor is the most commonly used component in an electronic circuit. Resistors are available with fixed or variable resistance values. They are available in a variety of shapes and sizes to meet specific circuit, space, and operating requirements (Figures 4–2 and 4–3). Resistors are drawn schematically as a series of jagged lines as shown in Figure 4–4.

A resistor's **tolerance** is the amount that the resistor may vary and still be acceptable. It is expensive for a manufacturer to hold a resistor to a certain value when an exact value is not needed. Therefore, the larger the tolerance, the cheaper it is to manufacture. Resistors are available with tolerances of ±20%, ±10%, ±5%, ±2%, and ±1%. Precision resistors are available with even smaller

HOT-MOLDED POTENTIOMETERS

TRIMMER POTENTIOMETERS

TRIMMING POTENTIOMETERS

Figure 4–3 Variable resistors come in many styles to meet the needs of manufacturers of electronic equipment. *(Courtesy of the Allen-Bradley Company.)*

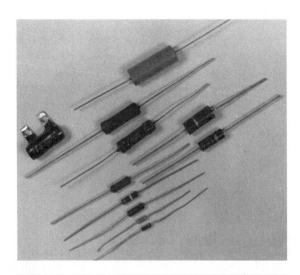

Figure 4–2 Fixed resistors come in various sizes and shapes.

Figure 4–4 Schematic diagram of a fixed resistor.

±5% TOLERANCE	±10% TOLERANCE	±20% TOLERANCE
1.0	1.0	1.0
1.1		
1.2	1.2	
1.3		
1.5	1.5	1.5
1.6		
1.8	1.8	
2.0		
2.2	2.2	2.2
2.4		
2.7	2.7	
3.0		
3.3	3.3	3.3
3.6		
3.9	3.9	
4.3		
4.7	4.7	4.7
5.1		
5.6	5.6	
6.2		
6.8	6.8	6.8
7.5		
8.2	8.2	
9.1		

Figure 4–5 Standard resistor values (exclusive of multiplier band). *(From De Guilmo,* Electricity/ Electronics, *by Delmar Publishers.)*

tolerances. In most electronic circuits, resistors of 10% tolerance are satisfactory.

EXAMPLE How much can a 1000-ohm resistor with a 20% tolerance vary and still be acceptable?

Solution: $1000 \times 0.20 = \pm 200$ ohms

or

$$\begin{array}{r} 1000 \\ \times\, 0.20 \\ \hline 200.00 \text{ ohms} \end{array}$$

The tolerance is ±200 ohms. Therefore, the 1000-ohm resistor may vary from 800 to 1200 ohms and still be satisfactory.

For the sake of uniformity, electronic manufacturers offer a number of standard resistor values. Figure 4–5 is a list of standard values for resistors with ±5%, ±10%, and ±20% tolerance. After the value on the chart is obtained, it is multiplied by the value associated with the color of the multiplier band.

Resistors fall into three major categories, named for the material they are made of: molded carbon composition resistors, wirewound resistors, and film resistors.

The *molded carbon composition* resistor is the most commonly used resistor in electronic circuits (Figure 4–6). It is inexpensive and is manufactured in the standard resistor values.

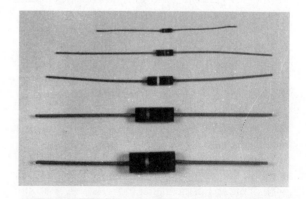

Figure 4–6 Carbon composition resistors are the most widely used resistors in electronic circuits.

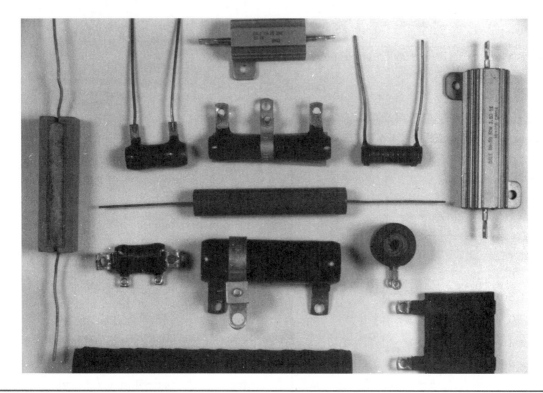

Figure 4–7 Wirewound resistors are available in many different styles.

The *wirewound* resistor is constructed of a nickel-chromium alloy (nichrome) wire wound on a ceramic form (Figure 4–7). Leads are attached and the entire resistor sealed with a coating. Wirewound resistors are used in high-current circuits where precision is necessary. The resistance range varies from a fraction of an ohm to several thousand ohms.

Film resistors are becoming increasingly popular (Figure 4–8). They offer the small size of the composition resistor with the accuracy of the wirewound resistor. A thin film of carbon or metal alloy is deposited on a cylindrical ceramic core and sealed in an epoxy or glass coating. The value of the resistor is set by cutting a spiral groove through the film, the length of the resistor. The closer the pitch of the spiral, the higher the resistance. Carbon film resistors are available from 10 ohms to 2.5 megohms at a ±1% tolerance. Metal film resistors are physically similar to carbon film resistors but

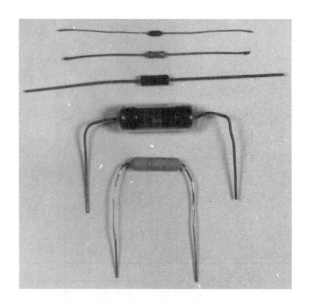

Figure 4–8 The film resistor offers the size of the carbon resistor with the accuracy of the wirewound resistor.

Figure 4–9 Tin oxide resistors.

are more expensive. They are available from 10 ohms to 1.5 megohms at a ±1% tolerance, although tolerances down to ±0.1% are available. Another type of film resistor is the tin oxide resistor (Figure 4–9). It consists of a tin oxide film on a ceramic substrate.

Variable resistors allow the resistance to vary. They have a resistive element of either carbon composition or wire that is connected to two terminals. A third terminal is attached to a movable wiper, which is connected to a shaft. The wiper slides along the resistive element when the shaft is rotated. As the shaft is rotated, the resistance between the center terminal and one outer terminal increases while the resistance between the center terminal and the other outer terminal decreases (Figure 4–10). Variable resistors are available with resistance that varies linearly (a linear taper) or logarithmically (an audio taper).

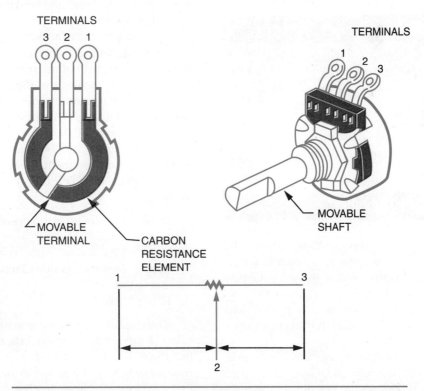

Figure 4–10 Variable resistors allow the resistance to increase or decrease at random.

Figure 4–11 A rheostat is a variable resistor used to control current.

A variable resistor used to control voltage is called a **potentiometer** or *pot*. A variable resistor used to control current is called a **rheostat** (Figure 4–11).

4—3 Questions

1. What is the purpose of specifying the tolerance of a resistor?
2. What are the three major types of fixed resistors?
3. What is the advantage of film resistors over carbon composition resistors?
4. Explain how a variable resistor works.

4—4 Resistor Identification

The small size of the resistor prevents printing the resistance value and tolerance on its case. Therefore, a color-coded strip system is used to display the resistor value. The strips can be seen and read in any position that the resistor is placed. The Electronic Industries Association (EIA) color code is shown in Figure 4–12.

The meaning of the colored bands on a resistor is as follows. The first band, closest to the end of the resistor, represents the first digit of the resis-

tor value. The second band represents the second digit of the resistor value. The third band represents the number of zeros to be added to the first two digits. The fourth band represents the tolerance of the resistor (Figure 4–13).

For example, the resistor shown in Figure 4–14 has a resistance value of 1500 ohms. The brown band (first band) represents the first digit (1). The green band (second band) represents the second digit (5). The red band (third band) represents the number of zeros (two zeros—00) to be added to the first two digits. The silver band (fourth band) indicates a resistance tolerance of ±10%. Therefore, this is a 1500-ohm resistor with a ±10% tolerance.

A resistor may have a fifth band (Figure 4–15). This band indicates the reliability of the resistor. It tells how many of the resistors (per thousand) will fail after 100 hours of operation. Generally, when there are five bands on a resistor, the same amount of body color shows at each end. In this case, look for the tolerance band (gold or silver), position it on the right, and read the resistor as described previously.

There are two instances where the third band does not mean the number of zeros. For resistor values of less than 10 ohms, the third band is gold. This means that the first two digits should be multiplied by 0.1. For resistor values of less than 1 ohm, the third band is silver. This means the first two digits are multiplied by 0.01.

A resistor may also be identified by a letter-and-number (alphanumeric) system (Figure 4–16). For example, RN60D5112F has the following meaning:

RN60	Resistor style (composition, wire-wound, film)
D	Characteristic (effects of temperature)
5112	Resistance value (2 represents number of zeros)
F	Tolerance

The resistor value is the primary concern. The value of the resistor may be indicated by three to five digits. In all cases, the last digit indicates the number of zeros to be added to the preceding digits. In the example given, the last digit (2) indicates

STANDARD RESISTOR COLOR CODE

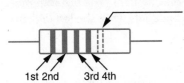

Note: A fifth band may be present, which represents reliability factors and may be ignored.

1st 2nd   3rd 4th

	1ST BAND 1ST DIGIT	2ND BAND 2ND DIGIT	3RD BAND NUMBER OF ZEROS	4TH BAND TOLERANCE
Black	0	0	—	—
Brown	1	1	0	1%
Red	2	2	00	2%
Orange	3	3	000	—
Yellow	4	4	0,000	—
Green	5	5	00,000	0.5%
Blue	6	6	000,000	0.25%
Violet	7	7		0.10%
Gray	8	8		0.05%
White	9	9		—
Gold			×0.1	5%
Silver			×0.01	10%
No Color				20%

Figure 4–12 The Electronic Industries Association (EIA) color code.

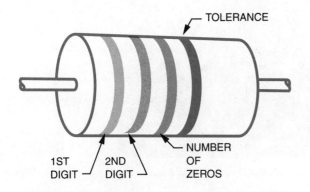

Figure 4–13 Meaning of the colored bands on a carbon composition resistor.

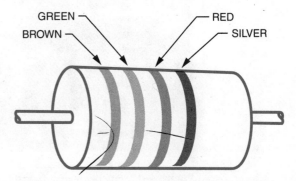

Figure 4–14 This resistor has a resistance value of 1500 ohms.

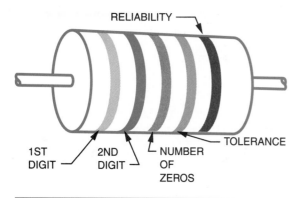

Figure 4–15 The fifth band on a resistor indicates the resistor's reliability.

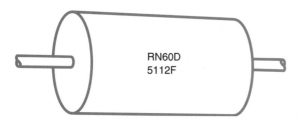

RN60D
5112F

Figure 4–16 Resistors may also be identified by a letter-and-number system.

the number of zeros to be added to the first three digits (511). So 5112 translates to 51,100 ohms.

In some cases an *R* may be inserted into the number. The R represents a decimal point and is used when the resistor value is less than 10 ohms. For example, 4R7 represents 4.7 ohms.

The five-digit numbering system is similar to the three- and four-digit systems. The first four digits represent significant digits while the last digit indicates the number of zeros to be added. For values of less than 1000 ohms, the R is used to designate a decimal point.

Potentiometers (variable resistors) are also imprinted with their values (Figure 4–17). These may be their actual values or an alphanumeric code. With the alphanumeric code system, the resistance value is determined from the last part of the code. For example, in MTC253L4, the number 253 means 25 followed by three zeros, or 25,000 ohms. The L4 indicates the resistor construction and body type.

4–4 Questions

1. Write the color code from memory.
2. What do the four bands on a carbon composition resistor represent?
3. Decode the following resistors:

	1st Band	2nd Band	3rd Band	4th Band
a.	Brown	Black	Red	Silver
b.	Blue	Green	Orange	Gold
c.	Orange	White	Yellow	(None)
d.	Red	Red	Red	Silver
e.	Yellow	Violet	Brown	Gold

4. What does a fifth band on a resistor indicate?
5. What does a gold or silver third band represent?

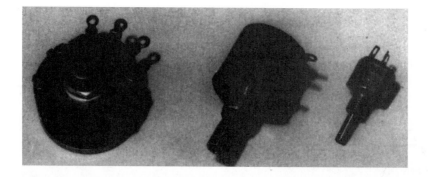

Figure 4–17 Potentiometers (variable resistors) are also labeled with their values. (*From Shimizu,* Electronic Fabrication, *by Delmar Publishers.*)

4—5 Connecting Resistors

There are three important types of resistive circuits: the **series circuit,** the **parallel circuit,** and the **series-parallel circuit** (Figure 4–18). A *series circuit* provides a single path for current flow. A *parallel circuit* provides two or more paths for current flow. A *series-parallel circuit* is a combination of a series circuit and a parallel circuit.

4—5 Question

1. What are the three basic types of circuit configuration?

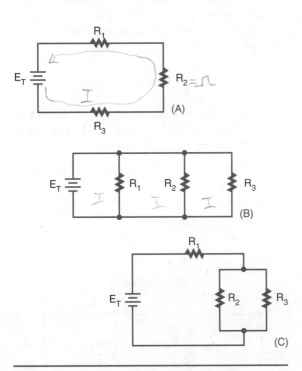

Figure 4–18 Three types of resistive circuits: (A) series circuit, (B) parallel circuit, (C) series-parallel circuit.

4—6 Connecting Resistors in Series

A series circuit contains two or more resistors and provides one path for current to flow. The current flows from the negative side of the voltage source through each resistor to the positive side of the voltage source. If there is only one path for current to flow between two points in a circuit, the circuit is a series circuit.

The more resistors connected in series, the more opposition there is to current flow. The more opposition there is to current flow, the higher the resistance in the circuit. In other words, when a resistor is added in series to a circuit, the total resistance in the circuit increases. The total resistance in a series circuit is the sum of the individual resistances in the circuit. This can be expressed as:

$$R_T = R_1 + R_2 + R_3 \ldots + R_n$$

The numerical subscripts refer to the individual resistors in the circuit. R_n is the last resistor in the circuit. The symbol R_T represents the total resistance in the circuit.

EXAMPLE What is the total resistance of the circuit shown in Figure 4–19?

Given:

$$R_T = ?$$
$$R_1 = 10 \text{ ohms}$$
$$R_2 = 20 \text{ ohms}$$
$$R_3 = 30 \text{ ohms}$$

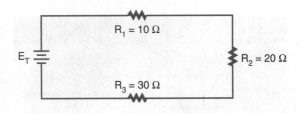

Figure 4–19

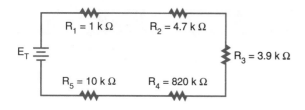

Figure 4–20

Solution:

$$R_T = R_1 + R_2 + R_3$$
$$R_T = 10 + 20 + 30$$
$$R_T = 60 \text{ ohms}$$

EXAMPLE Calculate the total resistance for the circuit shown in Figure 4–20.

Given:

$$R_T = ?$$
$$R_1 = 1 \text{ kilohm}$$
$$R_2 = 4.7 \text{ kilohms}$$
$$R_3 = 3.9 \text{ kilohms}$$
$$R_4 = 820 \text{ ohms}$$
$$R_5 = 10 \text{ kilohms}$$

Solution:

$$R_T = R_1 + R_2 + R_3 + R_4 + R_5$$
$$R_T = 1 \text{ k}\Omega + 4.7 \text{ k}\Omega + 3.9 \text{ k}\Omega + 0.82 \text{ k}\Omega$$
$$+ 10 \text{ k}\Omega$$
$$R_T = 1000 + 4700 + 3900 + 820 + 10,000$$
$$R_T = 20,420 \text{ ohms}$$

4–6 Questions

1. Write the formula for determining total resistance in a series circuit.
2. What is the total resistance of a circuit with three resistors—1500 ohms, 3300 ohms, and 4700 ohms—connected in series? (First draw the circuit.)

4–7 Connecting Resistors in Parallel

A parallel circuit contains two or more resistors and provides two or more paths for current to flow. Each current path in a parallel circuit is called a *branch*. The current flows from the negative side of the voltage source, through each branch of the parallel circuit, to the positive side of the voltage source. If there is more than one path for current to flow between two points in a circuit with two or more resistors, the circuit is a parallel circuit.

The more resistors are connected in parallel, the less opposition there is to current flow. The less opposition there is to current flow, the lower the resistance in the circuit. In other words, when a resistor is added in parallel to a circuit, the total resistance in the circuit decreases, because additional paths for current flow are provided. In a parallel circuit, the total resistance is always less than the resistance of any branch.

The total resistance in a parallel circuit is given by the formula:

$$\frac{1}{R_T} = \frac{1}{R_1} + \frac{1}{R_2} + \frac{1}{R_3} \cdots + \frac{1}{R_n}$$

Again, R_T is the total resistance, R_1, R_2, and R_3 are the individual (branch) resistors, and R_n is the number of the last resistor in the circuit.

EXAMPLE What is the total resistance of the circuit shown in Figure 4–21?

Given:

$$R_T = ?$$
$$R_1 = 10 \text{ ohms}$$
$$R_2 = 20 \text{ ohms}$$
$$R_3 = 30 \text{ ohms}$$

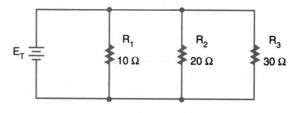

Figure 4–21

Solution:

$$\frac{1}{R_T} = \frac{1}{R_1} + \frac{1}{R_2} + \frac{1}{R_3}$$

$$\frac{1}{R_T} = \frac{1}{10} + \frac{1}{20} + \frac{1}{30} \quad \text{(common denominator is 60)}$$

$$\frac{1}{R_T} = \frac{6}{60} + \frac{3}{60} + \frac{2}{60}$$

$$\frac{1}{R_T} = \frac{11}{60}$$

$$\frac{1}{R_T} \diagup\!\!\!\!\times \frac{11}{60} \quad \text{(cross multiply)}$$

$$(11)(R_T) = (1)(60)$$

$$11R_T = 60$$

$$\frac{\cancel{11}R_T}{\cancel{11}} = \frac{60}{11} \quad \text{(divide both sides by 11)}$$

$$1R_T = \frac{60}{11}$$

$$R_T = 5.45 \text{ ohms}$$

Note that the total resistance is less than that of the smallest resistor. The circuit shown in Figure 4–21 could be replaced with one 5.45-ohm resistor.

EXAMPLE Calculate the total resistance for the circuit shown in Figure 4–22.

Given:

$$R_T = ?$$
$$R_1 = 1 \text{ kilohm (1000 ohms)}$$
$$R_2 = 4.7 \text{ kilohms (4700 ohms)}$$
$$R_3 = 3.9 \text{ kilohms (3900 ohms)}$$
$$R_4 = 820 \text{ ohms}$$
$$R_5 = 10 \text{ kilohms (10,000 ohms)}$$

Solution:

$$\frac{1}{R_T} = \frac{1}{R_1} + \frac{1}{R_2} + \frac{1}{R_3} + \frac{1}{R_4} + \frac{1}{R_5}$$

$$\frac{1}{R_T} = \frac{1}{1000} + \frac{1}{4700} + \frac{1}{3900} + \frac{1}{820} + \frac{1}{10,000}$$

It is too complicated to find a common denominator, so work with decimals.

$$\frac{1}{R_T} = 0.001 + 0.000213 + 0.000256$$
$$+ 0.00122 + 0.0001$$

$$\frac{1}{R_T} = 0.002789$$

$$\frac{1}{R_T} = \frac{0.002789}{1}$$

$$\frac{1}{R_T} \diagup\!\!\!\!\times \frac{0.002789}{1} \quad \text{(cross multiply)}$$

$$(0.002799)(R_T) = (1)(1)$$

$$0.002799 \, R_T = 1$$

$$\frac{\cancel{0.002799} \, R_T}{\cancel{0.002799}} = \frac{1}{0.002789} \quad \text{(divide both sides by 0.002789)}$$

$$1R_T = \frac{1}{0.002789}$$

$$R_T = 358.59 \, \Omega$$

Note that depending on how many places each number is rounded off to will significantly affect the accuracy of the final answer.

EXAMPLE What resistor value must be connected in parallel with a 47-ohm resistor to provide a total resistance of 27 ohms? See Figure 4–23.

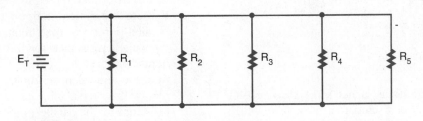

Figure 4–22

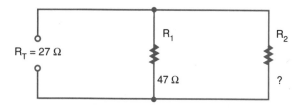

Figure 4–23

Given:

$$R_T = 27 \text{ ohms}$$
$$R_1 = 47 \text{ ohms}$$
$$R_2 = ?$$

Solution:

$$\frac{1}{R_T} = \frac{1}{R_1} + \frac{1}{R_2}$$

$$\frac{1}{27} = \frac{1}{47} + \frac{1}{R_2}$$

$$\frac{1}{27} - \frac{1}{47} = \frac{1}{47} - \frac{1}{47} + \frac{1}{R_2} \quad \text{(Subtract } \frac{1}{47} \text{ from both sides)}$$

$$\frac{1}{27} - \frac{1}{47} = \frac{1}{R_2} \quad \text{(easier to work with decimals)}$$

$$0.0370 - 0.0213 = \frac{1}{R_2}$$

$$0.0157 = \frac{1}{R_2}$$

$$\frac{0.0157}{1} \times \frac{1}{R_2} \quad \text{(cross multiply)}$$

$$(1)(1) = (0.0157)(R_2)$$

$$1 = 0.0157R_2$$

$$\frac{1}{0.0157} = \frac{0.0157R_2}{0.0157} \quad \text{(divide both sides by 0.0157)}$$

$$\frac{1}{0.0157} = 1R_2$$

$$63.69 \text{ ohms} = R_2$$

Note that 63.69 ohms is not a standard resistor value. Use the closest standard resistor value which is 62 ohms.

4–7 Questions

1. Write the formula for determining total resistance in a parallel circuit.
2. What is the total resistance of a circuit with three resistors—1500 ohms, 3300 ohms, and 4700 ohms—connected in parallel? (First draw the circuit.)

4–8 Connecting Resistors in Series and Parallel

A series-parallel circuit is a combination of a series and a parallel circuit. Figure 4–24 shows a simple series-parallel circuit with resistors. Notice that R_2 and R_3 are in parallel, and that this parallel combination is in series with R_1 and R_4. The current flows from the negative side of the voltage source through resistor R_4 and divides at point A to flow through the two branches, R_2 and R_3. At point B, the current recombines and flows through R_1.

The total resistance for a series-parallel circuit or compound circuit is computed using the series formula:

$$R_T = R_1 + R_2 + R_3 \ldots + R_n$$

and the parallel formula:

$$\frac{1}{R_T} = \frac{1}{R_1} + \frac{1}{R_2} + \frac{1}{R_3} \ldots + \frac{1}{R_n}$$

Most circuits can be broken down to a simple parallel or series circuit. The procedure is as follows:

1. Calculate the parallel portion of the circuit first to determine the equivalent resistance.
2. If there are series components within the parallel portion of the circuit, determine the equivalent resistance for the series components first.
3. After the equivalent resistance is determined, redraw the circuit, substituting the equivalent resistance for the parallel portion of the circuit.
4. Do final calculations.

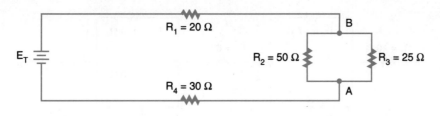

Figure 4–24

EXAMPLE What is the total resistance for the circuit shown in Figure 4–24?

The first step is to determine the equivalent resistance (R_A) for R_2 and R_3.

Given:

$$R_A = ?$$
$$R_2 = 50 \text{ ohms}$$
$$R_3 = 25 \text{ ohms}$$

Solution:

$$\frac{1}{R_A} = \frac{1}{R_2} + \frac{1}{R_3}$$
$$\frac{1}{R_A} = \frac{1}{50} + \frac{1}{25}$$
$$\frac{1}{R_A} = \frac{1}{50} + \frac{2}{50}$$
$$\frac{1}{R_A} = \frac{3}{50}$$
$$3R_A = 50$$
$$1R_A = \frac{50}{3}$$
$$R_A = 16.7 \text{ ohms}$$

Redraw the circuit, substituting the equivalent resistance for the parallel portion of the circuit. See Figure 4–25.

Now determine the total series resistance for the redrawn circuit.

Given:

$R_T = ?$

$R_1 = 20 \text{ ohms}$

$R_A = 16.7 \text{ ohms}$

$R_4 = 30 \text{ ohms}$

Solution:

$R_T = R_1 + R_A + R_4$

$R_T = 20 + 16.7 + 30$

$R_T = 66.7 \text{ ohms}$

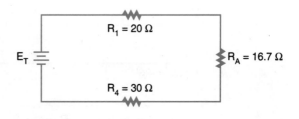

Figure 4–25

EXAMPLE Calculate the total resistance for the circuit shown in Figure 4–26.

First find the equivalent resistance (R_A) for parallel resistors R_2 and R_3. Then find the equivalent resistance (R_B) for resistors R_5, R_6, and R_7.

Given:

$$R_A = ?$$
$$R_2 = 47 \text{ ohms}$$
$$R_3 = 62 \text{ ohms}$$

Solution:

$$\frac{1}{R_A} = \frac{1}{R_2} + \frac{1}{R_3}$$
$$\frac{1}{R_A} = \frac{1}{47} + \frac{1}{62}$$
$$\frac{1}{R_A} = 0.0213 + 0.0161$$
$$\frac{1}{R_A} = 0.0374$$
$$\frac{1}{R_A} = \frac{0.0374}{1}$$
$$(0.0374)(R_A) = (1)(1)$$
$$(0.0374)(R_A) = 1$$

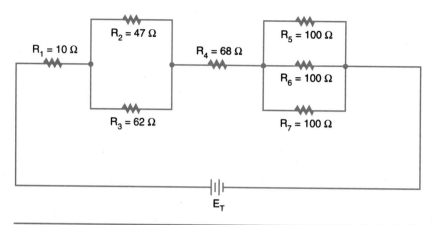

Figure 4–26

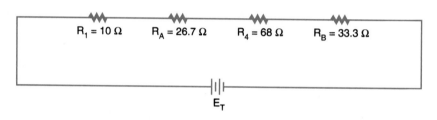

Figure 4–27

$$1R_A = \frac{1}{0.0374}$$

$$R_A = 26.7 \text{ ohms}$$

Given:

$$R_B = ?$$

$$R_5 = 100 \text{ ohms}$$

$$R_6 = 100 \text{ ohms}$$

$$R_7 = 100 \text{ ohms}$$

Solution:

$$\frac{1}{R_B} = \frac{1}{R_5} + \frac{1}{R_6} + \frac{1}{R_7}$$

$$\frac{1}{R_B} = \frac{1}{100} + \frac{1}{100} + \frac{1}{100}$$

$$\frac{1}{R_B} = \frac{3}{100}$$

$$3R_B = 100$$

$$R_B = \frac{100}{3}$$

$$R_B = 33.3 \text{ ohms}$$

Now redraw the circuit using equivalent resistance R_A and R_B and determine the total series resistance for the redrawn circuit. See Figure 4–27.

Given: *Solution:*

$R_T = ?$ $R_T = R_1 + R_A + R_4 + R_B$

$R_1 = 10 \text{ ohms}$ $R_T = 10 + 26.7 + 68 + 33.3$

$R_A = 26.7 \text{ ohms}$ $R_T = 138 \text{ ohms}$

$R_4 = 68 \text{ ohms}$

$R_B = 33.3 \text{ ohms}$

The circuit shown in Figure 4–26 could be replaced with a single resistor of 138 ohms (Figure 4–28).

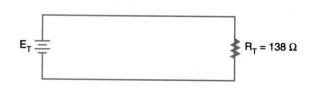

Figure 4–28

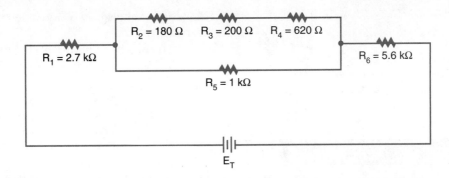

Figure 4–29

EXAMPLE Find the total resistance for the circuit shown in Figure 4–29.

The equivalent resistance of the series in the parallel portion of the circuit must be determined first. This is labeled R_S.

Given: *Solution:*

$R_S = ?$ $R_S = R_2 + R_3 + R_4$

$R_2 = 180$ ohms $R_S = 180 + 200 + 620$

$R_3 = 200$ ohms $R_S = 1000$ ohms

$R_4 = 620$ ohms

Redraw the circuit, substituting equivalent resistance R_S for the series resistors R_2, R_3, and R_4. See Figure 4–30.

Determine the equivalent parallel resistance R_A for R_S and R_5.

Given:

$R_A = ?$

$R_S = 1000$ ohms

$R_5 = 1000$ ohms

Solution:

$$\frac{1}{R_A} = \frac{1}{R_S} + \frac{1}{R_5}$$

$$\frac{1}{R_A} = \frac{1}{1000} + \frac{1}{1000}$$

$$\frac{1}{R_A} = \frac{2}{1000}$$

$$1R_A = \frac{1000}{2}$$

$$R_A = 500 \text{ ohms}$$

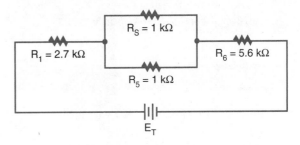

Figure 4–30

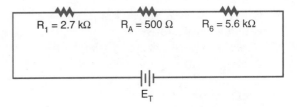

Figure 4–31

Redraw the circuit again, substituting equivalent resistance R_A for parallel resistors R_S and R_5, and determine the total series resistance for the redrawn circuit. See Figure 4–31.

Given: *Solution:*

$R_T = ?$ $R_T = R_1 + R_A + R_6$

$R_1 = 2700$ ohms $R_T = 2700 + 500 + 5600$

$R_A = 500$ $R_T = 8800$ ohms

$R_6 = 5600$ ohms

The circuit shown in Figure 4–29 can be replaced with a single resistor of 8800 ohms (Figure 4–32).

Figure 4–32

4–8 Question

1. What is the total resistance of a circuit with a 1500-ohm and a 3300-ohm resistor in parallel connected in series with a 4700-ohm resistor? (First draw the circuit.)

SUMMARY

■ Resistors are either fixed or variable.
■ The tolerance of a resistor is the amount that its resistance can vary and still be acceptable.
■ Resistors are either carbon composition, wirewound, or film.
■ Carbon composition resistors are the most commonly used resistors.
■ Wirewound resistors are used in high-current circuits that must dissipate large amounts of heat.
■ Film resistors offer small size with high accuracy.
■ Variable resistors used to control voltage are called potentiometers.
■ Variable resistors used to control current are called rheostats.

■ Resistor values may be identified by colored bands:
  — The first band represents the first digit.
  — The second band represents the second digit.
  — The third band represents the number of zeros to be added to the first two digits.
  — The fourth band represents the tolerance.
  — A fifth band may be added to represent reliability.
■ Resistor values of less than 100 ohms are shown by a black third band.
■ Resistors may be placed in three configurations—series, parallel, and compound.
■ Resistor values of less than 10 ohms are shown by a gold third band.
■ Resistor values of less than 1 ohm are shown by a silver third band.
■ Resistor values for one-percent tolerance resistors are shown with the fourth band as the multiplier.
■ Resistor values may also be identified by an alphanumerical system.
■ The total resistance in a series circuit can be found by the formula:

$$R_T = R_1 + R_2 + R_3 \ldots + R_n$$

■ The total resistance in a parallel circuit can be found by the formula:

$$\frac{1}{R_T} = \frac{1}{R_1} + \frac{1}{R_2} + \frac{1}{R_3} \ldots + \frac{1}{R_n}$$

■ The total resistance in a series-parallel circuit is determined by both series and parallel formulas.

Chapter 4 Self-Test

1. Describe how the resistance of material is determined.
2. What is the tolerance range of a 2200-ohm resistor with a 10% tolerance?
3. Write the color codes for the following resistors:
   a. 5600 ohms ±5%
   b. 1.5 megohms ±10%
   c. 2.7 ohms ±5%
   d. 100 ohms ±20%
   e. 470 kilohms ±10%
4. Determine the total resistance for the circuit shown.
5. Describe how current flows through a series-parallel circuit.

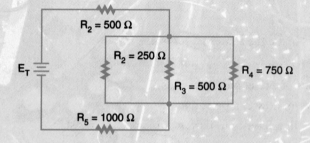

5

OHM'S LAW

Objectives

After completing this chapter, the student will be able to:

- Identify the three basic parts of a circuit.
- Identify three types of circuit configurations.
- Describe how current flow can be varied in a circuit.
- State Ohm's law with reference to current, voltage, and resistance.
- Solve problems using Ohm's law for current, resistance or voltage in series, parallel, and series-parallel circuits.
- Describe how the total current flow differs between series and parallel circuits.
- Describe how the total voltage drop differs between series and parallel circuits.
- Describe how the total resistance differs between series and parallel circuits.

Ohm's law defines the relationship among three fundamental quantities: current, voltage, and resistance. It states that current is directly proportional to voltage and inversely proportional to resistance.

This chapter examines Ohm's law and how it is applied to a circuit. Some of the concepts are introduced in previous chapters.

5–1 Electric Circuits

As stated earlier, current flows from a point with an excess of electrons to a point with a deficiency of electrons. The path that the current follows is called an *electric circuit*. All electric circuits consist of a voltage source, a load, and a conductor. The *voltage source* establishes a difference of potential that forces the current to flow. The source can be a battery, a generator, or another of the devices described in Chapter 3, Voltage. The *load* consists of some type of resistance to current flow. The resistance may be high or low depending on the purpose of the circuit. The current in the circuit flows through a *conductor* from the source to the load. The conductor must give up electrons easily. Copper is used for most conductors.

The path the electric current takes to the load may be through any of three types of circuits: a series circuit, a parallel circuit, or a series-parallel circuit. A series circuit (Figure 5–1) offers a single continuous path for the current flow, going from the source to the load. A parallel circuit (Figure 5–2) offers more than one path for current flow. It allows the source to apply voltage to more than one load. It also allows several sources to be connected to a single load. A series-parallel circuit (Figure 5–3) is a combination of the series and parallel circuit.

Current in an electric circuit flows from the negative side of the voltage source through the

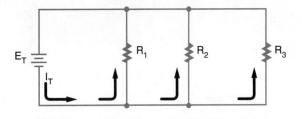

Figure 5–2 A parallel circuit offers more than one path for current flow.

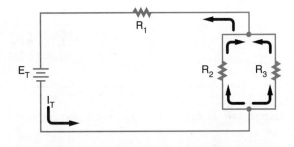

Figure 5–3 A series-parallel circuit is a combination of a series circuit and a parallel circuit.

load to the positive side of the voltage source (Figure 5–4). As long as the path is not broken, it is a closed circuit and current flows (Figure 5–5). However, if the path is broken, it is an open circuit and no current can flow (Figure 5–6).

The current flow in an electric circuit can be varied by changing either the voltage applied to the circuit or the resistance in the circuit. The current changes in exact proportions to the change in the voltage or resistance. If the voltage is increased, the current also increases. If the voltage is

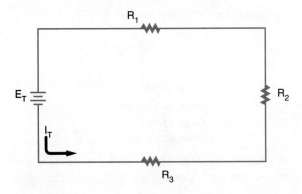

Figure 5–1 A series circuit offers a single path for current flow.

Figure 5–4 Current flow in an electric circuit flows from the negative side of the voltage source, through the load, and returns to the voltage source through the positive terminal.

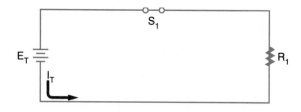

Figure 5–5 A closed circuit supports current flow.

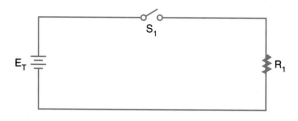

Figure 5–6 An open circuit does not support current flow.

Figure 5–7 Current flow in an electric circuit can be changed by varying the voltage.

Figure 5–8 Current flow in an electric circuit can also be changed by varying the resistance in the circuit.

decreased, the current also decreases (Figure 5–7). On the other hand, if the resistance is increased, the current decreases (Figure 5–8). This relationship of voltage, current, and resistance is called Ohm's law.

5–1 Questions

1. What are the three basic parts of an electric circuit?
2. Define the following:
   a. Series circuit
   b. Parallel circuit
   c. Series-parallel circuit
3. Draw a diagram of a circuit showing how current would flow through the circuit. (Use arrows to indicate current flow.)
4. What is the difference between an open circuit and a closed circuit?
5. What happens to the current in an electric circuit when the voltage is increased? When it is decreased? When the resistance is increased? When it is decreased?

5–2 Ohm's Law

Ohm's law, or the relationship among current, voltage, and resistance, was first observed by George Ohm in 1827. **Ohm's law** states that the current in an electric circuit is directly proportional to the voltage and inversely proportional to the resistance in a circuit. This may be expressed as:

$$\text{current} = \frac{\text{voltage}}{\text{resistance}}$$

or

$$I = \frac{E}{R}$$

where: I = current in amperes
E = voltage in volts
R = resistance in ohms

Whenever two of the three quantities are known, the third quantity can always be determined.

EXAMPLE How much current flows in the circuit shown in Figure 5–9?

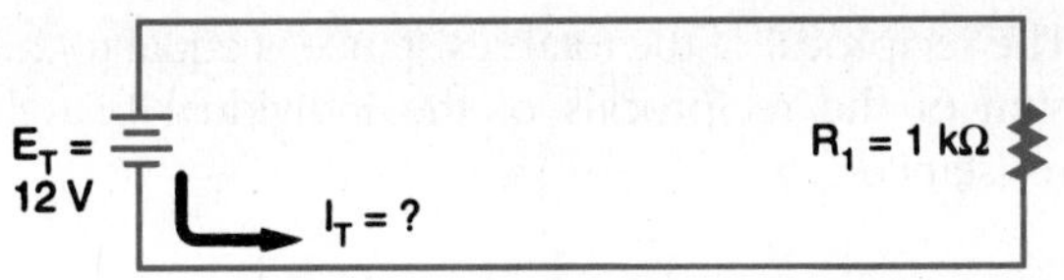

Figure 5–9

Given:

$$I_T = ?$$
$$E_T = 12 \text{ volts}$$
$$R_T = 1000 \text{ ohms}$$

Solution:

$$I_T = \frac{E_T}{R_T}$$
$$I_T = \frac{12}{1000}$$
$$I_T = 0.012 \text{ amp or } 12 \text{ milliamps}$$

EXAMPLE In the circuit shown in Figure 5–10, how much voltage is required to produce 20 milliamps of current flow?

Given:

$$I_T = 20 \text{ mA} = 0.02 \text{ amp}$$
$$E_T = ?$$
$$R_T = 1.2 \text{ k}\Omega = \text{ohms}$$

Solution:

$$I_T = \frac{E_T}{R_T}$$
$$0.02 = \frac{E_T}{1200}$$
$$E_T = (0.02)(1200)$$
$$E_T = 24 \text{ volts}$$

EXAMPLE What resistance value is needed for the circuit shown in Figure 5–11 to draw 2 amperes of current?

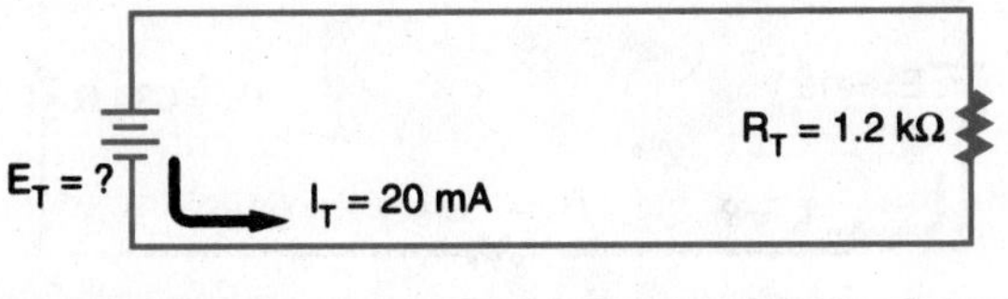

Figure 5–10

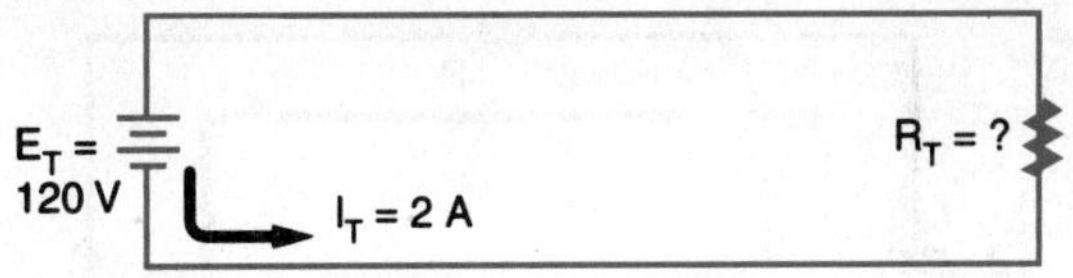

Figure 5–11

Given:

$$I_T = 2 \text{ amps}$$
$$E_T = 120 \text{ volts}$$
$$R_T = ?$$

Solution:

$$I_T = \frac{E_T}{R_T}$$
$$2 = \frac{120}{R_T}$$
$$(1)(120) = (2)(R_T)$$
$$\frac{120}{2} = 1R_T$$
$$60 \text{ ohms} = R_T$$

5–2 Questions

1. State Ohm's law as a formula.
2. How much current flows in a circuit with 12 volts applied and a resistance of 2400 ohms?
3. How much resistance is required to limit current flow to 20 milliamperes with 24 volts applied?
4. How much voltage is needed to produce 3 amperes of current flow through a resistance of 100 ohms?

5–3 Application of Ohm's Law

In a series circuit (Figure 5–12), the same current flows throughout the circuit.

$$I_T = I_{R_1} = I_{R_2} = I_{R_3} \ldots = I_{R_n}$$

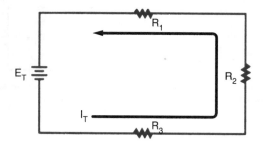

Figure 5–12 In a series circuit, the current flow is the same throughout the circuit.

The total voltage in a series circuit is equal to the voltage drop across the individual loads (resistance) in the circuit.

$$E_T = E_{R_1} + E_{R_2} + E_{R_3} \ldots + E_{R_n}$$

The total resistance in a series circuit is equal to the sum to the individual resistances in the circuit.

$$R_T = R_1 + R_2 + R_3 \ldots + R_n$$

In a parallel circuit (Figure 5–13), the same voltage is applied to each branch in the circuit.

$$E_T = E_{R_1} = E_{R_2} = E_{R_3} \ldots = E_{R_n}$$

The total current in a parallel circuit is equal to the sum of the individual branch currents in the circuit.

$$I_T = I_{R_1} + I_{R_2} + I_{R_3} \ldots + I_{R_n}$$

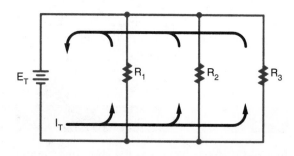

Figure 5–13 In a parallel circuit, the current divides among the branches of the circuit and recombines on returning to the voltage source.

The reciprocal of the total resistance is equal to the sum of the reciprocals of the individual branch resistances.

$$\frac{1}{R_T} = \frac{1}{R_1} + \frac{1}{R_2} + \frac{1}{R_3} \ldots + \frac{1}{R_n}$$

The total resistance in a parallel circuit will always be smaller than the smallest branch resistance.

Ohm's law states that the current in a circuit (series, parallel, or series-parallel) is directly proportional to the voltage and inversely proportional to the resistance.

$$I = \frac{E}{R}$$

In determining unknown quantities in a circuit, follow these steps:

1. Draw a schematic of the circuit and label all known quantities.
2. Solve for equivalent circuits and redraw the circuit.
3. Solve for the unknown quantities.

Remember: Ohm's law is true for any point in a circuit and can be applied at any time. The same current flows throughout a series circuit and the same voltage is present at any branch of a parallel circuit.

EXAMPLE What is the total current flow in the circuit shown in Figure 5–14?

Given:

$$I_T = ?$$
$$E_T = 12 \text{ volts}$$
$$R_T = ?$$
$$R_1 = 560 \text{ ohms}$$

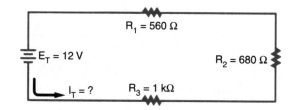

Figure 5–14

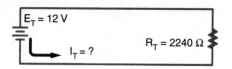

Figure 5–15

$R_2 = 680$ ohms

$R_3 = 1\ k\Omega = 1000$ ohms

Solution:

First solve for the total resistance of the circuit:

$$R_T = R_1 + R_2 + R_3$$
$$R_T = 560 + 680 + 1000$$
$$R_T = 2240 \text{ ohms}$$

Draw an equivalent circuit. See Figure 5–15. Now solve for the total current flow:

$$I_T = \frac{E_T}{R_T}$$
$$I_T = \frac{12}{2240}$$
$$I_T = 0.0054 \text{ amp or } 5.4 \text{ milliamps}$$

EXAMPLE How much voltage is dropped across resistor R_2 in the circuit shown in Figure 5–16?

Given:

$$I_T = ?$$
$$E_T = 48 \text{ volts}$$
$$R_T = ?$$
$$R_1 = 1.2\ k\Omega = 1200 \text{ ohms}$$
$$R_2 = 3.9\ \Omega = 3900 \text{ ohms}$$
$$R_3 = 5.6\ k\ \Omega = 5600 \text{ ohms}$$

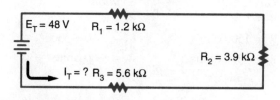

Figure 5–16

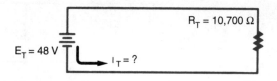

Figure 5–17

Solution:

First solve for the total circuit resistance:

$$R_T = R_1 + R_2 + R_3$$
$$R_T = 1200 + 3900 + 5600$$
$$R_T = 10,700 \text{ ohms}$$

Draw the equivalent circuit. See Figure 5–17. Solve for the total current in the circuit:

$$I_T = \frac{E_T}{R_T}$$
$$I_T = \frac{48}{10,7000}$$
$$I_T = 0.0045 \text{ amp or } 4.5 \text{ milliamps}$$

Remember, in a series circuit, the same current flows throughout the circuit. Therefore, $I_{R_2} = I_T$.

$$I_{R_2} = \frac{E_{R_2}}{R_2}$$
$$0.0045 = \frac{E_{R_2}}{3900}$$
$$E_2 = (0.0045)(3900)$$
$$E_2 = 17.55 \text{ volts}$$

EXAMPLE What is the value of R_2 in the circuit shown in Figure 5–18?

First solve for the current that flows through R_1 and R_3. Because the voltage is the same in each branch of a parallel circuit, each branch voltage is equal to the source voltage of 120 volts.

Given:

$$I_{R_1} = ?$$
$$E_{R_1} = 120 \text{ volts}$$
$$R_1 = 1000 \text{ ohms}$$

Solution:

$$I_{R_1} = \frac{E_{R_1}}{R_1}$$
$$I_{R_1} = \frac{120}{1000}$$
$$I_{R_1} = 0.12 \text{ amp}$$

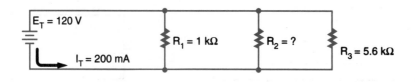

Figure 5–18

Given:

$I_{R_3} = ?$

$E_{R_3} = 120$ volts

$R_3 = 5600$ ohms

Solution:

$$I_{R_3} = \frac{E_{R_3}}{R_3}$$

$$I_{R_3} = \frac{120}{5600}$$

$$I_{R_3} = 0.021 \text{ amp}$$

In a parallel circuit, the total current is equal to the sum of the currents in the branch currents.

Given:

$$I_T = 0.200 \text{ amp}$$

$$I_{R_1} = 0.120 \text{ amp}$$

$$I_{R_2} = ?$$

$$I_{R_3} = 0.021 \text{ amp}$$

Solution:

$$I_T = I_{R_1} + I_{R_2} + I_{R_3}$$

$$0.200 = 0.120 + I_{R_2} + 0.021$$

$$0.200 = 0.141 + I_{R_2}$$

$$0.200 - 0.141 = I_{R_2}$$

Resistor R_2 can now be determined using Ohm's law.

Given:

$$I_{R_2} = 0.059 \text{ amp}$$

$$E_{R_2} = 120 \text{ volts}$$

$$R_2 = ?$$

Solution:

$$I_{R_2} = \frac{E_{R_2}}{R_2}$$

$$0.059 = \frac{120}{R_2}$$

$$(0.059)(R_2) = (120)(1)$$

$$0.059\,R_2 = 120$$

$$1R_2 = \frac{120}{0.059}$$

$$R_2 = 2033.9 \text{ ohms}$$

EXAMPLE What is the current through R_3 in the circuit shown in Figure 5–19?

First determine the equivalent resistance (R_A) for resistors R_1 and R_2.

Given:

$$R_A = ?$$

$$R_1 = 1000 \text{ ohms}$$

$$R_2 = 2000 \text{ ohms}$$

Solution:

$$\frac{1}{R_A} = \frac{1}{R_1} + \frac{1}{R_2} \quad \text{(adding fractions requires a common denominator)}$$

$$\frac{1}{R_A} = \frac{1}{1000} + \frac{1}{2000}$$

$$\frac{1}{R_A} = \frac{2}{2000} + \frac{1}{2000}$$

$$\frac{1}{R_A} = \frac{3}{2000}$$

$$3R_A = 2000$$

$$1R_A = \frac{2000}{3}$$

$$R_A = 666.67 \text{ ohms}$$

Then determine the equivalent resistance (R_B) for resistors R_4, R_5, and R_6. First, find the total series resistance (R_S) for resistors R_5 and R_6.

Given:

$R_S = ?$

$R_5 = 1500 \text{ ohms}$

$R_6 = 3300 \text{ ohms}$

Solution:

$R_S = R_5 + R_6$

$R_S = 1500 + 3300$

$R_S = 4800 \text{ ohms}$

Given:

$$R_B = ?$$

$$R_4 = 4700 \text{ ohms}$$

$$R_S = 4800 \text{ ohms}$$

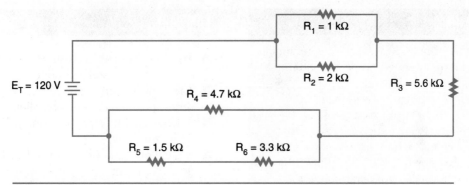

Figure 5–19

Solution:

$$\frac{1}{R_B} = \frac{1}{R_4} + \frac{1}{R_S}$$

$$\frac{1}{R_B} = \frac{1}{4700} + \frac{1}{4800}$$

(In this case, a common denominator would be too large. Use decimals!)

$$\frac{1}{R_B} = 0.000213 + 0.000208$$

$$\frac{1}{R_B} = 0.000421$$

$$(0.000421)(R_B) = (1)(1)$$

$$0.000421\,R_B = 1$$

$$1R_B = \frac{1}{0.000421}$$

$$R_B = 2375.30 \text{ ohms}$$

Redraw the equivalent circuit substituting R_A and R_B, and find the total series resistance of the equivalent circuit. See Figure 5–20.

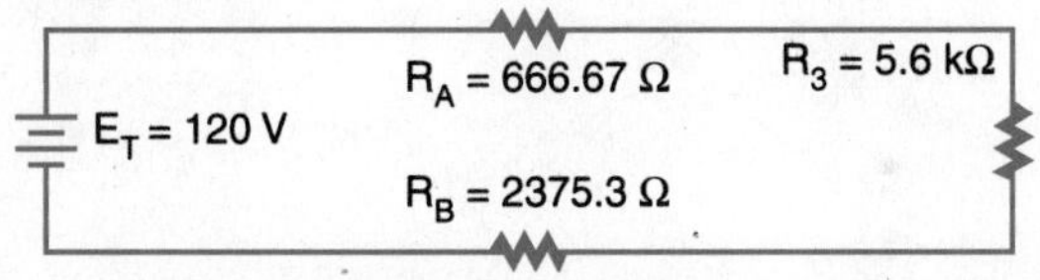

Given:

$$R_T = ?$$
$$R_A = 666.67 \text{ ohms}$$
$$R_3 = 5600 \text{ ohms}$$
$$R_B = 2375.30 \text{ ohms}$$

Solution:

$$R_T = R_A + R_3 + R_B$$
$$R_T = 666.67 + 5600 + 2375.30$$
$$R_T = 8641.97 \text{ ohms}$$

Now solve for the total current through the equivalent circuit using Ohm's law.

Given:

$$I_T = ?$$
$$E_T = 120 \text{ volts}$$
$$R_T = 8641.97 \text{ ohms}$$

Solution:

$$I_T = \frac{E_T}{R_T}$$

$$I_T = \frac{120}{8641.97}$$

$$I_T = 0.0139 \text{ amp or } 13.9 \text{ milliamps}$$

In a series circuit, the same current flows throughout the circuit. Therefore, the current flowing through R_3 is equal to the total current in the circuit.

$$I_{R_3} = I_T$$
$$I_{R_3} = 13.9 \text{ milliamps}$$

Figure 5–20

5–3 Questions

1. State the formulas necessary for determining total current in series and parallel circuits when the current flow through the individual components is known.
2. State the formulas necessary for determining total voltage in series and parallel circuits when the individual voltage drops are known.
3. State the formulas for determining total resistance in series and parallel circuits when the individual resistances are known.
4. State the formula to solve for total current, voltage, or resistance in a series or parallel circuit when at least two of the three values (current, voltage, and resistance) are known.
5. What is the total circuit current in Figure 5–21?

$I_T = ?$

$E_T = 12$ volts

$R_1 = 500$ ohms

$R_2 = 1200$ ohms

$R_3 = 2200$ ohms

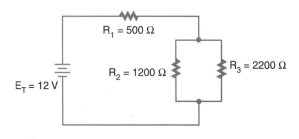

$R_1 = 500 \ \Omega$

$R_2 = 1200 \ \Omega$

$R_3 = 2200 \ \Omega$

$E_T = 12 \ V$

Figure 5–21

SUMMARY

- An electric circuit consists of a voltage source, a load, and a conductor.
- The current path in an electric circuit can be: series, parallel, or series-parallel.
- A series circuit offers only one path for current to flow.
- A parallel circuit offers several paths for the flow of current.
- A series-parallel circuit provides a combination of series and parallel paths for the flow of current.
- Current flows from the negative side of the voltage source through the load to the positive side of the voltage source.
- Current flow in an electric circuit can be varied by changing either the voltage or the resistance.
- The relationship of current, voltage, and resistance is given by Ohm's law.
- Ohm's law states that the current in an electric circuit is directly proportional to the voltage applied and inversely proportional to the resistance in the circuit.

$$I = \frac{E}{R}$$

- Ohm's law applies to all series, parallel, and series-parallel circuits.
- To determine unknown quantities in a circuit:
  — Draw a schematic of the circuit and label all quantities.
  — Solve for equivalent circuits and redraw the circuit.
  —Solve for all unknown quantities.

Chapter 5 Self-Test

Using Ohm's law, find the unknown value for the following:

1. $I = ?$ $E = 9$ V $R = 4500$ ohms
2. $I = 250$ mA $E = ?$ $R = 470$ ohms
3. $I = 10$ A $E = 240$ V $R = ?$
4. Find the total current for the circuits shown.

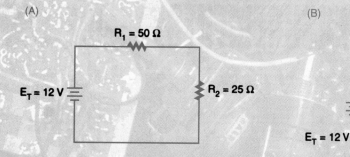

(A)

(B)

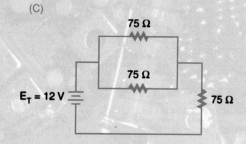

(C)

Chapter

6

ELECTRICAL MEASUREMENTS—METERS

Objectives

After completing this chapter, the student will be able to:

- Identify the two types of meter movements available.
- Describe how a voltmeter is used in a circuit.
- Describe how an ammeter is used in a circuit.
- Describe how an ohmmeter is used for measuring resistance.
- Identify the functions of a multimeter.
- Describe how to use a multimeter to measure voltage, current, and resistance.
- Describe how to measure current using an ammeter.
- Describe how to connect an ammeter into a circuit.
- List safety precautions for using an ammeter.
- Describe how to connect a voltmeter to an electrical circuit.
- List safety precautions for connecting a voltmeter to a circuit.
- Describe how resistance values are measured using an ohmmeter.
- Define *continuity check*.
- Describe how an ohmmeter is used to check open, short, or closed circuits.

In the field of electricity, accurate quantitative measurements are essential. A technician commonly works with current, voltage, and resistance. Ammeters, voltmeters, and ohmmeters are used to provide the essential measurements. A good understanding of the design and operation of electrical measuring meters is important.

This chapter describes the more commonly used analog meters, including the multimeter or multi-function meter.

6–1 Introduction to Meters

Meters are the means by which the invisible action of electrons can be detected and measured. Meters are indispensable in examining the operation of a circuit. Two types of meters are available. One type is the **analog meter,** which uses a graduated scale with a pointer (Figure 6–1). The other type is the **digital meter,** which provides a reading in numbers (Figure 6–2). Digital meters are easier to read and provide a more accurate reading than analog meters. However, analog meters provide a better graphic display of rapid changes in current or voltage.

Most meters are housed in a protective case. Terminals are provided for connecting the meter to the circuit. The polarity of the terminals must be observed for proper connection. A red terminal is positive and a black terminal is negative.

Figure 6–2 A digital meter. *(Courtesy of Protek.)*

Prior to use of an analog meter, the pointer should be adjusted to zero. A small screw is located on the front of the meter to permit this adjustment (Figure 6–3). To zero the meter, place the meter in the position where it is to be used. If the needle does not point to zero, use a screwdriver to turn the screw until it does. The meter should not be connected to a circuit while this adjustment is being made.

6–1 Questions

1. What is the purpose of a meter?
2. What are the two types of meters available?
3. What colors identify the positive and negative terminals of a meter?
4. What adjustments should be made before using an analog meter?

Figure 6–1 Analog meter. *(Courtesy of Protek.)*

Figure 6–3 Location of the zero adjustment screw on an analog meter. *(Courtesy of Protek.)*

6–2 Types of Meters

An **ammeter** is used to measure current in a circuit. An *ammeter* (schematic symbol shown in Figure 6–4) can be considered a flow meter. It measures the number of electrons flowing past a given point in a circuit. The electrons must flow through the ammeter to obtain a reading. As shown in Figure 6–5 this is accomplished by opening the circuit and inserting the ammeter.

A **voltmeter** is used to measure the voltage (difference of potential) between two points in a circuit. A voltmeter can be considered a pressure gauge, used to measure the electrical pressure in a circuit (Figure 6–6).

Resistance is measured with an **ohmmeter.** To measure resistance, a voltage is placed across the device to be measured, inducing a current flow

Figure 6–4 Schematic symbol for an ammeter.

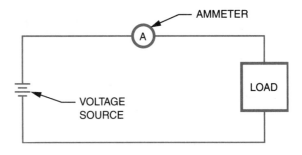

Figure 6–5 The placement of an ammeter in a circuit.

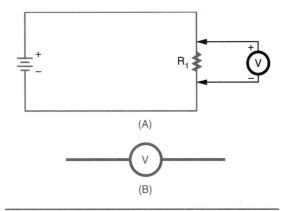

Figure 6–6 (A) A voltmeter is connected in parallel in a circuit. (B) Schematic symbol for a voltmeter.

through the device (Figure 6–7). When there is little resistance, a large current flows and the ohmmeter registers a low resistance. When there is great resistance, a small current flows and the ohmmeter registers a high resistance.

6–2 Questions

1. What is used to measure current?
2. What is used to measure voltage?
3. What meter is used to measure resistance?
4. Describe how to measure current with an ammeter.
5. Describe how to measure voltage with a voltmeter.
6. Describe how to measure resistance with an ohmmeter.

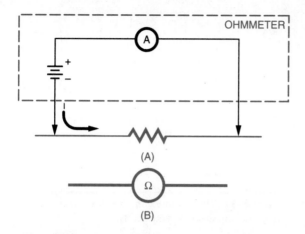

Figure 6–7 (A) An ohmmeter applies a voltage across the component being measured and monitors the current flowing through it. (B) Schematic symbol for an ohmmeter.

6–3 Measuring Current

To use an ammeter to measure current, the circuit must be opened and the meter inserted into the circuit in series (Figure 6–8).

When placing the ammeter in the circuit, polarity must be observed. The two terminals on an ammeter are marked: red for positive and black for negative (or common) (Figure 6–9).

Caution: Always turn off the power before you connect an ammeter to a circuit.

The negative terminal must be connected to the more negative point in the circuit, and the positive terminal to the more positive point in the cir-

cuit (Figure 6–10). When the ammeter is connected, the needle (pointer) of the meter moves from left to right. If the needle moves in the opposite direction, reverse the leads.

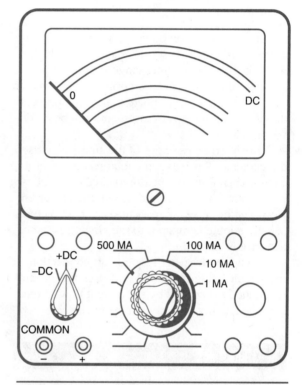

Figure 6–9 An ammeter is one part of this meter. A black negative lead plugs into the common or negative jack. A red positive lead plugs into the jack with the plus symbol.

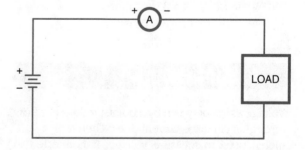

Figure 6–10 Connect the positive terminal of the ammeter to the more positive point in the circuit. Connect the negative terminal to the more negative point in the circuit.

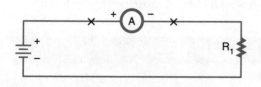

Figure 6–8 An ammeter is connected in series in a circuit.

Caution: An analog ammeter must never be connected in parallel with any of the circuit components. If connected in parallel, the fuse in the ammeter will blow and seriously damage the meter or the circuit. Also, never connect an ammeter directly to a voltage source.

Before turning on power to the circuit after the ammeter is installed, set the meter to its highest scale (its highest ammeter range prior to applying power). After the power is applied, the ammeter can be set to the appropriate scale. This prevents the needle of the meter movement from being driven into its stop.

The internal resistance of the ammeter adds to the circuit and increases the total resistance of the circuit. Therefore, the measured circuit current can be slightly lower than the actual circuit current. However, because the resistance of an ammeter is usually minute compared to the circuit resistance, the error is ignored.

A clip-on ammeter requires no connection to the circuit being measured and uses the electromagnetic field created by the current flow to measure the amount of current in the circuit.

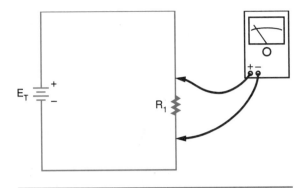

Figure 6–11 When connecting a voltmeter into a circuit, be sure to observe polarity.

Polarity is important. The negative terminal of the voltmeter must be connected to the more negative point in the circuit, and the positive terminal to the more positive point in the circuit (Figure 6–11). If the connections are reversed, the pointer deflects toward the left side of the meter, registering no measurement. If this occurs, reverse the meter leads.

A good practice is to remove power from the circuit, connect the voltmeter, and then reapply power. Initially, set the voltmeter for its highest scale. After the voltage is applied to the circuit, set the meter down to the proper scale.

The voltmeter's internal resistance is connected in parallel with the component being measured. The total resistance of resistors in parallel is always less than that of the smallest resistor. As a result, the voltage read by the voltmeter is smaller than the actual voltage across the component. In most cases, the internal resistance of a voltmeter is high, so that the error is small and can be ignored. However, if voltage is being measured in a high-resistance circuit, meter resistance may have a noticeable effect. Some voltmeters are designed with extra-high internal resistance for such purposes.

6—3 Questions

1. How is an ammeter connected to a circuit?
2. What is the first step prior to connecting an ammeter in a circuit?
3. What should be done if the ammeter reads in the reverse direction?
4. What scale should the ammeter be placed on prior to applying power?

6—4 Measuring Voltage

Voltage exists between two points; it does not flow through a circuit as current does. Therefore, a voltmeter, used to measure voltage, is connected in parallel with the circuit.

Caution: If an analog voltmeter is connected in series with a circuit, a large current can flow through the meter and might damage it.

6—4 Questions

1. How is a voltmeter connected to a circuit?
2. What is the recommended practice for connecting a voltmeter to a circuit?

3. What should be done if the voltmeter reads in the reverse direction?

4. What caution must be taken when measuring a high resistance circuit?

6—5 Measuring Resistance

An ohmmeter measures the resistance of a circuit or component by applying a known voltage. The voltage is supplied by batteries. When a constant voltage is applied to the meter circuit through the component under test, the pointer is deflected based on the current flow. The meter deflection varies with the resistance being measured. To measure the resistance of a circuit or component, the ohmmeter is connected in parallel with the circuit or component.

Caution: Prior to connecting an ohmmeter to a circuit, make sure the power is turned off.

When measuring a component in a circuit, disconnect one end of the component from the circuit. This eliminates parallel paths, which result in an incorrect resistance reading. The device must be removed from the circuit to obtain an accurate reading. Then, the ohmmeter leads are connected across the device (Figure 6–12).

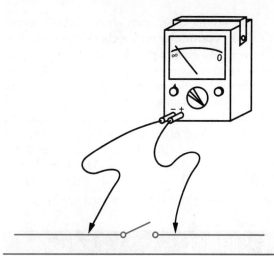

Figure 6–13 An ohmmeter can be used to determine if a circuit is open. An open circuit indicates a high resistance.

Since the primary purpose of an ohmmeter is to measure resistance, it can be used in determining whether a circuit is open, shorted, or closed. An **open circuit** has infinite resistance because no current flows through it (Figure 6–13). A **short circuit** has zero ohms of resistance because current flows through it without developing a voltage drop. A **closed circuit** is a complete path for current flow. Its resistance varies depending on the components in the circuit (Figure 6–14).

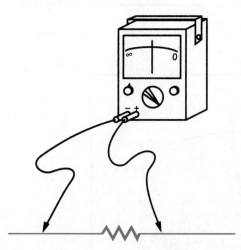

Figure 6–12 When using an ohmmeter to measure resistance, the component being measured must be removed from the circuit.

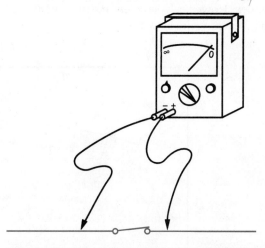

Figure 6–14 An ohmmeter can also be used to determine whether a circuit offers a complete path for current flow. A closed circuit indicates a low resistance.

The testing for an open, short, or closed circuit is called a **continuity test** or check. It is a check to determine if a current path is continuous. To determine whether a circuit is open or closed, the lowest scale on the ohmmeter should be used. First ensure that there are no components in the circuit that may be damaged by the current flow from the ohmmeter. Then place the leads of the ohmmeter across the points in the circuit to be measured. If a reading occurs, the path is closed or shorted. If no reading occurs, the path is open. This test is useful to determine why a circuit does not work.

6—5 Questions

1. How does an ohmmeter work?
2. What caution must be observed prior to connecting an ohmmeter to a circuit?
3. What is the primary purpose of an ohmmeter?
4. For what other purpose may an ohmmeter be used?

6—6 Reading Meter Scales

Voltmeter and ammeter scales are read in the same manner. However, voltmeters measure volts and ammeters measure amperes.

The maximum value indicated by a meter is called the **full scale value.** In other words, the maximum voltage or current that a meter can read is its full scale value.

The measured value of voltage or current is read on the scale under the pointer. For example, the pointer in Figure 6–15 is shown deflected one major division, indicating a voltage of 1 volt or a current of 1 ampere. In Figure 6–16, the meter is shown deflected seven major divisions, indicating a current of 7 amperes or a voltage of 7 volts.

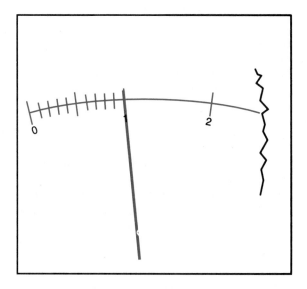

Figure 6–15 The reading indicates 1 volt or ampere.

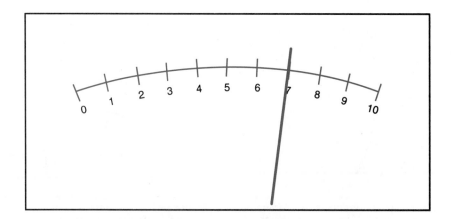

Figure 6–16 The reading indicates 7 volts or amperes.

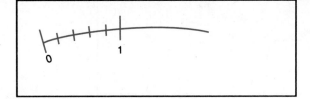

Figure 6–17 Each small division represents 0.2 volt or ampere.

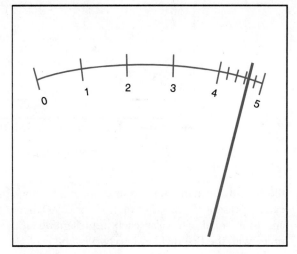

Figure 6–19 The reading indicates 4.65 volts or amperes.

If the pointer of the meter rests between the major divisions of the scale, the smaller divisions are read. Figure 6–17 shows four small lines between each major division of the scale, creating five equally spaced intervals. Each of these small intervals represents one-fifth of the major interval, or 0.2 unit.

If the pointer falls between the small lines on a meter scale, the value must be estimated. In Figure 6–18 the pointer falls between the ⅖ (0.4) and ⅗ (0.6) marks. This indicates a value of approximately 2.5 volts or amperes. In Figure 6–19, the pointer is one-fourth of the distance between the ⅗ (0.6) and ⅘ (0.8) marks. Each small interval represents 0.2. One-fourth of 0.2 is 0.05. Therefore, the pointer indicates a value of about 4.65 volts or amperes.

The number of major and minor divisions on a meter scale depends on the range of the voltage or current that the meter is designed to measure. In all cases, the value of the small intervals can be found by dividing the value of the major interval by the number of spaces it contains.

The ohm scale on a meter is different from most voltage and current scales (Figure 6–20). It is read from right to left instead of from left to right. Also, it is a nonlinear scale, so the number of small spaces between the major intervals is not the same throughout the scale. Between 0 and 1 there are five small spaces, which equal 0.2 unit each. There are four intervals between 6 and 10, representing 1 unit each, and between each of these there is a minor division that represents 0.5 unit. Between

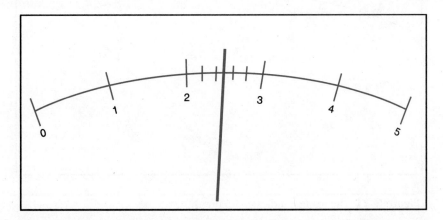

Figure 6–18 The reading indicates 2.5 volts or amperes.

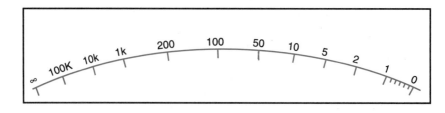

Figure 6–20 The ohmmeter scale is read from right to left.

the 50 and 100 marks are five small intervals, which each represent 10 units. Between 100 and 500 there are four small intervals, each representing 100 units, with the first divided into two intervals of 50 each. The last mark on the left is labeled infinity (∞). If the pointer deflects to this mark, the resistance is beyond the range of the meter. The pointer normally rests on the infinity mark when no resistance is being measured. Figure 6–21 shows the pointer deflected to 1.5 ohms. Figure 6–22 shows the pointer indicating 200 ohms.

Prior to using an ohmmeter, the test leads are shorted together and the zero control is adjusted so the pointer rests on the zero mark. This cali-

brates the meter and compensates for battery deterioration.

6—6 Questions

1. What determines the maximum value an analog meter can measure?
2. What are the differences between an ohmmeter scale and a voltmeter or ammeter scale?
3. Estimate the reading of the voltmeter scales in Figure 6–23.

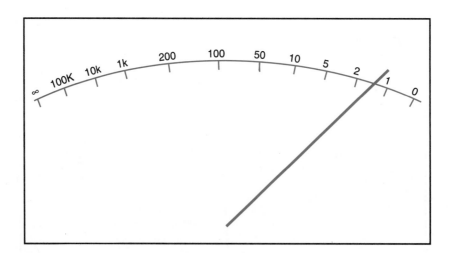

Figure 6–21 The reading indicates 1.5 ohms.

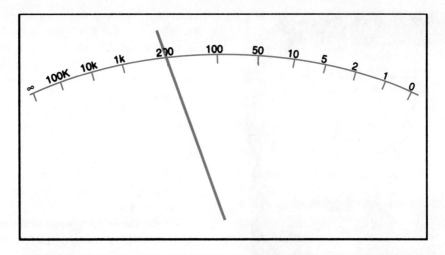

Figure 6–22 The reading indicates 200 ohms.

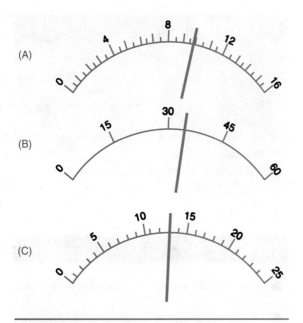

Figure 6–23

6–7 Multimeters

When working on a piece of equipment, many different measurements must be taken. To eliminate the need for several meters, the voltmeter, amme-ter, and ohmmeter can be combined into a single instrument called a **multimeter.** A multimeter is also referred to as a *volt-ohm-milliammeter (VOM)*, Figure 6–24.

The meter in Figure 6–24 has five voltage positions, four current positions, and three resistance positions. There are five scales on the meter to accommodate the various ranges and functions. The technician selects the switch on the multimeter for the desired voltage, current, or resistance range. The zero ohm control adjusts the ohmmeter circuit to compensate for variations in the voltage of the internal batteries. The function switch has three positions: –DC, +DC, and AC. To measure current, DC voltage, and resistance, the function switch is placed at –DC or +DC according to the polarity of the applied current or voltage. The function switch permits reversal of the test lead connections without removing the leads from the circuit being tested.

To measure DC voltage, set the function switch to +DC. With the function switch set at +DC, the common jack is negative and the plus jack is positive. The voltmeter is connected in parallel with the circuit. When measuring an unknown voltage, always set the meter to the highest range (500 volts). If the measured voltage is lower, a lower position can be selected. This procedure protects the meter

Figure 6–24 Analog multimeter.

from damage. Read the voltage on the scale marked DC. For the 2.5-volt range, use the 0–250 scale and divide by 100. For the 10-volt, 50-volt, and 250-volt ranges, use the scales directly. For the 500-volt range, use the 0–50 scale and multiply by 10.

To measure current, the selector switch is set for the desired current position and the meter is connected in series with the circuit. The DC scale on the meter is used. For the 1-mA range, use the 0–10 scale and divide by 10. For the 10-mA range, use the 0–10 scale directly. For the 100-mA range, use the 0–10 scale and multiply by 10. For the 500-mA range, use the 0–50 scale and multiply by 10.

To measure resistance, set the selector switch to the desired resistance range. Short the test leads together. Rotate the zero ohm control until the pointer indicates zero ohms. Separate the test leads and connect them across the component being

measured. For measuring resistance between 0 and 200 ohms, use the R × 1 range. To measure resistance between 200 and 20,000 ohms, use the R × 100 range. For resistance above 20,000 ohms, use the R × 10,000. When measuring resistance in digital and semiconductor circuits, do not use the R × 1 scale. This scale is energized by a 9V battery and could damage the circuitry. Observe the reading on the ohm scale at the top of the meter scales. Note that the ohm scale reads from right to left. To determine the actual resistance value, multiply the reading by the factor at the switch position. The letter *K* equals 1000.

To use the other voltage and current jacks located on the multimeter, refer to the operator's manual.

6–7 Questions

1. What is a multimeter?
2. Describe how a voltage measurement is taken with a multimeter.
3. Describe how a current measurement is taken with a multimeter.
4. Explain how to use the ohmmeter portion of a multimeter.
5. What precautions should be observed when using a multimeter?

SUMMARY

■ Analog meters use a graduated scale with a pointer.
■ Digital meters provide a direct readout.
■ On both analog and digital meters, the red terminal is positive and the black terminal is negative.
■ Before using an analog meter, check the mechanical zero adjustment.
■ An ammeter must be connected in series with a circuit.
■ A voltmeter is connected in parallel with a circuit.

■ An ohmmeter measures resistance by the amount of current flowing through the resistor being measured.

■ The maximum value of a meter scale is called the full scale value.

■ The number of divisions on the meter scale depends on the range the meter is designed to measure.

■ Ammeters and voltmeters are read from left to right and have a linear-scale.

■ Ohmmeters are read from right to left and have nonlinear scales.

■ An analog ohmmeter must be calibrated before use to compensate for battery deterioration.

■ A multimeter combines a voltmeter, ammeter, and ohmmeter into one package.

■ A VOM is a multimeter that measures volts, ohms, and milliamperes.

■ On a multimeter, the range selector switch selects the function to be used.

Chapter 6 Self-Test

1. Which type of meter, analog or digital, would you use for an accurate reading?

2. Which type of meter, analog or digital, would you use to gauge rapid changes in a source?

3. Draw a meter scale for each of the following and show where the needle would point for the following readings.
   a. 23 V
   b. 220 mA
   c. 2700 ohms

4. What are the advantages of using a multimeter?

Chapter

7

POWER

Objectives

After completing this chapter, the student will be able to:

- Define *power* as it relates to electric circuits.
- State the relationship of current and voltage.
- Solve for power consumption in an electrical circuit.
- Determine the total power consumption in a series, parallel, or series-parallel circuit.

In addition to current, voltage, and resistance, a fourth quantity is important in circuit analysis. This quantity is called power.

Power is the rate at which work is done. Power is expended every time a circuit is energized. Power is directly proportional to both current and voltage.

This chapter looks at circuit applications involving power.

7-1 Power

Electrical or mechanical power relates to the rate at which work is being done. Work is done whenever a force causes motion. If a mechanical force is used to lift or move a weight, work is being done. However, force exerted without causing motion, such as a force of a compressed spring between two fixed objects, does not constitute work.

Voltage is an electrical force that creates current to flow into a closed circuit. When voltage exists between two points and current cannot flow, no work is done. This is similar to a spring under tension that produces no motion. When voltage causes electrons to move in a circuit, work is being done. The instantaneous rate at which work is done is called the *electric power rate* and is measured in watts.

The total amount of work done may be accomplished in different lengths of time. For example, a given quantity of electrons may be moved from one location to another in one second, one minute, or one hour, depending on the rate in which they were moved. In all cases, the total amount of work done is the same. However, when the work is done in a short period of time, the instantaneous power rate (wattage) is greater than when the same amount of work is done over a longer period of time.

As mentioned, the basic unit of power is the watt. A watt is equal to the voltage across a circuit multiplied by the current through the circuit. It represents the rate at any given instant in which work is being done, moving electrons through the circuit. The symbol P represents electrical power. The relationship of power, current, and voltage may be expressed as follows:

$$P = I \times E$$

I represents the current through the circuit and E represents the voltage applied to the circuit being measured. The amount of power will differ with any change in the voltage or current in the circuit.

EXAMPLE Calculate the power consumed in the circuit shown in Figure 7–1.

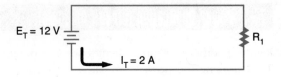

Figure 7–1

Given:	Solution:
P = ?	P = IE
I = 2 amps	P = (2)(12)
E = 12 volts	P = 24 watts

EXAMPLE What voltage is required to deliver 2 amperes of current at 200 watts?

Given:
P = 200 watts
I = 2 amps
E = ?

Solution:
$$P = IE$$
$$200 = 2(E)$$
$$\frac{200}{2} = \frac{\cancel{2}E}{\cancel{2}} \text{ (divide both sides by 2)}$$
$$\frac{200}{2} = 1E$$
$$100 \text{ volts} = E$$

EXAMPLE How much current does a 100-watt, 120-volt light bulb use?

Given:
$$P = 100 \text{ watts}$$
$$I = ?$$
$$E = 120 \text{ volts}$$

Solution:
$$P = IE$$
$$100 = (I)(120)$$
$$\frac{100}{120} = 1I$$
$$0.83 \text{ amp} = I$$

7-1 Questions

1. Define *power* as it relates to electricity.
2. What unit is used to measure power?
3. Calculate the missing value:
   a. P = ?, E = 12 V, I = 1 A
   b. P = 1000 W, E = ?, I = 10 A
   c. P = 150 W, E = 120 V, I = ?

7-2 Power Application (Circuit Analysis)

Resistive components in a circuit consume power. To determine the power dissipated by a component, multiply the voltage drop across the component by the current flowing through the component.

$$P = IE$$

The total power dissipated in a series or parallel circuit is equal to the sum of the power dissipated by the individual components. This can be expressed as:

$$P_T = P_{R_1} + P_{R_2} + P_{R_3} \ldots + P_{R_n}$$

The power dissipated in a circuit is often less than 1 watt. To facilitate the use of these smaller numbers, the milliwatt (mW) and the microwatt (µW) are used.

$$1000 \text{ milliwatts} = 1 \text{ watt}$$

$$1 \text{ milliwatt} = \frac{1}{1000} \text{ watt}$$

$$1,000,000 \text{ microwatts} = 1 \text{ watt}$$

$$1 \text{ microwatt} = \frac{1}{1,000,000} \text{ watt}$$

EXAMPLE How much power is consumed in the circuit shown in Figure 7–2?

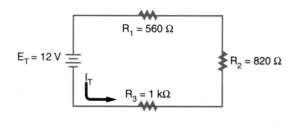

Figure 7–2

First determine the total resistance for the circuit.

Given:	*Solution:*
$R_T = ?$	$R_T = R_1 + R_2 + R_3$
$R_1 = 560$ ohms	$R_T = 560 + 820 + 1000$
$R_2 = 820$ ohms	$R_T = 2380$ ohms
$R_3 = 1000$ ohms	

Now determine the total current flowing in the circuit, using Ohm's law.

Given: $I_T = ?$
$E_T = 12$ volts
$R_T = 2380$ ohms

Solution:
$$I_T = \frac{E_T}{R_T}$$
$$I_T = \frac{12}{2380}$$
$$I_T = 0.005 \text{ amp}$$

The total power consumption can now be determined using the power formula.

Given: $P_T = ?$
$I_T = 0.005$ amp
$E_T = 12$ volts

Solution:
$$P_T = I_T E_T$$
$$P_T = (0.005)(12)$$
$$P_t = 0.06 \text{ watt or } 60 \text{ mW}$$

EXAMPLE What is the value of resistor R_2 in the circuit shown in Figure 7–3?

First determine the voltage drop across resistor R_1. In a parallel circuit the voltage is the same in all branches.

Given:
$$P_{R_1} = 0.018 \text{ watt}$$
$$I_{R_1} = 0.0015 \text{ amp}$$
$$E_{R_1} = ?$$

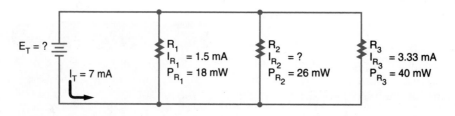

Figure 7–3

Solution:

$$P_{R_1} = I_{R_1} E_{R_1}$$
$$0.018 = (0.0015)(E_{R_1})$$
$$\frac{0.018}{0.0015} = 1E_{R_1}$$
$$12 \text{ volts} = E_{R_1}$$

Now the current through resistor R_2 can be determined.

Given:

$$P_{R_2} = 0.026 \text{ watt}$$
$$I_{R_2} = ?$$
$$E_{R_2} = 12 \text{ volts}$$

Solution:

$$P_{R_2} = I_{R_2} E_{R_2}$$
$$0.026 = (I_{R_2})(12)$$
$$\frac{0.026}{12} = 1 I_{R_2}$$
$$0.00217 \text{ amp} = I_{R_2}$$

The resistance value of R_2 can now be determined using Ohm's law.

Given:

$$I_{R_2} = 0.00217 \text{ amp}$$
$$E_{R_2} = 12 \text{ volts}$$
$$R_2 = ?$$

Solution:

$$I_{R_2} = \frac{E_{R_2}}{R_2}$$
$$0.00217 = \frac{12}{R_2}$$
$$(R_2)(0.00217) = (12)(1)$$
$$R_2 = \frac{12}{0.00217}$$
$$R_2 = 5530 \text{ ohms}$$

EXAMPLE If a current of 0.05 ampere flows through a 22-ohm resistor, how much power is dissipated by the resistor?

The voltage drop across the resistor must first be determined using Ohm's law.

Given:

$$I_R = 0.05 \text{ amp}$$
$$E_R = ?$$
$$R = 22 \text{ ohms}$$

Solution:

$$I_R = \frac{E_R}{R}$$
$$0.05 = \frac{E_R}{22}$$
$$\frac{0.05}{1} = \frac{E_R}{22}$$
$$(1)(E_R) = (0.05)(22)$$
$$E_R = 1.1 \text{ volts}$$

The power consumed by the resistor can now be determined using the power formula.

Given:

$$P_R = ?$$
$$E_R = 1.1 \text{ volts}$$
$$I_R = 0.05 \text{ amp}$$

Solution:

$$P_R = I_R E_R$$
$$P_R = (0.05)(1.1)$$
$$P_R = 0.055 \text{ watt or 55 mW}$$

7–2 Questions

1. What is the formula for power when both the current and voltage are known?
2. What is the formula for determining the total power in a series circuit? A parallel circuit?
3. Convert the following:
   a. 100 mW = _____ µW
   b. 10 W = _____ mW
   c. 10 µW = _____ W
   d. 1000 µW = _____ mW
   e. 0.025 W = _____ mW
4. What is the power consumed by each resistor in the circuit shown in Figure 7–4?
5. What is the power consumed by each resistor in the circuit shown in Figure 7–5?
6. What is the power consumed by each resistor in the circuit shown in Figure 7–6?

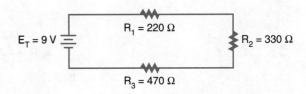

Figure 7–4

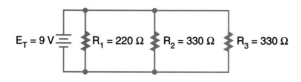

Figure 7–5

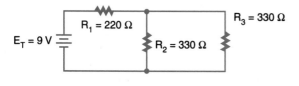

Figure 7–6

SUMMARY

■ Power is the rate at which energy is delivered to a circuit.

■ Power is also the rate at which energy (heat) is dissipated by the resistance in a circuit.

■ Power is measured in watts.

■ Power is the product of current and voltage:

$$P = IE$$

■ The total power dissipated in a series or parallel circuit is equal to the sum of the power dissipated by the individual components.

$$P_T = P_1 + P_2 + P_3 \ldots + P_n$$

Chapter 7 Self-Test

Using Watt's law, find the missing value for the following:

1. P = ? E = 30 V I = 40 mA

2. P = 1 W E = ? I = 10 mA

3. P = 12.3 W E = 30 V I = ?

4. What is the total power consumption for the following circuits?

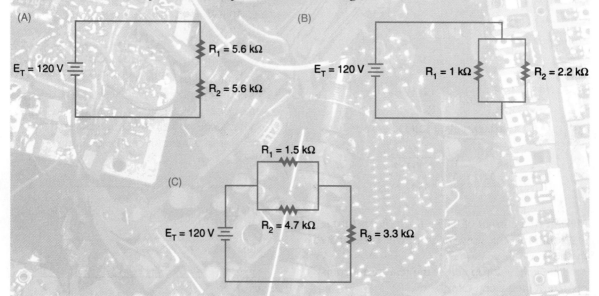

Chapter

8

DC CIRCUITS

Objectives

After completing this chapter, the student will be able to:

- Solve for all unknown values (current, voltage, resistance, and power) in a series, parallel, or series-parallel circuit.

In the study of electronics, certain circuits appear again and again. The most commonly used circuits are the series circuit, the parallel circuit, and the series-parallel circuit.

This chapter applies information from the last few chapters to the solving of all unknowns in these three basic types of circuits.

8-1 Series Circuits

A series circuit (Figure 8–1) provides only one path for current flow. The factors governing the operation of a series circuit are:

 1. The same current flows through each component in a series circuit.

$$(I_T = I_{R_1} = I_{R_2} = I_{R_3} \ldots = I_{R_n})$$

 2. The total resistance in a series circuit is equal to the sum of the individual resistances.

$$(R_T = R_1 + R_2 + R_3 \ldots + R_n)$$

 3. The total voltage across a series circuit is equal to the sum of the individual voltage drops.

$$(E_T = E_{R_1} = E_{R_2} = E_{R_3} \ldots + E_{R_n})$$

 4. The voltage drop across a resistor in a series circuit is proportional to the size of the resistor. $(I = E/R)$

 5. The total power dissipated in a series circuit is equal to the sum of the individual power dissipations. $(P_T = P_{R_1} + P_{R_2} + P_{R_3} \ldots + P_{R_n})$

EXAMPLE Three resistors, 47 ohms, 100 ohms, and 150 ohms, are connected in series with a battery rated at 12 volts. Solve for all values in the circuit.

The first step is to draw a schematic of the circuit and list all known variables. See Figure 8–2.

Given:

$I_T = ?$	$R_1 = 47$ ohms	$E_{R_1} = ?$
$E_T = 12$ volts	$R_2 = 100$ ohms	$E_{R_2} = ?$
$R_T = ?$	$R_3 = 150$ ohms	$E_{R_3} = ?$
$P_T = ?$		$P_{R_1} = ?$
		$P_{R_2} = ?$
		$P_{R_3} = ?$

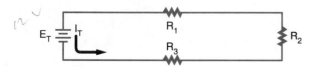

Figure 8–1 Series circuit.

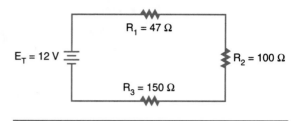

Figure 8–2

In solving for all values in a circuit, the total resistance must be found first. Then the total circuit current can be determined. Once the current is known, the voltage drops and power dissipation can be determined.

Given: *Solution:*

$R_T = ?$ $R_T = R_1 + R_2 + R_3$

$R_1 = 47$ ohms $R_T = 47 + 100 + 150$

$R_2 = 100$ ohms $R_T = 297$ ohms

$R_3 = 150$ ohms

Using Ohm's law, the current is:

Given: *Solution:*

$I_T = ?$ $I_T = \dfrac{E_T}{R_T}$

$E_T = 12$ volts

$R_T = 297$ ohms $I_T = \dfrac{12}{297}$

                          $I_T = 0.040$ amp

Since $I_T = I_{R_1} = I_{R_2} = I_{R_3}$, the voltage drop (E_{R_1}) across resistor R_1 is:

Given: *Solution:*

$I_{R_1} = 0.040$ amp $I_{R_1} = \dfrac{E_{R_1}}{R_1}$

$E_{R_1} = ?$

$R_1 = 47$ ohms $0.040 = \dfrac{E_{R_1}}{47}$

                          $\dfrac{0.040}{1} = \dfrac{E_{R_1}}{47}$

                          $E_{R_1} = (0.040)(47)$

                          $E_{R_1} = 1.88$ volts

The voltage drop (E_{R_2}) across resistor R_2 is:

Given:

$I_{R_2} = 0.040$ amp

$E_{R_2} = ?$

$R_2 = 100$ ohms

Solution:

$$I_{R_2} = \frac{E_{R_2}}{R_2}$$

$$0.040 = \frac{E_{R_2}}{100}$$

$$\frac{0.040}{1} = \frac{E_{R_2}}{100}$$

$$E_{R_2} = (0.040)(100)$$

$$E_{R_2} = 4 \text{ volts}$$

The voltage drop (E_{R_3}) across resistor R_3 is:

Given:

$I_{R_3} = 0.040$ amp

$E_{R_3} = ?$

$R_3 = 150$ ohms

Solution:

$$I_{R_3} = \frac{E_{R_3}}{R_3}$$

$$0.040 = \frac{E_{R_3}}{150}$$

$$\frac{0.040}{1} = \frac{E_{R_3}}{150}$$

$$E_{R_3} = (0.040)(150)$$

$$E_{R_3} = 6 \text{ volts}$$

Verify that the sum of the individual voltages is equal to the total voltage.

Given:

$E_T = 12$ volts

$E_{R_1} = 1.88$ volts

$E_{R_2} = 4$ volts

$E_{R_3} = 6$ volts

Solution:

$$E_T = E_{R_1} + E_{R_2} + E_{R_3}$$

$$E_T = 1.88 + 4 + 6$$

$$E_T = 11.88 \text{ volts}$$

There is a difference between the calculated and the total given voltage due to the rounding of the total current to three decimal places.

The power dissipated across resistor R_1 is:

Given:

$P_{R_1} = ?$

$I_{R_1} = 0.040$ amp

$E_{R_1} = 1.88$ volts

Solution:

$$P_{R_1} = I_{R_1} E_{R_1}$$

$$P_{R_1} = (0.040)(1.88)$$

$$P_{R_1} = 0.075 \text{ watt}$$

The power dissipated across resistor R_2 is:

Given:

$P_{R_2} = ?$

Solution:

$$P_{R_2} = I_{R_2} E_{R_2}$$

$I_{R_2} = 0.040$ amp

$E_{R_2} = 4$ volts

$$P_{R_2} = (0.040)(4)$$

$$P_{R_2} = 0.16 \text{ watt}$$

The power dissipated across resistor R_3 is:

Given:

$P_{R_3} = ?$

$I_{R_3} = 0.040$ amp

$E_{R_3} = 6$ volts

Solution:

$$P_{R_3} = I_{R_3} E_{R_3}$$

$$P_{R_3} = (0.040)(6)$$

$$P_{R_3} = 0.24 \text{ watt}$$

The total power dissipated is:

Given:

$P_T = ?$

$P_{R_1} = 0.075$ watt

$P_{R_2} = 0.16$ watt

$P_{R_3} = 0.24$ watt

Solution:

$$P_T = P_{R_1} + P_{R_2} + P_{R_3}$$

$$P_T = 0.075 + 0.16 + 0.24$$

$$P_T = 0.475 \text{ watt or } 475 \text{ mW}$$

8–1 Question

1. Four resistors, 270 ohms, 560 ohms, 1200 ohms, and 1500 ohms, are connected in series with a battery rated at 28 volts. Solve for all values of the circuit.

8–2 Parallel Circuits

A *parallel circuit* (Figure 8–3) is defined as a circuit having more than one current path. The factors governing the operations of a parallel circuit are:

1. The same voltage exists across each branch of the parallel circuit and is equal to that of the voltage source. ($E_T = E_{R_1} = E_{R_2} = E_{R_3} \ldots = E_{R_n}$)
2. The current through each branch of a parallel circuit is inversely proportional to the amount of resistance of the branch. (I = E/R)

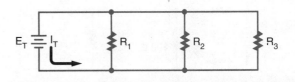

Figure 8–3 Parallel circuit.

3. The total current in a parallel circuit is the sum of the individual branch currents.

$$(I_T = I_{R_1} + I_{R_2} + I_{R_3} \ldots + I_{R_n})$$

4. The reciprocal of the total resistance in a parallel circuit is equal to the sum of the reciprocals of the individual resistances.

$$\left(\frac{1}{R_T} = \frac{1}{R_1} + \frac{1}{R_2} + \frac{1}{R_3} \ldots + \frac{1}{R_n} \right)$$

5. The total power consumed in a parallel circuit is equal to the sum of the power consumed by the individual resistors.

$$(P_T = P_{R_1} + P_{R_2} + P_{R_3} \ldots + P_{R_n})$$

EXAMPLE Three resistors, 100 ohms, 220 ohms, and 470 ohms, are connected in parallel with a battery rated at 48 volts. Solve for all values of the circuit.

First draw a schematic of the circuit and list all known variables (Figure 8–4).

Given:

$I_T = ?$	$R_1 = 100$ ohms	$I_{R_1} = ?$
$E_T = 48$ volts	$R_2 = 220$ ohms	$I_{R_2} = ?$
$R_T = ?$	$R_3 = 470$ ohms	$I_{R_3} = ?$
$P_T = ?$		$P_{R_1} = ?$
		$P_{R_2} = ?$
		$P_{R_3} = ?$

In solving the circuit for all values, the total resistance must be found first. Then the individual currents flowing through each branch of the circuit can be found. Once the current is known, the power dissipation across each resistor can be determined.

Given:

$$R_T = ?$$
$$R_1 = 100 \text{ ohms}$$
$$R_2 = 220 \text{ ohms}$$
$$R_3 = 470 \text{ ohms}$$

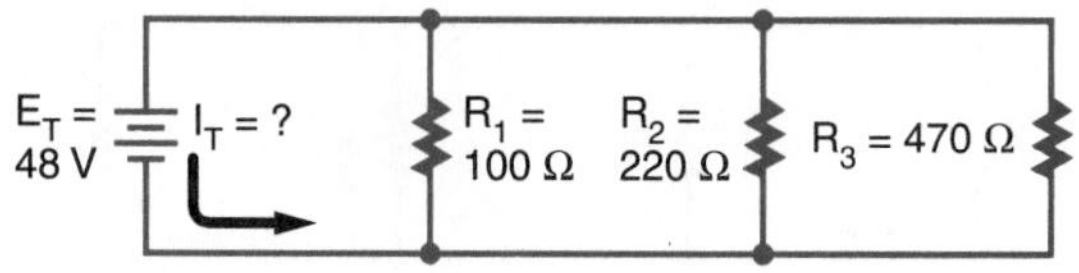

Figure 8–4

Solution:

$$\frac{1}{R_T} = \frac{1}{R_1} + \frac{1}{R_2} + \frac{1}{R_3}$$

(A common denominator would be too large. Change to decimals.)

$$\frac{1}{R_T} = \frac{1}{100} + \frac{1}{220} + \frac{1}{470}$$

$$\frac{1}{R_T} = 0.01 + 0.004 + 0.002$$

$$\frac{1}{R_T} = 0.016$$

$$\frac{1}{R_T} = \frac{0.016}{1}$$

$$(0.017)(R_T) = (1)(1)$$

$$R_T = \frac{1}{0.016}$$

$$R_T = 62.5 \text{ ohms}$$

The current (I_{R_1}) through resistor R_1 is:

Given:

$I_{R_1} = ?$

$E_{R_1} = 48$ volts

$R_1 = 100$ ohms

Solution:

$$I_{R_1} = \frac{E_{R_1}}{R_1}$$

$$I_{R_1} = \frac{48}{100}$$

$$I_{R_1} = 0.48 \text{ amp}$$

The current (I_{R_2}) through resistor R_2 is:

Given:

$I_{R_2} = ?$

$E_{R_2} = 48$ volts

$R_2 = 220$ ohms

Solution:

$$I_{R_2} = \frac{E_{R_2}}{R_2}$$

$$I_{R_2} = \frac{48}{220}$$

$$I_{R_2} = 0.218 \text{ amp}$$

The current (I_{R_3}) through resistor R_3 is:

Given:

$I_{R_3} = ?$

$E_{R_3} = 48$ volts

$R_3 = 470$ ohms

Solution:

$$I_{R_3} = \frac{E_{R_3}}{R_3}$$

$$I_{R_3} = \frac{48}{470}$$

$$I_{R_3} = 0.102 \text{ amp}$$

The total current is:

Given:
$I_T = ?$
$I_{R_1} = 0.48$ amp
$I_{R_2} = 0.218$ amp
$I_{R_3} = 0.102$ amp

Solution:
$I_T = I_{R_1} + I_{R_2} + I_{R_3}$
$I_T = 0.48 + 0.218 + 0.102$
$I_T = 0.800$ amp

The total current can also be found using Ohm's law:

Given:
$I_T = ?$
$E_T = 48$ volts
$R_T = 62.5$ ohms

Solution:
$$I_T = \frac{E_T}{R_T}$$
$$I_T = \frac{48}{62.5}$$
$I_T = 0.768$ amps

Again a difference occurs because of rounding. The power dissipated by resistor R_1 is:

Given:
$P_{R_1} = ?$
$I_{R_1} = 0.48$ amp
$E_{R_1} = 48$ volts

Solution:
$P_{R_1} = I_{R_1} E_{R_1}$
$P_{R_1} = (0.48)(48)$
$P_{R_1} = 23.04$ watts

The power dissipated by resistor R_2 is:

Given:
$P_{R_2} = ?$
$I_{R_2} = 0.218$ amp
$E_{R_2} = 48$ volts

Solution:
$P_{R_2} = I_{R_2} E_{R_2}$
$P_{R_2} = (0.218)(48)$
$P_{R_2} = 10.46$ watts

The power dissipated by resistor R_3 is:

Given:
$P_{R_3} = ?$
$I_{R_3} = 0.102$ amp
$E_{R_3} = 48$ volts

Solution:
$P_{R_3} = I_{R_3} E_{R_3}$
$P_{R_3} = (0.102)(48)$
$P_{R_3} = 4.90$ watts

The total power is:

Given:
$P_T = ?$
$P_{R_1} = 23.04$ watts
$P_{R_2} = 10.46$ watts
$P_{R_3} = 4.90$ watts

Solution:
$P_T = P_{R_1} + P_{R_2} + P_{R_3}$
$P_T = 23.04 + 10.46 + 4.90$
$P_T = 38.40$ watts

The total power can also be determined using Ohm's law:

Given:
$P_T = ?$
$I_T = 0.80$ amp
$E_T = 48$ volts

Solution:
$P_T = I_T E_T$
$P_T = (0.80)(48)$
$P_T = 38.4$ watts

8–2 Question

1. Four resistors, 2220 ohms, 2700 ohms, 3300 ohms, and 5600 ohms, are connected in parallel with a battery rated at 9 volts. Solve for all values of the circuit.

8–3 Series-Parallel Circuits

Most circuits consist of both series and parallel circuits. Circuits of this type are referred to as *series-parallel circuits* (Figure 8–5). The solution of most series-parallel circuits is simply a matter of applying the laws and rules discussed earlier. Series formulas are applied to the series part of the circuit, and parallel formulas are applied to the parallel parts of the circuit.

EXAMPLE Solve for all unknown quantities in Figure 8–6. Given:

$$I_T = ?, E_T = 48 \text{ volts}, R_T = ?, P_T = ?$$

$R_1 = 820$ ohms	$I_{R_1} = ?$	$E_{R_1} = ?$	$P_{R_1} = ?$
$R_2 = 330$ ohms	$I_{R_2} = ?$	$E_{R_2} = ?$	$P_{R_2} = ?$
$R_3 = 680$ ohms	$I_{R_3} = 3$	$E_{R_3} = ?$	$P_{R_3} = ?$
$R_4 = 470$ ohms	$I_{R_4} = ?$	$E_{R_4} = ?$	$P_{R_4} = ?$
$R_5 = 120$ ohms	$I_{R_5} = ?$	$E_{R_5} = ?$	$P_{R_5} = ?$
$R_6 = 560$ ohms	$I_{R_6} = ?$	$E_{R_6} = ?$	$P_{R_6} = ?$

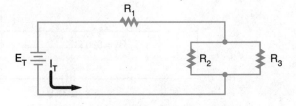

Figure 8–5 Series-parallel circuit.

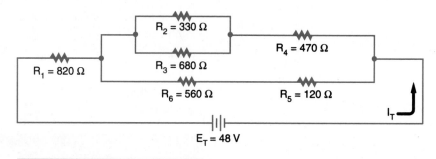

Figure 8–6

To solve for total resistance (R_T), first find the equivalent resistance (R_A) for parallel resistors R_2 and R_3. Then solve for the equivalent resistance of resistors R_A and R_4 (identified as R_{S_1}) and R_5 and R_6 (identified as R_{S_2}). Then the equivalent resistance ($R_{B_{TS}}$) can be determined for R_{S_1} and R_{S_2}. Finally find the total series resistance for R_1 and R_B.

Given:

$$R_A = ?$$
$$R_2 = 330 \text{ ohms}$$
$$R_3 = 680 \text{ ohms}$$

Solution:

$$\frac{1}{R_A} = \frac{1}{R_2} + \frac{1}{R_3}$$

$$\frac{1}{R_A} = \frac{1}{330} + \frac{1}{680}$$ (A common denominator would be too large. Change to decimals.)

$$\frac{1}{R_A} = 0.00303 + 0.00147$$

$$\frac{1}{R_A} = 0.0045$$

$$(1)(1) = (0.0045)(R_A)$$
$$1 = 0.0045 \, R_A$$

$$\frac{1}{0.0045} = R_A$$

$$222.22 \text{ ohms} = R_A$$

Redraw the circuit, substituting resistor R_A for resistors R_2 and R_3. See Figure 8–7.

Now determine series resistance R_{S_1} for resistors R_A and R_4.

Given:

$R_{S_1} = ?$

$R_A = 222.22$ ohms

$R_4 = 470$ ohms

Solution:

$R_{S_1} = R_A + R_4$

$R_{S_1} = 222.22 + 470$

$R_{S_1} = 692.22$ ohms

Determine series resistance R_{S_2} for resistors R_5 and R_6.

Given:

$R_{S_2} = ?$

$R_5 = 120$ ohms

$R_6 = 560$ ohms

Solution:

$R_{S_2} = R_5 + R_6$

$R_{S_2} = 120 + 560$

$R_{S_2} = 680$ ohms

Redraw the circuit with resistors R_{S_1} and R_{S_2}. See Figure 8–8.

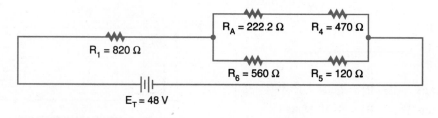

Figure 8–7

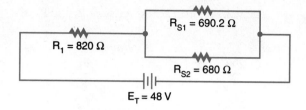

Figure 8–8

Now determine the parallel resistance (R_B) for resistors R_{S_1} and R_{S_2}.

Given:

$$R_B = ?$$
$$R_{S_1} = 692.22 \text{ ohms}$$
$$R_{S_2} = 680 \text{ ohms}$$

Solution:

$$\frac{1}{R_B} = \frac{1}{R_{S_1}} + \frac{1}{R_{S_2}}$$

$$\frac{1}{R_B} = \frac{1}{692.22} + \frac{1}{680}$$

$$\frac{1}{R_B} = 0.00144 + 0.00147$$

$$\frac{1}{R_B} = 0.00291$$

$$\frac{1}{R_B} = \frac{0.00291}{1}$$

$$0.00291 \, R_B = (1)(1)$$

$$R_B = \frac{1}{0.00291}$$

$$R_B = 343.64 \text{ ohms}$$

Redraw the circuit using resistor R_B. See Figure 8–9.

Now determine the total resistance in the circuit.

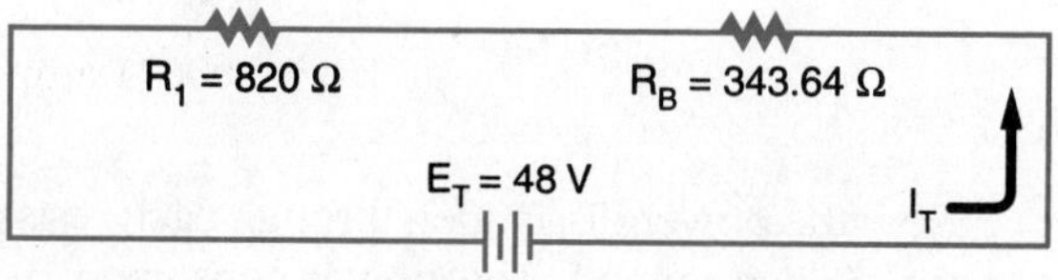

R₁ = 820 Ω   R_B = 343.64 Ω   E_T = 48 V   I_T

Figure 8–9

Given:

$$R_T = ?$$
$$R_1 = 820 \text{ ohms}$$
$$R_B = 343.64 \text{ ohms}$$

Solution:

$$R_T = R_1 + R_B$$
$$R_T = 820 + 343.64$$
$$R_T = 1163.64 \text{ ohms}$$

The total current in the circuit can now be determined using Ohm's law.

Given:

$$I_T = ?$$
$$E_T = 48 \text{ volts}$$
$$R_T = 1163.64 \text{ ohms}$$

Solution:

$$I_T = \frac{E_T}{R_T}$$

$$I_T = \frac{48}{1163.64}$$

$$I_T = 0.0412 \text{ amp}$$
$$\text{or } 41.2 \text{ mA}$$

The voltage drop across resistor R_1 can now be determined.

Given:

$$I_{R_1} = 0.0412 \text{ amp}$$
$$E_{R_1} = ?$$
$$R_1 = 820 \text{ ohms}$$

Solution:

$$I_{R_1} = \frac{E_{R_1}}{R_1}$$

$$0.0412 = \frac{E_{R_1}}{820}$$

$$\frac{0.0412}{1} = \frac{E_{R_1}}{820}$$

$$(1)(E_{R_1}) = (0.0412)(820)$$

$$E_{R_1} = 33.78 \text{ volts}$$

The voltage drop across equivalent resistance R_B is:

Given:

$$I_{R_B} = 0.0412 \text{ amp}$$
$$E_{R_B} = ?$$
$$R_B = 343.64 \text{ ohms}$$

Solution:

$$I_{R_B} = \frac{E_{R_B}}{R_B}$$

$$0.0412 = \frac{E_{R_B}}{343.64}$$

$$E_{R_B} = (343.64)(0.0412)$$
$$E_{R_B} = 14.158 \text{ volts}$$

The current through each branch in a parallel circuit has to be calculated separately.

The current through branch resistance R_{S_1} is:

Given:
$I_{R_{S_1}} = ?$
$E_{R_{S_1}} = 14.158 \text{ volts}$
$R_{S_1} = 692.22 \text{ ohms}$

Solution:
$$I_{R_{S_1}} = \frac{E_{S_1}}{R_{S_1}}$$
$$I_{R_{S_1}} = \frac{14.158}{692.22}$$
$$I_{R_{S_1}} = 0.0204 \text{ amps}$$

The current through branch resistance R_{S_2} is:

Given:
$I_{R_{S_2}} = ?$
$E_{R_{S_2}} = 14.158 \text{ volts}$
$R_{S_2} = 680 \text{ ohms}$

Solution:
$$I_{R_{S_2}} = \frac{E_{R_{S_2}}}{R_{S_2}}$$
$$I_{R_{S_2}} = \frac{14.158}{680}$$
$$I_{R_{S_2}} = 0.0208 \text{ amp}$$

The voltage drop across resistors R_A and R_4 can now be determined.

Given:
$I_{R_A} = 0.0205 \text{ amp}$
$E_{R_A} = ?$
$R_A = 222.22 \text{ ohms}$

Solution:
$$I_{R_A} = \frac{E_{R_A}}{R_A}$$
$$0.0205 = \frac{E_{R_A}}{222.22}$$
$$E_{R_A} = (0.0205)(222.22)$$
$$E_{R_A} = 4.56 \text{ volts}$$

Given:
$I_{R_4} = 0.0205 \text{ amp}$
$E_{R_4} = ?$
$R_4 = 470 \text{ ohms}$

Solution:
$$I_{R_4} = \frac{E_{R_4}}{R_4}$$
$$0.0205 = \frac{E_{R_4}}{470}$$
$$E_{R_4} = (0.0205)(470)$$
$$E_{R_4} = 9.64 \text{ volts}$$

The voltage drop across resistors R_5 and R_6 is:

Given:
$I_{R_5} = 0.0208 \text{ amp}$
$E_{R_5} = ?$
$R_5 = 120 \text{ ohms}$

Solution:
$$I_{R_5} = \frac{E_{R_5}}{R_5}$$
$$0.0208 = \frac{E_{R_5}}{120}$$
$$E_{R_5} = (0.0208)(120)$$
$$E_{R_5} = 2.50 \text{ volts}$$

Given:
$I_{R_6} = 0.0208 \text{ amp}$
$E_{R_6} = ?$
$R_6 = 560 \text{ ohms}$

Solution:
$$I_{R_6} = \frac{E_{R_6}}{R_6}$$
$$0.0208 = \frac{E_{R_6}}{560}$$
$$E_{R_6} = (0.0208)(560)$$
$$E_{R_6} = 11.65 \text{ volts}$$

The current through equivalent resistance R_A splits through parallel branches with resistors R_2 and R_3. Current through each resistor has to be calculated separately.

The current through resistor R_2 is:

Given:
$I_{R_2} = ?$
$E_{R_2} = 4.56 \text{ volts}$
$R_2 = 330 \text{ ohms}$

Solution:
$$I_{R_2} = \frac{E_{R_2}}{R_2}$$
$$I_{R_2} = \frac{4.56}{330}$$
$$I_{R_2} = 0.0138 \text{ amp}$$

Given:
$I_{R_3} = ?$
$E_{R_3} = 4.56 \text{ volts}$
$R_3 = 680 \text{ ohms}$

Solution:
$$I_{R_3} = \frac{E_{R_3}}{R_3}$$
$$I_{R_3} = \frac{4.56}{680}$$
$$I_{R_3} = 0.00671 \text{ amp}$$

Now the power dissipation through each resistor can be determined. The power consumed by resistor R_1 is:

Given:

$P_{R_1} = ?$

$I_{R_1} = 0.0412$ amp

$E_{R_1} = 33.78$ volts

Solution:

$P_{R_1} = I_{R_1} E_{R_1}$

$P_{R_1} = (0.0412)(33.78)$

$P_{R_1} = 1.39$ watts

The power consumed by resistor R_2 is:

Given:

$P_{R_2} = ?$

$I_{R_2} = 0.0138$ amp

$E_{R_2} = 4.56$ volts

Solution:

$P_{R_2} = I_{R_2} E_{R_2}$

$P_{R_2} = (0.0138)(4.56)$

$P_{R_2} = 0.063$ watt or 63 mW

The power consumed by resistor R_3 is:

Given:

$P_{R_3} = ?$

$E_{R_3} = 4.56$ volts

$I_{R_3} = 0.00670$ amp

Solution:

$P_{R_3} = I_{R_3} E_{R_3}$

$P_{R_3} = (0.00670)(4.56)$

$P_{R_3} = 0.031$ watt or 31 mW

The power consumed by resistor R_4 is:

Given:

$P_{R_4} = ?$

$E_{R_4} = 9.64$ volts

$I_{R_4} = 0.0205$ amp

Solution:

$P_{R_4} = I_{R_4} E_{R_4}$

$P_{R_4} = (0.0205)(9.64)$

$P_{R_4} = 0.20$ watt or 200 mW

The power consumed by resistor R_5 is:

Given:

$P_{R_5} = ?$

$E_{R_5} = 2.50$ volts

$I_{R_5} = 0.0208$ amp

Solution:

$P_{R_5} = I_{R_5} E_{R_5}$

$P_{R_5} = (0.0208)(2.50)$

$P_{R_5} = 0.052$ watt or 52 mW

The power consumed by resistor R_6 is:

Given:

$P_{R_6} = ?$

$E_{R_6} = 11.65$ volts

$I_{R_6} = 0.0208$ amp

Solution:

$P_{R_6} = I_{R_6} E_{R_6}$

$P_{R_6} = (0.0208)(11.65)$

$P_{R_6} = 0.24$ watt or 240 mW

The total power consumption of the circuit is:

Given:

$P_T = ?$

$P_{R_1} = 1.39$ watts

$P_{R_2} = 0.063$ watt

$P_{R_3} = 0.031$ watt

$P_{R_4} = 0.20$ watt

$P_{R_5} = 0.052$ watt

$P_{R_6} = 0.24$ watt

Solution:

$P_T = P_{R_1} + P_{R_2} + P_{R_3} + P_{R_4} + P_{R_5} + P_{R_6}$

$P_T = 1.39 + 0.063 + 0.031 + 0.20 + 0.052 + 0.24$

$P_T = 1.98$ watts

The total power consumption could also be determined by the use of the power formula.

Given:

$P_T = ?$

$E_T = 48$ volts

$I_T = 0.0413$ amp

Solution:

$P_T = E_T I_T$

$P_T = (48)(0.0413)$

$P_T = 1.98$ watts

8–3 Question

1. Solve for all unknown values in Figure 8–10.

SUMMARY

■ A series circuit provides only one path for current flow.

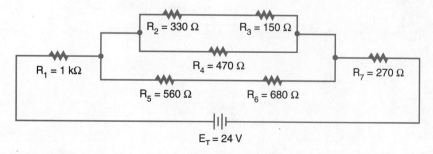

Figure 8–10

■ Formulas governing the operation of a series circuit include:

$$I_T = I_{R_1} = I_{R_2} = I_{R_3} \ldots = I_{R_n}$$

$$R_T = R_1 + R_2 + R_3 \ldots + R_n$$

$$E_T = E_{R_1} + E_{R_2} + E_{R_3} \ldots + E_{R_n}$$

$$I = E/R$$

$$P_T = P_{R_1} + P_{R_2} + P_{R_3} \ldots + P_{R_n}$$

■ A parallel circuit provides more than one path for current flow.

■ Formulas governing the operation of a parallel circuit include:

$$I_T = I_{R_1} + I_{R_2} + I_{R_3} \ldots + I_{R_n}$$

$$\frac{1}{R_T} = \frac{1}{R_1} + \frac{1}{R_2} + \frac{1}{R_3} \ldots + \frac{1}{R_n}$$

$$E_T = E_{R_1} = E_{R_2} = E_{R_3} \ldots = E_{R_n}$$

$$I = E/R$$

$$P_T = P_{R_1} + P_{R_2} + P_{R_3} \ldots + P_{R_n}$$

■ Series-parallel circuits are solved by using series formulas for the series parts of the circuit and parallel formulas for the parallel parts of the circuit.

Chapter 8 Self-Test

1. Solve for all unknown quantities in the circuits shown.

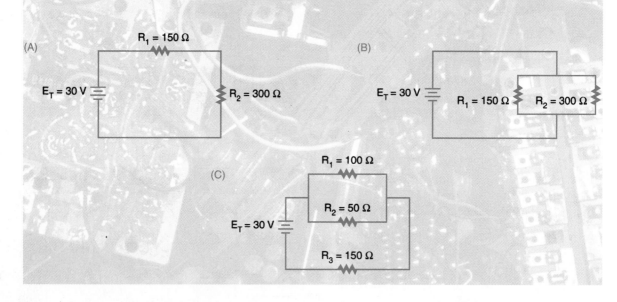

9

MAGNETISM

Objectives

After completing this chapter, the student will be able to:

- Identify three types of magnets.
- Describe the basic shapes of magnets.
- Describe the differences between permanent magnets and temporary magnets.
- Describe how the earth functions as a magnet.
- State the laws of magnetism.
- Explain magnetism based on the theory of atoms and electron spin.
- Explain magnetism based on the domain theory.
- Identify flux lines and their significance.
- Define *permeability*.
- Describe the magnetic effects of current flowing through a conductor.
- Describe the principle of an electromagnet.
- Describe how to determine the polarity of an electromagnet using the left-hand rule.
- Define *magnetic induction*.
- Define *retentivity* and *residual magnetism*.
- Define a *magnetic shield*.
- Describe how magnetism is used to generate electricity.
- State the basic law of electromagnetism.
- Describe how the left-hand rule for generators can be used to determine the polarity of induced voltage.
- Describe how AC and DC generators convert mechanical energy into electrical energy.
- Describe how a relay operates as an electromechanical switch.

■ Discuss the similarities between a doorbell and a relay.

■ Discuss the similarities between a solenoid and a relay.

■ Describe how a magnetic phonograph cartridge works.

■ Describe how a loudspeaker operates.

■ Describe how information can be stored and retrieved using magnetic recording.

■ Describe how a DC motor operates.

Electricity and magnetism are inseparable. To understand electricity means to understand the relationship that exists between magnetism and electricity.

Electric current always produces some form of magnetism, and magnetism is the most common method for generating electricity. In addition, electricity behaves in specific ways under the influence of magnetism.

This chapter looks at magnetism, electromagnetism, and the relationship between magnetism and electricity.

9–1 Magnetic Fields

The word **magnet** is derived from *magnetite,* the name of a mineral found in Magnesia, a part of Asia Minor. This mineral is a natural magnet. Another type of magnet is the **artificial magnet.** This magnet is created by rubbing a piece of soft iron with a piece of magnetite. A third type of magnet is the **electromagnet.** It is created by current flowing through a coil of wire.

Magnets come in various shapes (Figure 9–1). Among the more common shapes are the horseshoe, the bar or rectangle, and the ring.

Magnets that retain their magnetic properties are called **permanent magnets.** Magnets that retain only a small portion of their magnetic properties are called **temporary magnets.**

Magnets are made of metallic or ceramic materials. Alnico (*al*uminum, *ni*ckel, and *co*balt) and Cunife (copper [*Cu*], *ni*ckel, and Iron [*Fe*]) are two metallic alloys used for magnets.

The earth itself is a huge magnet (Figure 9–2). The earth's magnetic North and South Poles are situated close to the geographic North and South Poles. If a bar magnet is suspended, the magnet aligns in a north–south direction, with one end pointing toward the North Pole of the earth and the other toward the South Pole. This is the principle

behind the compass. It is also the reason the two ends of the magnet are called the North Pole and the South Pole.

Magnets align in a north–south direction because of laws similar to those of positive and negative charges: Unlike magnetic poles attract each other and like magnetic poles repel each other. The color code for a magnet is red for the North Pole and blue for the South Pole.

Figure 9–1 Magnets come in various sizes and shapes.

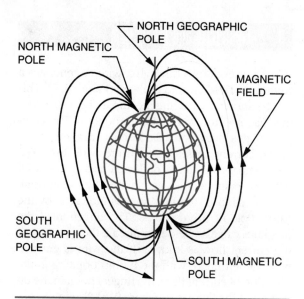

Figure 9–2 The earth's magnetic North and South Poles are situated close to the geographic North and South Poles.

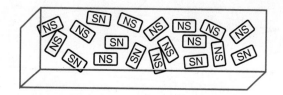

Figure 9–3 The domains in unmagnetized material are randomly arranged with no overall magnetic effect.

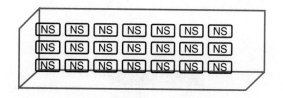

Figure 9–4 When material becomes magnetized, all the domains align in a common direction.

Magnetism, the property of the magnet, can be traced to the atom. As electrons orbit around the nucleus of the atom, they spin on their axis, like the earth as it orbits the sun. This moving electrostatic charge produces a magnetic field. The direction of the magnetic field depends on the electron's direction of spin. Iron, nickel, and cobalt are the only naturally magnetic elements. Each of these materials has two valence electrons that spin in the same direction. Electrons in other materials tend to spin in opposite directions and this cancels their magnetic characteristics.

Ferromagnetic materials are materials that respond to magnetic fields. In ferromagnetic materials, the atoms combine into *domains,* or groups of atoms arranged in the form of magnets. In an unmagnetized material, the magnetic domains are randomly arranged, and the net magnetic effect is zero (Figure 9–3). When the material becomes magnetized, the domains align in a common direction and the material becomes a magnet (Figure 9–4). If the magnetized material is divided into smaller pieces, each piece becomes a magnet with its own poles.

Evidence of this "domain theory" is that if a magnet is heated, or hit repeatedly with a hammer, it loses its magnetism (the domains are jarred back into a random arrangement). Also, if an artificial magnet is left by itself, it slowly loses its magnetism. To prevent this loss, bar magnets should be stacked on top of each other with opposite poles together; keeper bars should be placed across horseshoe magnets (Figure 9–5). Both methods maintain the magnetic field.

A *magnetic field* consists of invisible lines of force that surround a magnet. These lines of force are called *flux lines.* They can be "seen" by placing a sheet of paper over a magnet and sprinkling iron filings on the paper. When the paper is lightly tapped, the iron filings arrange themselves into a definite pattern that reflects the forces attracting them (Figure 9–6).

Flux lines have several important characteristics: They have polarity from North to South. They always form a complete loop. They do not cross each other because like polarities repel. They tend to form the smallest loop possible, because unlike poles attract and tend to pull together.

The characteristic that determines whether a substance is ferromagnetic or not is called permeability. *Permeability* is the ability of a material to accept magnetic lines of force. A material with great permeability offers less resistance to flux lines than air.

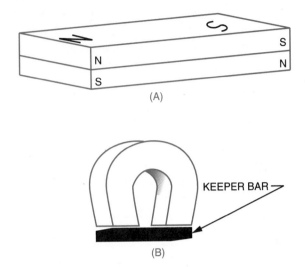

(A)

(B)

Figure 9–5 To prevent loss of magnetism, (A) bar magnets are stacked on top of each other, and (B) keeper bars are placed across horseshoe magnets.

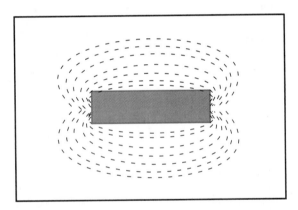

Figure 9–6 Magnetic lines of flux can be seen in patterns of iron filings.

9–1 Questions

1. What are the three types of magnets?
2. What are the basic shapes of magnets?
3. How are the ends of a magnet identified?
4. What are the two theories of magnetism?
5. What are flux lines?

9–2 Electricity and Magnetism

When current flows through a wire, it generates a **magnetic field** around the wire (Figure 9–7). This can be shown by placing a compass next to a wire that has no current flowing through it. The compass aligns itself with the earth's magnetic field. When current passes through the wire, however, the compass needle is deflected and aligns itself with the magnetic field generated by the current. The direction of the flux lines is indicated by the North Pole of the compass. The direction of the flux lines can also be determined if the direction of the current flow is known. If the wire is grasped with the left hand, with the thumb pointing in the direction of current flow, the fingers point in the direction of the flux lines (Figure 9–8). When the polarity of the voltage source is reversed, the flux lines are also reversed.

If two wires are placed next to each other with current flowing in opposite directions, they create opposing magnetic fields that repel each other (Figure 9–9). If the two wires carry current flowing in the same direction the fields combine (Figure 9–10).

A single piece of wire produces a magnetic field with no North or South Pole and little strength or practical value. If the wire is twisted into a loop, three things occur. One, the flux lines are brought together. Two, the flux lines are concentrated at

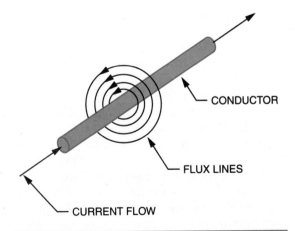

Figure 9–7 A current flowing through a conductor creates a magnetic field around the conductor.

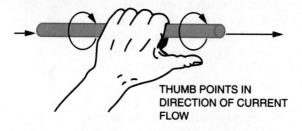

FINGERS INDICATE
DIRECTION OF FLUX LINES

THUMB POINTS IN
DIRECTION OF CURRENT
FLOW

Figure 9–8 Determining the direction of the flux lines around a conductor when the direction of the current flow is known (left hand rule for conductors).

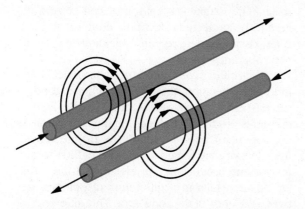

Figure 9–9 When current flows in opposite directions through two conductors placed next to each other, the resulting magnetic fields repel each other.

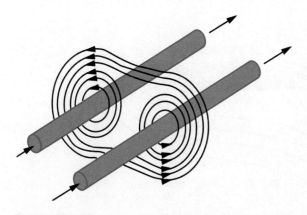

Figure 9–10 When current flows in the same direction through two conductors placed next to each other, the magnetic fields combine.

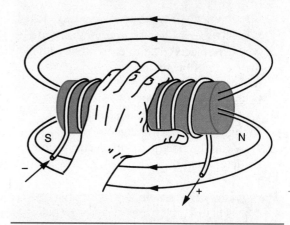

Figure 9–11 Determining the polarity of an electromagnet.

the center of the loop. Three, a North and South Pole are established. This is the principle of the electromagnet.

An electromagnet is composed of many turns of wire close together. This allows the flux lines to add together when a current flows through the wire. The more turns of wire, the more flux lines are added together. Also, the greater the current, the greater the number of flux lines generated. The strength of the magnetic field, then, is directly proportional to the number of turns in the coil and the amount of current flowing through it.

A third method of increasing the strength of the magnetic field is to insert a ferromagnetic core into the center of the coil. An iron core is typically used because it has a higher permeability (can support more flux lines) than air.

To determine the polarity of the electromagnet, grasp the coil with the left hand, with the fingers pointing in the direction of the current flow. The thumb then points in the direction of the North Pole (Figure 9–11).

9–2 Questions

1. How can a magnetic field be shown to exist when a current flows through a wire?
2. How can the direction of flux lines around a wire be determined?

3. What happens when two current-carrying wires are placed next to each other with the current flowing in:
a. The same direction?
b. Opposite directions?
4. What are three ways to increase the strength of an electromagnetic field?
5. How can the polarity of an electromagnet be determined?

9—3 Magnetic Induction

Magnetic induction is the effect a magnet has on an object without physical contact. For example, a magnet may induce a magnetic field in an iron bar (Figure 9–12). In passing through the iron bar, the magnetic lines of force cause the domains in the iron bar to align in one direction. The iron bar is now a magnet. The domains in the iron bar align themselves with their South Pole toward the North Pole of the magnet, because opposite poles attract. For the same reason, the iron bar is drawn toward the magnet. The flux lines now leave from the end of the iron bar; the iron bar is an extension of the magnet. This is an effective way to increase the length or change the shape of a magnet without physically altering it.

When the magnet and iron bar are separated, the domains in the iron bar return to their random pattern although a few domains remain in North–South alignment, giving the iron bar a weak magnetic field. This remaining magnetic field is called *residual magnetism*. The ability of a material to retain its magnetic field after the magnetizing force is removed is called *retentivity*. Soft iron has low retentivity. Alnico, an alloy made of aluminum, nickel, and cobalt, has high retentivity.

Flux lines can be bent by inserting a low-reluctance material in front of the magnetic field. Low-reluctance materials are called *magnetic shields*. An example is the material called Mu-metal. The magnetic shield is placed around the item to be protected. Electronic equipment, especially oscilloscopes, require shielding from magnetic flux lines.

Electromagnetic induction is the principle behind the generation of electricity: When a conductor passes or is passed by a magnetic field, a current is produced (induced) in the conductor. As the conductor passes through the magnetic field, free electrons are forced to one end of the conductor, leaving a deficiency of electrons at the other end. This results in a difference of potential between the ends of the conductor. This difference of potential exists only when the conductor is passing through a magnetic field. When the conductor is

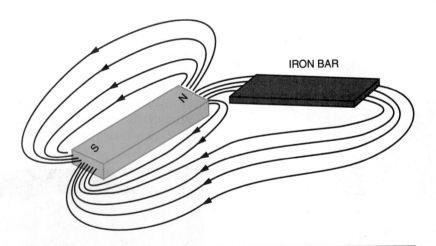

Figure 9–12 Placing an iron bar in a magnetic field extends the magnetic field and magnetizes the iron bar.

removed from the magnetic field, the free electrons return to their parent atoms.

For electromagnetic induction to occur, either the conductor must move or the magnetic field must move. The voltage produced in the conductor is called *induced voltage*. The amount of induced voltage is determined by the strength of the magnetic field, the speed with which the conductor moves through the magnetic field, the angle at which the conductor cuts the magnetic field, and the length of the conductor.

The stronger the magnetic field, the greater the induced voltage. The faster the conductor moves through the field, the greater the induced voltage. Motion between the conductor and the magnetic field can be produced by moving the conductor, the magnetic field, or both. The maximum voltage is induced when the conductor moves at right angles to the field. Angles less than 90 degrees induce less voltage. If a conductor is moved parallel to the flux lines, no voltage is induced. The longer the conductor, the greater the induced voltage.

Faraday's law, the basic law of electromagnetism, states: The induced voltage in a conductor is directly proportional to the rate at which the conductor cuts the magnetic lines of force.

The polarity of an induced voltage can be determined by the following left-hand rule for generators: The thumb, index finger, and middle finger are held at right angles to each other (Figure 9–13).

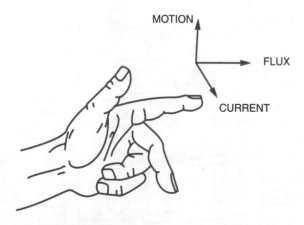

MOTION

FLUX

CURRENT

Figure 9–13 The left-hand rule for generators can be used to determine the direction of the induced current flow in a generator.

The thumb points in the same direction as the motion of the conductor, the index finger points in the direction of the flux lines (North to South), and the middle finger points toward the negative end of the conductor, the direction of the current flow.

9–3 Questions

1. How can the length of a magnet be increased without physically altering the magnet?
2. What is residual magnetism?
3. How does a magnetic shield work?
4. How is electromagnetic induction used to generate electricity?

9–4 Magnetic and Electromagnetic Applications

An alternating current (AC) generator converts mechanical energy to electrical energy by utilizing the principle of electromagnetic induction. The mechanical energy is needed to produce motion between the magnetic field and the conductor.

Figure 9–14 shows a loop of wire (conductor) being rotated (moved) in a magnetic field. The loop has a light and dark side for ease of explanation. At the point shown in part A, the dark half is parallel to the lines of force, as is the light half. No voltage is induced. As the loop is rotated toward the position shown in part B, the lines of force are cut and a voltage is induced. The induced voltage is greatest at this position, when the motion is at right angles to the magnetic field. As the loop is rotated to position C, fewer lines of force are cut and the induced voltage decreases from the maximum value to zero volts. At this point the loop has rotated 180 degrees, or half a circle.

The direction of current flow can be determined by applying the left-hand rule for generators. The direction of current flow at position B is shown by the arrow. When the loop is rotated to position D, the action reverses. As the dark half moves up through the magnetic lines of force and the light half moves down, applying the left-hand rule for generators shows that the induced voltage

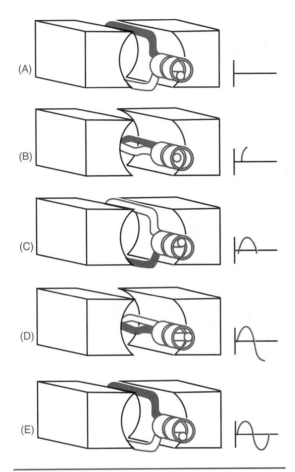

Figure 9–14 Induced voltage in an AC generator.

changes polarities. The voltage reaches a peak at position D and decreases until the loop reaches the original position. The induced voltage has completed one cycle of two alternations.

The rotating loop is called an *armature* and the source of the electromagnetic field is called the *field*. The armature can have any number of loops. The term "armature" refers to the part that rotates in the magnetic field, regardless of whether it consists of one loop or multiple loops. The *frequency* of the alternating current or voltage is the number of complete cycles completed per second. The speed of rotation determines the frequency. An AC generator is often called an *alternator* because it produces alternating current.

A direct current (DC) generator also converts mechanical energy into electrical energy. It functions like an AC generator with the exception that it converts the AC voltage to DC voltage. It does this with a device called a *commutator*, as shown in Figure 9–15. The output is taken from the commutator, a split ring. When the loop is rotated from position A to position B, a voltage is induced. The induced voltage is greatest as the motion is at right angles to the magnetic field. As the loop is rotated to position C, the induced voltage decreases from the maximum value to zero. As the loop continues to rotate to position D, a voltage is induced, but the commutator reverses the output polarity so that it remains the same as before. The loop then returns to the original position, E. The voltage generated

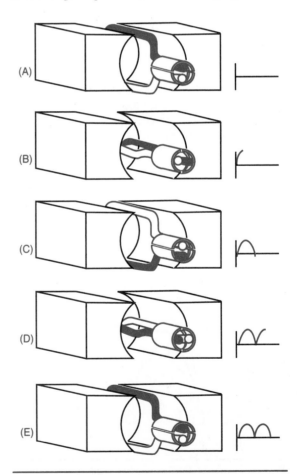

Figure 9–15 Induced voltage in a DC generator.

Figure 9–16 Examples of various types of relays. *(Courtesy of Struthers-Dunn, Inc.)*

from the commutator pulsates, but in one direction only, varying twice during each revolution between zero and maximum.

A **relay** is an electromechanical switch that opens and closes with an electromagnetic coil (Figure 9–16). As a current flows through the coil, it generates a magnetic field that attracts the plunger, pulling it down. As the plunger is pulled down, it closes the switch contacts. When the current through the coil stops, a spring pulls the armature back to its original position and opens the switch.

A relay is used where it is desirable to have one circuit control another circuit. It electrically isolates the two circuits. A small voltage or current can control a large voltage or current. A relay can also be used to control several circuits some distance away.

A doorbell is an application of the relay. The striker to ring the bell is attached to the plunger. As the doorbell is pressed, the relay coil is energized, pulling down the plunger and striking the bell. As the plunger moves down, it opens the circuit, deenergizing the relay. The plunger is pulled back by the spring closing the switch contacts, energizing the circuit again, and the cycle repeats until the button is released.

A **solenoid** is similar to a relay (Figure 9–17). A coil, when energized, pulls a plunger that does some mechanical work. This is used on some door chimes where the plunger strikes a metal bar. It is also used on automotive starters. The plunger pulls

the starter gear in to engage the flywheel to start the engine.

Phonograph pickups use the electromagnetic principle. A magnetic field is produced by a permanent magnet that is attached to a stylus (needle). The permanent magnet is placed inside a small coil. As the stylus is tracked through the groove of a record, it moves up and down and from side to side in response to the audio signal recorded. The movement of the magnet in the coil induces a small voltage that varies at the audio signal response. The induced voltage is then amplified and used to drive a loudspeaker, reproducing the audio signal.

Loudspeakers are used for all types of audio amplification. Most speakers today are constructed of a moving coil around a permanent magnet. The magnet produces a stationary magnetic field. As current is passed through the coil, it produces a magnetic field that varies at the rate of the audio signal. The varying magnetic field of the coil is attracted and repelled by the magnetic field of the permanent magnet. The coil is attached to a cone that moves back and forth in response to the audio signal. The cone moving back and forth moves the air, reproducing the audio sound.

Magnetic recording is used for cassette recorders, video recorders, reel-to-reel recorders, floppy disk drives, and hard disk drives. All these mediums use the same electromagnetic principle to store information. A signal is stored on the tape

A. A magnetic field surrounds single conductor when electric current flows through it

B. An enlarged magnetic field surrounds coil of wire with electric current flowing through it

C. Addition of metal frame further increases magnetic force

D. Seated plunger provides all metallic path for magnetic field of maximum strength

E. Position of armature under load before being energized. Current flowing through coil sets up magnetic field that pulls plunger into seated position

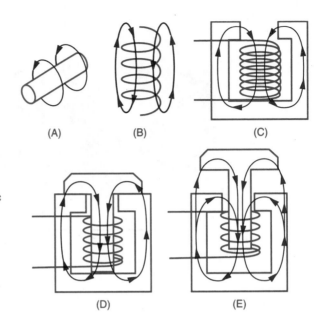

(A)　　(B)　　(C)

(D)　　(E)

Figure 9–17 Example of a solenoid. *(Courtesy of Dormeyer Industries, Division of A. F. Dormeyer Manufacturing Co., Inc.)*

or disk with a record head, to be read back later with a playback head. In some equipment the record and playback head are combined in one package or they may be the same head. The record and playback head are a coil of wire with a ferromagnetic core. In a tiny gap between the ends of the core is a magnetic field. As the storage medium, a piece of material covered with iron oxide, is pulled across the record head, the magnetic field penetrates the type, magnetizing it. The information is written on it in a magnetic pattern that corresponds to the original information. To play back or read the information, the medium is moved past the gap in the playback head. The changing magnetic field induces a small voltage into the coil winding. When this voltage is amplified, the original information is reproduced.

The operation of a DC motor depends on the principle that a current-carrying conductor, placed in and at right angles to a magnetic field, tends to move at right angles to the direction of the field. Figure 9–18A shows a magnetic field extending between a north pole and south pole. Figure 9–18B shows the

magnetic field that exists around a current-carrying conductor. The plus sign implies that the current flows inward. The direction of the flux lines can be determined using the left-hand rule. Figure 9–18C shows the conductor placed in the magnetic field. Note that both fields become distorted. Above the wire, the field is weakened, and the conductor tries to move upward. The amount of force exerted upward depends on the strength of the field between the poles and the amount of current flowing through the conductor. If the current through the conductor is reversed (Figure 9–18D), the direction of the magnetic flux around the conductor is reversed. If this occurs, the field below the conductor is weakened and the conductor tends to move downward.

A method of determining the direction of a current-carrying conductor in a magnetic field uses the right-hand motor rule: When the thumb, index finger, and middle finger of the right hand are extended at right angles to each other, the middle finger points in the direction of current flow in the conductor; the index finger indicates the magnetic field from the north pole to the south pole; the

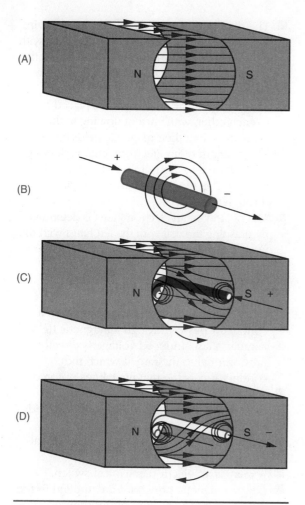

(A)

(B)

(C)

(D)

Figure 9–18 Operation of a DC motor.

thumb points in the direction of the motion of the conductor.

The force acting on a current-carrying conductor in a magnetic field depends on the strength of the magnetic field, the length of the conductor, and the amount of current flowing in the conductor.

If a loop of wire, free to rotate horizontally, is placed between the two poles of a magnet, the loop spins as the poles repel each other. The current flows in one direction on one side of the loop and in the other direction on the other side of the

loop. One side of the loop moves down and the other side of the loop moves upward. The loop rotates in a counterclockwise direction around its axis. A commutator reverses the direction of current flow in the loop every time it reaches the top or zero torque position. This is how a DC motor rotates. The loop or armature rotates in a magnetic field. Permanent magnets or electromagnets may produce this field. The commutator reverses the direction of current through the armature. Note the resemblance between a DC motor and a DC generator.

The basic meter movement uses the principle of the DC motor. It consists of a stationary permanent magnet and a moveable coil. When the current flows through the coil, the resulting magnetic field reacts with the field of the permanent magnet and causes the coil to rotate. The greater the current flow through the coil, the stronger the magnetic field produced. The stronger the magnetic field produced, the greater the rotation of the coil. To determine the amount of current flow, a pointer is attached to the rotating coil. As the coil turns, the pointer also turns. The pointer moves across a graduated scale and indicates the amount of current flow. This type of meter movement is used for analog meters such as voltmeters, ammeters, and ohmmeters.

A conductor carrying current can be deflected (moved) by a magnetic field. It is not the conductor that is deflected, but the electrons traveling in the conductor. The electrons are restricted to the conductor, so the conductor also moves. Electrons can travel through other mediums. In the case of television picture tubes, electrons travel through a vacuum to strike a phosphor screen where they emit light. The electrons are produced by an electron gun. By varying the electron beam over the surface of the picture screen, a picture can be created. To move the beam back and forth across the screen, two magnetic fields deflect the beam. One magnetic field moves the beam up and down the screen and the other magnetic field moves the beam from side to side. This method is used in television, radar, oscilloscopes, computer terminals and other applications where a picture is desired on a screen.

9—4 Questions

1. What are the differences between an AC and a DC generator?
2. Why are relays important?
3. How does a loudspeaker produce sound?
4. What is the principle behind DC motors and meter movements?
5. How does an electromagnetic field produce an image on a screen?

SUMMARY

■ The word *magnet* is derived from the name of magnetite, a mineral that is a natural magnet.
■ A magnet can be created by rubbing a piece of soft iron with another magnet.
■ An electromagnet is created by current flowing in a coil of wire.
■ Horseshoe, bar, rectangular, and ring are the most common shapes of magnets.
■ Unlike poles attract and like poles repel.
■ One theory of magnetism is based on the spin of electrons as they orbit around an atom.
■ Another theory of magnetism is based on the alignment of domains.
■ Flux lines are invisible lines of force surrounding a magnet.
■ Flux lines form the smallest loop possible.
■ Permeability is the ability of a material to accept magnetic lines of force.
■ A magnetic field surrounds a wire when current flows through it.

■ The direction of the flux lines around a wire can be determined by grasping the wire with the left hand, with the thumb pointing in the direction of current flow. The fingers then point in the direction of the flux lines.
■ If two current-carrying wires are placed next to each other, with current flowing in the same direction, their magnetic fields combine.
■ The strength of an electromagnet is directly proportional to the number of turns in the coil and the amount of current flowing through the coil.
■ The polarity of an electromagnet is determined by grasping the coil with the left hand with the fingers in the direction of current flow. The thumb then points toward the North Pole.
■ *Retentivity* is the ability of a material to retain a magnetic field.
■ Electromagnetic induction occurs when a conductor passes through a magnetic field.
■ Faraday's law: Induced voltage is directly proportional to the rate at which the conductor cuts the magnetic lines of force.
■ The left-hand rule for generators can be used to determine the direction of induced voltage.
■ AC and DC generators convert mechanical energy into electrical energy.
■ A relay is an electromechanical switch.
■ Electromagnetic principles are applied in the design and manufacture of doorbells, solenoids, phonograph pickups, loudspeakers, and magnetic recordings.
■ DC motors and meters use the same principles.
■ Electron beams can be deflected by an electromagnetic field to produce images on television, radar, and oscilloscope screens.

Chapter 9 Self-Test

1. How can the domain theory of magnetism be verified?
2. What three methods can be used to increase the strength of an electromagnet?
3. Explain how a DC generator operates through one cycle.

10

INDUCTANCE

Objectives

After completing this chapter, the student will be able to:

■ Explain the principles of inductance.

■ Identify the basic units of inductance.

■ Identify different types of inductors.

■ Determine the total inductance in series and parallel circuits.

■ Explain L/R time constants and how they relate to inductance.

When a current flows through a conductor, a magnetic field builds up around the conductor. This field contains energy and is the foundation for inductance.

This chapter examines inductance and its application to DC circuits. Inductance is covered in more detail in Chapter 16.

10—1 Inductance

Inductance is the characteristic of an electrical conductor that opposes a change in current flow. An inductor is a device that stores energy in a magnetic field.

Inductance exhibits the same effect on current in an electric circuit as inertia does on velocity of a mechanical object. It takes more work to start a load moving than it does to keep it moving because the load possesses the property of inertia. Inertia is the characteristic of mass that opposes a change in velocity. Once current is moving through a conductor, inductance helps to keep it moving. The effects of inductance are sometimes desirable and other times undesirable.

The basic principle behind inductance that states, *when a current flows through a conductor, it generates a magnetic field around the conductor,* was covered in Chapter 9. As the magnetic flux lines build up, they create an opposition to the flow of current.

When the current changes direction or stops, or the magnetic field changes, an electromotive force (emf) is induced back into the conductor through the collapsing of the magnetic field. The opposition to the changes in current flow is identified as counter electromotive force (counter emf). This effect is summarized by Lenz's law—an induced emf in any circuit is always in a direction to oppose the effect that produced it. The amount of counter emf is in proportion to the rate of change. The faster the rate of change, the greater the counter emf.

All conductors have some inductance. The amount of inductance depends on the conductor and the shape of it. Straight wire has small amounts of inductance whereas coils of wire have much more inductance.

The unit by which inductance is measured is the henry (H), named for Joseph Henry (1797–1878), an American physicist. A henry is the amount of inductance required to induce an EMF (electromotive force) of 1 volt when the current in a conductor changes at the rate of 1 ampere per second. The henry is a large unit; the millihenry (mH) and microhenry (μH) are more commonly used. The symbol for inductance is L.

10—1 Questions

1. Define *inductance*.
2. What is the unit for measuring inductance?
3. Define a *henry*.
4. What letter is used to represent inductance?

10—2 Inductors

Inductors are devices designed to have a specific **inductance.** They consist of a conductor coiled around a core and are classified by the type of core material, magnetic or nonmagnetic. Figure 10–1 shows the symbol used for inductors.

Inductors can also be fixed or variable. Figure 10–2 shows the symbol for a variable inductor. Variable inductors are created with adjustable core material. Figure 10–3 shows several types of inductors used for adjusting the core material. Maximum inductance occurs when the core material is in line with the coil of the wire.

Air-core inductors, or inductors without core material, are used for up to 5 millihenries of inductance. They are wrapped on a ceramic or phenolic core (Figure 10–4).

Ferrite and powdered iron cores are used for up to 200 millihenries. The symbol used for an iron-core inductor is shown in Figure 10–5.

Toroid cores are donut-shaped and offer a high inductance for a small size (Figure 10–6). The magnetic field is contained within the core.

Shielded inductors have a shield made of magnetic material to protect them from the influence of other magnetic fields (Figure 10–7).

Figure 10–1 Schematic symbol for an inductor.

Figure 10–2 Schematic symbol for a variable inductor.

Figure 10–3 Several types of inductors used to vary the inductance.

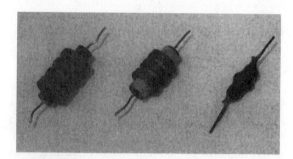

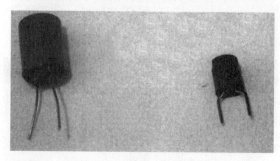

Figure 10–4 Types of air-core inductors. *(From Shimizu,* Electronic Fabrication, *by Delmar Publishers).*

Figure 10–5 Schematic symbol for an iron-core inductor.

Figure 10–6 Toroid-core inductor.

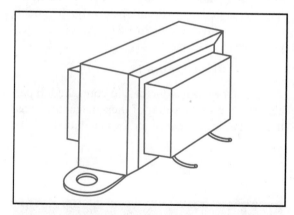

Figure 10–7 Shielded inductor.

Laminated iron-core inductors are used for all large inductors (Figure 10–8). These inductors vary from 0.1 to 100 henries, the inductance depending on the amount of current flowing through the inductor. These inductors are sometimes referred to as *chokes.* They are used in the filtering circuits of power supplies to remove AC components from the DC output. They will be discussed further later on.

Inductors typically have tolerances of ±10%, but tolerances of less than 1% are available. Inductors, like resistors, can be connected in series, parallel, or series-parallel combinations. The total

Figure 10–8 Laminated iron-core inductor.

inductance of several inductors connected in series (separated to prevent their magnetic fields from interacting) is equal to the sum of the individual inductances.

$$L_T = L_1 + L_2 + L_3 \ldots + L_n$$

If two or more inductors are connected in parallel (with no interaction of their magnetic fields) the total inductance is found by using the formula:

$$\frac{1}{L_T} = \frac{1}{L_1} + \frac{1}{L_2} + \frac{1}{L_3} \ldots + \frac{1}{L_n}$$

10–2 Questions

1. What are inductors?
2. Draw the symbols used to represent fixed and variable inductors.
3. What is another name for a laminated iron-core inductor?
4. What are the formulas for determining total inductance in:
   a. Series circuits?
   b. Parallel circuits?
5. What is the total inductance of a circuit with three inductors, 10 H, 3.5 H, and 6 H, connected in parallel?

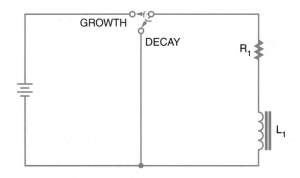

Figure 10–9 Circuit used to determine L/R time constant.

10–3 L/R Time Constants

A **time constant** is the time required for current through a conductor to increase to 63.2% or decrease to 36.8% of the maximum current. An RL circuit is shown in Figure 10–9. *L/R* is the symbol used for the time constant of an RL circuit. This can be expressed as:

$$t = \frac{L}{R}$$

where: t = time in seconds
        L = inductance in henries
        R = resistance in ohms

Figure 10–10 charts the growth and decay (or increase and decrease) of a magnetic field, in terms of time constants. It takes five time constants to fully transfer all the energy into the magnetic field, or to build up the maximum magnetic field. It takes five full-time constants to completely collapse the magnetic field.

10–3 Questions

1. What is a time constant for an inductor?
2. How is a time constant determined?
3. How many time constants are required to fully build up a magnetic field for an inductor?

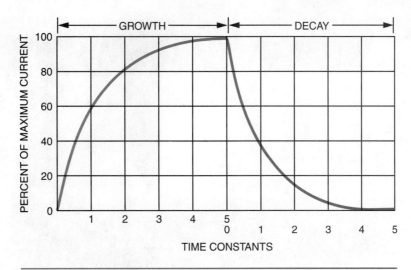

Figure 10–10 Time constants required to build up or collapse the magnetic field in an inductor.

4. How many time constants are required to fully collapse a magnetic field for an inductor?

5. How long does it take to fully build up a magnetic field for a 0.1-henry inductor in series with a 100,000-ohm resistor?

SUMMARY

■ Inductance is the ability to store energy in a magnetic field.

■ The unit for measuring inductance is the henry (H).

■ The letter *L* represents inductance.

■ Inductors are devices designed to have specific inductances.

■ The symbol for fixed inductance is:

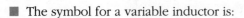

■ The symbol for a variable inductor is:

■ Types of inductors include: air core, ferrite or powdered iron core, toroid core, shielded, and laminated iron core.

■ The total inductance for inductors connected in series is calculated by the formula:

$$L_T = L_1 + L_2 + L_3 \ldots + L_n$$

■ The total inductance for inductors connected in parallel is:

$$\frac{1}{L_T} = \frac{1}{L_1} + \frac{1}{L_2} + \frac{1}{L_3} \ldots + \frac{1}{L_n}$$

■ A time constant is the time required for current to increase to 63.2% or decrease to 36.8% of the maximum current.

■ A time constant can be determined by the formula:

$$t = \frac{L}{R}$$

■ It takes five time constants to fully build up or collapse the magnetic field of an inductor.

Chapter 10 Self-Test

1. How can the magnetic field be increased for a particular inductance?
2. What is the total inductance for the circuit shown?

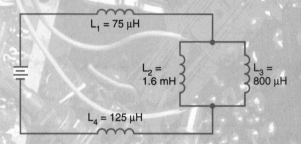

3. A 500-mH inductor and a 10-kilohm resistor are connected in series to a 25-volt source. What will be the voltage across the inductor 100 microseconds after energizing the circuit?

C h a p t e r

11

CAPACITANCE

Objectives

After completing this chapter, the student will be able to:

- Explain the principles of capacitance.
- Identify the basic units of capacitance.
- Identify different types of capacitors.
- Determine total capacitance in series and parallel circuits.
- Explain RC time constants and how they relate to capacitance.

Capacitance allows the storage of energy in an electrostatic field. Capacitance is present whenever two conductors are separated by an insulator.

This chapter focuses on capacitance and its application to DC circuits. Capacitance is covered in more detail in Chapter 15.

11–1 Capacitance

Capacitance is the ability of a device to store electrical energy in an electrostatic field. A **capacitor** is a device that possesses a specific amount of capacitance. A capacitor is made of two conductors separated by an insulator (Figure 11–1). The conductors are called *plates* and the insulator is called a *dielectric*. Figure 11–2 shows the symbols for capacitors.

When a voltage source is connected to a capacitor, a current flows until the capacitor is charged. The capacitor is charged with an excess of electrons on one plate (negative charge) and a deficiency of electrons on the other plate (positive charge). The dielectric prevents electrons from moving between the plates. Once the capacitor is charged, all current stops. The capacitor's voltage is equal to the voltage of the voltage source.

A charged capacitor can be removed from the voltage source and used as an energy source. However, as energy is removed from the capacitor, the voltage diminishes rapidly. In a DC circuit, a capacitor acts as an open circuit after its initial charge. An *open circuit* is a circuit with an infinite resistance.

Caution: Because a capacitor can be removed from a voltage source and hold the potential of the voltage source indefinitely, treat all capacitors as though they were charged. Never touch both leads of a capacitor with your hand until you have discharged it by shorting both leads together. A capacitor in a circuit can hold a potential indefinitely if it does not have a discharge path.

The amount of energy stored in a capacitor is in proportion to the size of the capacitor. The capacitors used in classrooms are usually small and deliver only a small shock if discharged through the body. If

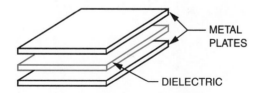

Figure 11–1 A capacitor consists of two plates (conductors) separated by a dielectric (insulator or nonconductor).

Figure 11–2 Schematic symbols for capacitors.

a capacitor is large and charged with a high voltage, however, the shock can be fatal. Charged capacitors should be treated like any other voltage source.

The basic unit of capacitance is the **farad** (F). A *farad* is the amount of capacitance that can store 1 coulomb (C) of charge when the capacitor is charged to 1 volt. The farad is too large for ordinary use, so the microfarad (μF) and picofarad (pF) are used. The letter C stands for capacitance.

$$1 \text{ microfarad} = 0.000,001 \text{ or } \frac{1}{1,000,000} \text{farad}$$

$$1 \text{ picofarad} = 0.000,000,000,001 \text{ or}$$

$$\frac{1}{1,000,000,000,000} \text{farad}$$

11–1 Questions

1. What is capacitance?
2. Draw the symbol for capacitance.
3. What precautions should be observed when handling capacitors?
4. What is the unit for measuring capacitance?
5. What units are normally associated with capacitors?

11–2 Capacitors

Four factors affect the capacitance of a capacitor. They are:

1. Area of the plate
2. Distance between the plates
3. Type of dielectric material
4. Temperature

A capacitor is either fixed or variable. A **fixed capacitor** has a definite value that cannot be changed. A **variable capacitor** is one whose capacitance can be changed either by varying the space between plates (trimmer capacitor) or by varying the amount of meshing between two sets of plates (tuning capacitor).

Capacitance is directly proportional to the area of the plate. For example, doubling the plate area doubles the capacitance if all other factors remain the same.

Capacitance is inversely proportional to the distance between the plates. In other words, as the plates move apart, the strength of the electric field between the plates decreases.

The ability of a capacitor to store electrical energy depends on the electrostatic field between the plates and the distortion of the orbits of the electrons in the dielectric material. The degree of distortion depends on the nature of dielectric material and is indicated by its dielectric constant. A *dielectric constant* is a measure of the effectiveness of a material as an insulator. The constant compares the material's ability to distort and store energy in an electric field to the ability of air, which has a dielectric constant of 1. Paper has a dielectric constant of 2 to 3; mica has a dielectric constant of 5 to 6; and titanium has a dielectric constant of 90 to 170.

The temperature of a capacitor is the least significant of the four factors. It need not be considered for most general-purpose applications.

Capacitors come in many types and styles to meet the needs of the electronics industry. *Electrolytic capacitors* offer large capacitance for small size and weight (Figure 11–3). Electrolytic capacitors consist of two metal foils separated by fine gauze or other absorbent material that is saturated with a chemical paste called an electrolyte. The

Figure 11–3 Electrolytic capacitors.

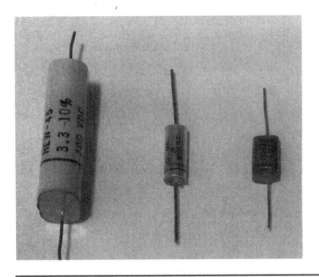

Figure 11–4 Paper and plastic capacitors. *(From Shimizu,* Electronic Fabrication, *by Delmar Publishers.)*

electrolyte is a good conductor and serves as part of the negative plate. The dielectric is formed by oxidation of the positive plate. The oxidized layer is thin and a good insulator. An electrolytic capacitor is polarized, having a positive and negative lead. Polarity must be observed when connecting an electrolytic capacitor in a circuit.

Paper and plastic capacitors are constructed by a rolled foil technique (Figure 11–4). A paper dielectric has less resistance than a plastic film dielectric, and plastic film is now being used more as a result. The plastic film allows a metallized film to be deposited directly on it. This reduces the distance between plates and produces a smaller capacitor.

The ceramic disk capacitor is popular because it is inexpensive to produce (Figure 11–5). It is used for capacitors of 0.1 microfarad and smaller. The ceramic material is the dielectric. This is a tough, reliable, general-purpose capacitor.

Variable capacitors also come in all sizes and shapes (Figure 11–6). Types include padders, trimmers, and tuning capacitors. Padder and trimmer capacitors must be adjusted by a technician. Tuning capacitors can be adjusted by the user.

Like resistors and inductors, capacitors can be connected in series, parallel, and series-parallel

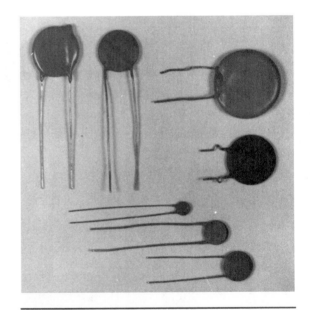

Figure 11–5 Ceramic disk capacitors.

combinations. Placing capacitors in series effectively increases the thickness of the dielectric. This decreases the total capacitance, because capacitance is inversely proportional to the distance

Figure 11–6 Variable capacitors. *(Courtesy of E. F. Johnson Company.)*

between the plates. The total capacitance of capacitors in series is calculated like the total resistance of parallel resistors:

$$\frac{1}{C_T} = \frac{1}{C_1} + \frac{1}{C_2} + \frac{1}{C_3} \ldots + \frac{1}{C_n}$$

When capacitors of different values are connected in series, the smaller capacitors charge up to the highest voltage.

Placing capacitors in parallel effectively adds to the plate area. This makes the total capacitance equal to the sum of the individual capacitances:

$$C_T = C_1 + C_2 + C_3 \ldots + C_n$$

Capacitors in parallel all charge to the same voltage.

11–2 Questions

1. What four factors affect capacitance?
2. What are the advantages of electrolytic capacitors?
3. What are other names for variable capacitors?
4. What is the formula for total capacitance in series circuits?
5. What is the formula for total capacitance in parallel circuits?

11–3 RC Time Constants

An RC time constant reflects the relationship between time, resistance, and capacitance. An RC circuit is shown in Figure 11–7. The time it takes a capacitor to charge and discharge is directly proportional to the amount of the resistance and capacitance. The time constant reflects the time required for a capacitor to charge up to 63.2% of the applied voltage or to discharge down to 63.2%. The time constant is expressed as:

$$t = RC$$

where: t = time in seconds
R = resistance in ohms
C = capacitance in farads

EXAMPLE What is the time constant of a 1-microfarad capacitor and a 1-megohm resistor?

Given:	*Solution:*
$C = 1\ \mu F$	$t = RC$
$R = 1\ M\Omega$	$t = (1{,}000{,}000)(0.000001)$
$t = ?$	$t = 1$ second

The time constant is not the time required to charge or discharge a capacitor fully. Figure 11–8 shows the time constants needed to charge and discharge a capacitor. Note that it takes approximately five time constants to fully charge or discharge a capacitor.

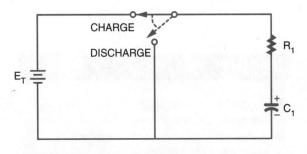

Figure 11–7 Circuit used to determine RC time constant.

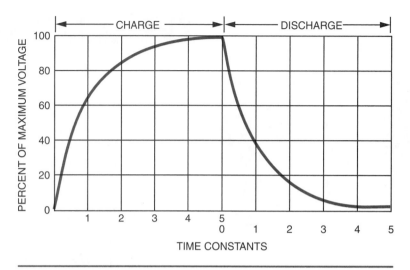

Figure 11–8 Chart of time constants required to charge and discharge a capacitor.

11—3 Questions

1. What is a time constant for a capacitor?
2. How is the time constant determined for a capacitor?
3. How many time constants are required to fully charge or discharge a capacitor?
4. A 1-microfarad and a 0.1-microfarad capacitor are connected in series. What is the total capacitance for the circuit?
5. A 0.015-microfarad capacitor is charged to 25 volts. What is the voltage 25 milliseconds after placing a 2-megohm resistor across its terminals?

SUMMARY

- Capacitance is the ability to store electrical energy in an electrostatic field.
- A capacitor consists of two conductors separated by an insulator.
- The symbol for a fixed capacitor is:

- The symbol for a variable capacitor is:

- The unit of capacitance is the farad (F).
- Because the farad is large, microfarads (μF) and picofarads (pF) are more often used.
- The letter C represents capacitance.
- Capacitance is affected by:
  1. Area of the capacitor plates
  2. Distance between the plates
  3. Types of dielectric materials
  4. Temperature
- Capacitor types include: electrolytic, paper, plastic, ceramic, and variable.
- The formula for total capacitance in a series circuit is:

$$\frac{1}{C_T} = \frac{1}{C_1} + \frac{1}{C_2} + \frac{1}{C_3} \ldots + \frac{1}{C_n}$$

- The formula for total capacitance in a parallel circuit is:

$$C_T = C_1 + C_2 + C_3 \ldots + C_n$$

- The formula for the RC circuit time constant is:

$$t = RC$$

- It takes five time constants to fully charge and discharge a capacitor.

Chapter 11 Self-Test

1. Where is the charge stored in a capacitor?
2. Four capacitors are connected in series, 1.5 μF, 0.05 μF, 2000 pF, and 25 pF. What is the total capacitance of the circuit?
3. Four capacitors are connected in parallel, 1.5 μF, 0.05 μF, 2000 pF, and 25 pF. What is the total capacitance of the circuit?

2

AC Circuits

12

ALTERNATING CURRENT

Objectives

After completing this chapter, the student will be able to:

- Describe how an AC voltage is produced with an AC generator.
- Define *alternation, cycle, hertz, sine wave, period,* and *frequency.*
- Identify the parts of an AC generator.
- Define *peak, peak-to-peak, effective,* and *rms.*
- Explain the relationship between time and frequency.
- Identify and describe three basic nonsinusoidal waveforms.
- Describe how nonsinusoidal waveforms consist of the fundamental frequency and harmonics.

Alternating current (AC) is the type of electricity most widely used today. Unlike direct current (DC), which flows in only one direction, alternating current changes direction periodically. Alternating current flows in one direction and then reverses and flows in the opposite direction.

Alternating current is easier to generate and transmit than direct current. Alternating-current generators are simpler, can be constructed in larger sizes, and are more economical to operate. An AC voltage can be increased or decreased, using a transformer, with very little power loss. Alternating current can also be converted to direct current easily. Alternating current can be used to transfer information from one point to another over transmission lines and can also be converted into electromagnetic waves and transmitted and received through an antenna system.

This chapter examines why alternating current is important, how alternating current is produced, and the important electrical characteristics of alternating current.

12–1 Generating Alternating Current

An alternating-current generator converts mechanical energy into electrical energy. The AC generator is capable of producing an alternating voltage by utilizing the principles of electromagnetic induction. Electromagnetic induction is the process of inducing a voltage in a conductor by passing it through a magnetic field.

As described in Chapter 9, the left-hand rule for generators can be used to determine the direction of current flow in a conductor that is being passed through a magnetic field: When the thumb is pointed in the direction of the conductor movement, the index finger (extended at right angles to the thumb) indicates the direction of the magnetic lines of flux from north to south, and the middle finger (extended at a right angle to the index finger) indicates the direction of current flow in the conductor. Maximum voltage is induced when the conductor is moved perpendicular to the lines of flux. When the conductor is moved parallel to the lines of flux, no voltage is induced.

Figure 12–1 shows a wire loop being rotated through a magnetic field. At position A, the loop is parallel to the lines of force. As previously stated, no voltage is induced when the conductor is moved parallel to the lines of force. As the loop is rotated to position B, it passes through more lines of force, and the maximum voltage is induced when the loop is at right angles to the lines of force. As the loop continues to rotate to position C, fewer lines of force are cut and the induced voltage decreases. Movement from position A to position C represents a rotation of 180 degrees. Rotating the loop to position D, results in a reversal of current flow. Again, the maximum voltage is induced when the loop is at right angles to the lines of force. As the loop returns to its original position at E, the induced voltage returns to zero.

Each time an AC generator moves through one complete revolution, it is said to complete one *cycle*. Its output voltage is referred to as one cycle of output voltage. Similarly, the generator produces one cycle of output current in a complete circuit. The two halves of a cycle are referred to as **alter-**

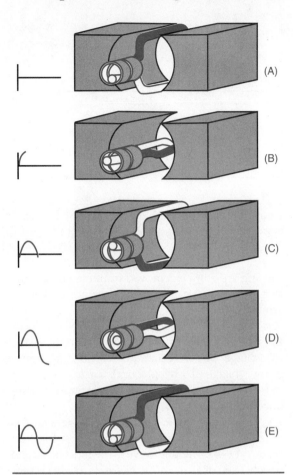

Figure 12–1 AC generator inducing a voltage output.

nations (Figure 12–2). Each alternation is produced by a change in the polarity of the voltage. The voltage exhibits one polarity during half the cycle (one alternation) and the opposite polarity during half the cycle (second alternation). Generally, one alternation is referred to as the positive alternation and the other as the negative alternation. The positive alternation produces output with a positive polarity; the negative alternation produces output with a negative polarity. Two complete alternations of voltage with no reference to time make up a **cycle.** One cycle per second is defined as a **hertz** (Hz).

The rotating loop of wire is referred to as the armature. The AC voltage that is induced in the

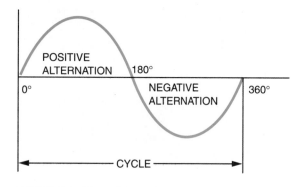

Figure 12–2 Each cycle consists of a positive and a negative alternation.

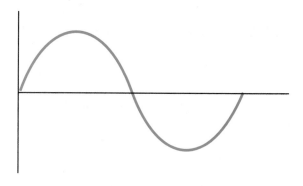

Figure 12–4 The sinusoidal waveform, the most basic of the AC waveforms.

rotating armature is removed from the ends of the loop by a sliding contact located at each end of the armature (Figure 12–3). Two metal rings, called slip rings, are attached to each end of the loop. Brushes slide against the slip rings to remove the AC voltage. The AC generator described generates a low voltage. To be practical, an AC generator must be made of many loops to increase the induced voltage.

The waveform produced by an AC generator is called a *sinusoidal waveform* or *sine wave* for short (Figure 12–4). The sine wave is the most basic and widely used of all the AC waveforms. It can be produced by both mechanical and electronic methods. The sine wave is identical to the trigonometric sine function. The values of a sine wave vary from zero to maximum in the same manner as the sine value. Both voltage and current exist in sine-wave form.

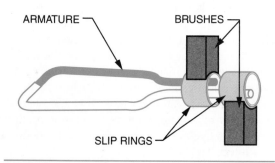

Figure 12–3 Voltage is removed from the armature of an AC generator through slip rings.

12–1 Questions

1. What is the function of an AC generator?
2. Explain how an AC generator operates.
3. Define the following terms:
   a. Cycle
   b. Alternation
   c. Hertz
   d. Sine wave
4. Identify the major parts of an AC generator.
5. What is the difference between a positive and a negative alternation?

12–2 AC Values

Each point on a sine wave has a pair of numbers associated with it. One value expresses the degree of rotation of the waveform. The **degree of rotation** is the angle to which the armature has turned. The second value indicates its amplitude. **Amplitude** is the maximum departure of the value of an alternating current or wave from the average value. There are several methods of expressing these values.

The **peak value** of a sine wave is the absolute value of the point on the waveform with the greatest amplitude (Figure 12–5). There are two peak values, one for the positive and one for the negative alternation. These peak values are equal.

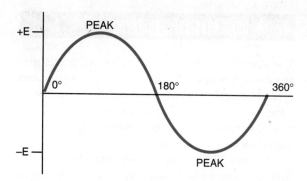

Figure 12–5 The peak value of a sine wave is the point on the AC waveform having the greatest amplitude. The peak value occurs during both the positive and the negative alternations of the waveform.

The **peak-to-peak value** refers to the vertical distance between the two peaks (Figure 12–6). The peak-to-peak value can be determined by adding the absolute values of each peak.

The **effective value** of alternating current is the amount that produces the same degree of heat in a given resistance as an equal amount of direct current. The effective value can be determined by a mathematical process called the root-mean-square (rms) process. Therefore, the effective value is also referred to as the **rms value.** Using the rms process shows that the effective value of a sine

wave is equal to 0.707 times the peak v[...] an AC voltage or current is given or me[...] assumed to be the effective value unle[...] indicated. Most meters are calibrated to indicate effective, or rms, value of voltage and current.

$$E_{rms} = 0.707E_p$$

where: E_{rms} = rms or effective voltage value
E_p = maximum voltage of one alternation

$$I_{rms} = 0.707I_p$$

where: I_{rms} = rms or effective current value
I_p = maximum current of one alternation

EXAMPLE A current sine wave has a peak value of 10 amperes. What is the effective value?

Given: Solution:
I_{rms} = ? $I_{rms} = 0.707I_p$
I_p = 10 amps $I_{rms} = (0.707)(10)$
                                             $I_{rms} = 7.07$ amperes

EXAMPLE A voltage sine wave has an effective value of 40 volts. What is the peak value of the sine wave?

Given: Solution:
E_p = ? $E_{rms} = 0.707E_p$
E_{rms} = 40 volts $40 = 0.707E_p$

$$\frac{40}{0.707} = \frac{0.707E_p}{0.707}$$

$$\frac{40}{0.707} = 1E_p$$

$$56.58 \text{ volts} = E_p$$

The time required to complete one cycle of a sine wave is called the **period.** The period is usually measured in seconds. The letter t is used to represent the period.

The number of cycles that occurs in a specific period of time is called the **frequency.** The frequency of an AC sine wave is usually expressed in terms of cycles per second. The unit of frequency is the *hertz*. One hertz equals one cycle per second.

The period of a sine wave is inversely proportional to its frequency. The higher the frequency, the shorter the period. The relationship between

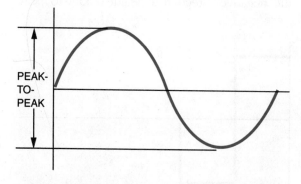

Figure 12–6 The peak-to-peak value can be determined by adding the peak values of the two alternations.

the frequency and the period of a sine wave is expressed as:

$$f = \frac{1}{t}$$

$$t = \frac{1}{f}$$

where: f represents the frequency, and t represents the period.

EXAMPLE What is the frequency of a sine wave with a period of 0.05 second?

Given:
f = ?
t = 0.05 second

Solution:
$$f = \frac{1}{t}$$

$$f = \frac{1}{0.05}$$

$$f = 20 \text{ hertz}$$

EXAMPLE If a sine wave has a frequency of 60 hertz, what is its period?

Given:
f = 60 hertz
t = ?

Solution:
$$t = \frac{1}{f}$$

$$t = \frac{1}{60}$$

$$t = 0.0167 \text{ second}$$

$$\text{or } 16.7 \text{ msec}$$

12–2 Questions

1. Define the following values:
   a. peak
   b. peak-to-peak
   c. effective
   d. rms
2. A voltage sine wave has a peak value of 125 volts. What is its effective value?
3. What is the relationship between time and frequency?
4. A current sine wave has an effective value of 10 amperes. What is its peak value?
5. What is the period of a 400-hertz sine wave?

12–3 Nonsinusoidal Waveforms

The sine wave is the most common AC waveform. However, it is not the only type of waveform used in electronics. Waveforms other than sinusoidal (sine wave) are classified as nonsinusoidal. Nonsinusoidal waveforms are generated by specially designed electronic circuits.

Figures 12–7 through 12–9 show three of the basic nonsinusoidal waveforms. These waveforms can represent either current or voltage. Figure 12–7 shows a square wave, so designated because its positive and negative alternations are square in shape. This indicates that the current or voltage immediately reaches its maximum value and remains there for the duration of the alternation. When the polarity changes, the current or voltage immediately reaches the opposite peak value and remains there for the next duration. The pulse width is equal to one-half the period. The **pulse width** is the duration the voltage is at its maximum or peak value until it drops to its minimum value. The amplitude can be any value. The square wave is useful as an electronic signal because its characteristics can be changed easily.

Figure 12–8 shows a triangular waveform. It consists of positive and negative ramps of equal slope. During the positive alternation the waveform rises at a linear rate from zero to its peak value and then decreases back to zero. During the negative alternation the waveform declines in the negative direction at a linear rate to peak

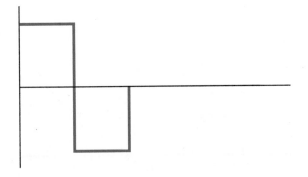

Figure 12–7 Square waveform.

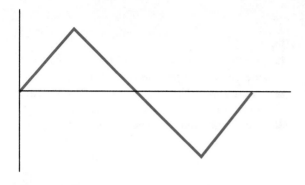

Figure 12–8 Triangular waveform.

value and then returns to zero. The period of this waveform is measured from peak to peak. Triangular waveforms are used primarily as electronic signals.

Figure 12–9 shows a sawtooth waveform. A sawtooth waveform is a special case of the triangular wave. First it rises at a linear rate and then rapidly declines to its negative peak value. The positive ramp is of relatively long duration and has a smaller slope than the short ramp. Sawtooth waves are used to trigger the operations of electronic circuits. In television sets and oscilloscopes they are used to sweep the electron beam across the screen, creating the image.

Pulse waveforms and other nonsinusoidal waveforms can be considered in two ways. One method considers the waveform as a momentary change in voltage followed, after a certain time interval, by another momentary change in voltage.

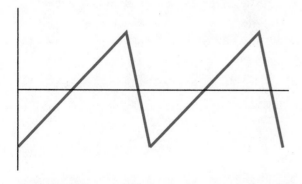

Figure 12–9 Sawtooth waveform.

The second method considers the waveform as the algebraic sum of many sine waves having different frequencies and amplitudes. This method is useful in the design of amplifiers. If the amplifier cannot pass all the sine-wave frequencies, distortion results.

Nonsinusoidal waves are composed of the fundamental frequency and harmonics. The **fundamental frequency** represents the repetition rate of the waveform. **Harmonics** are higher frequency sine waves that are exact multiples of the fundamental frequency. *Odd harmonics* are frequencies that are odd multiples of the fundamental frequency. *Even harmonics* are frequencies that are even multiples of the fundamental frequency.

Square waves are composed of the fundamental frequency and all odd harmonics.

Triangular waveforms are also composed of the fundamental frequency and all odd harmonics, but unlike the square wave, the odd harmonics are 180 degrees out of phase with the fundamental frequency.

Sawtooth waveforms are composed of both even and odd harmonics. The even harmonics are 180 degrees out of phase with the odd harmonics.

12–3 Questions

1. What are nonsinusoidal waveforms?
2. Draw two cycles of a:
   a. Square wave
   b. Triangular wave
   c. Sawtooth wave
3. What are the applications of these nonsinusoidal waveforms?
4. Describe the fundamental frequency and harmonics of three different nonsinusoidal waveforms.

SUMMARY

■ AC is the most commonly used type of electricity.
■ AC consists of current flowing in one direction and then reversing and flowing in the opposite direction.

- One revolution of an AC generator is called a cycle.
- The two halves of a cycle are referred to as alternations.
- Two complete alternations with no reference to time make up a cycle.
- One cycle per second is defined as a hertz.
- The waveform produced by an AC generator is called a sinusoidal waveform or sine wave.
- The peak value of a sine wave is the absolute value of the point on the waveform with the greatest amplitude.
- The peak-to-peak value is the vertical distance from one peak to the other peak.
- The effective value of AC is the amount of current that produces the same degree of heat in a given resistance as an equal amount of direct current.
- The effective value can be determined by a mathematical process called the root-mean-square (rms) process.
- The rms value of a sine wave is equal to 0.707 times the peak value.

$$E_{rms} = 0.707E_p$$
$$I_{rms} = 0.707I_p$$

- The time required to complete one cycle of a sine wave is called the period (t).
- The number of cycles occurring in a specific period of time is called frequency (f).
- The relationship between frequency and period is:

$$f = \frac{1}{t}$$

- Square waves are composed of the fundamental frequency and all odd harmonics.
- Triangular waveforms are composed of the fundamental frequency and all odd harmonics 180 degrees out of phase with the fundamental frequency.
- Sawtooth waveforms are composed of both even and odd harmonics, the even harmonics 180 degrees out of phase with the odd harmonics.

Chapter 12 Self-Test

1. What causes magnetic induction to occur?
2. Explain how the left-hand rule applies to AC generators.
3. Explain how the peak-to-peak value of a waveform is determined.
4. How is the effective value of alternating current determined?
5. Draw examples of three nonsinusoidal waveforms that can represent current and voltage.
6. Why are harmonics important in the study of waveforms?

13

AC MEASUREMENTS

Objectives

After completing this chapter, the student will be able to:

■ Identify the types of meters available for AC measurements.

■ Identify the types of meter movements used to make AC measurements.

■ Explain the function of an oscilloscope.

■ Identify the basic parts of an oscilloscope and explain their functions.

■ Demonstrate the proper setup of an oscilloscope.

■ Describe how to use an oscilloscope to take a measurement.

■ Explain how a counter works.

■ Identify the basic parts of a counter.

AC measurements of current, voltage, resistance, power, and frequency are necessary to operate and repair AC circuits and equipment.

This chapter covers some of the important test equipment used to take various AC measurements.

13–1 AC Meters

The *moving coil* meter movement, is referred to as the d'Arsonval meter movement. The analog meter (Figure 13–1) is a moving coil meter. The easy-to-read digital meter (Figure 13–2) has recently replaced the analog meter.

Moving coil meters are designed to measure DC current. To measure AC current on this meter the AC current must first be converted to DC current. This is accomplished with diodes called *rectifiers*. The process of converting AC to DC is called **rectification.** The rectifiers are placed between the meter input and the meter movement and allow current to flow in only one direction (Figure 13–3). The rectifiers convert the sine wave into a pulsating DC current that is applied to the meter movement.

A second type of AC meter uses the iron-vane meter movement (Figure 13–4). This does not require the conversion of AC to DC. It consists of

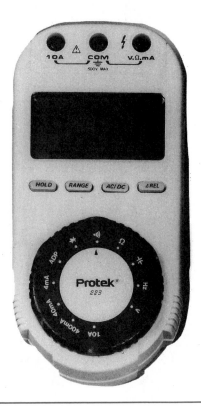

Figure 13–2 Digital meter used for AC measurements.

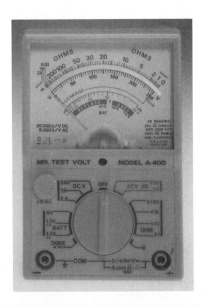

Figure 13–1 Analog meter used for AC measurements.

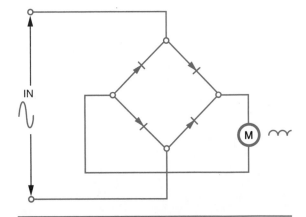

Figure 13–3 Rectifiers are used to convert the AC signal to a DC level prior to applying it to the meter movement.

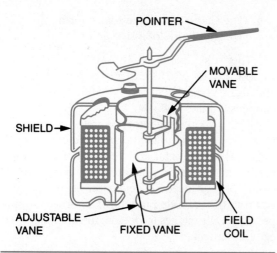

Figure 13–4 Iron-vane meter movements do not require the conversion of AC to DC. *(From Duff,* AC Fundamentals, *by Delmar Publishers.)*

Figure 13–5 A clamp-on meter operates on the principle that current flowing through a wire generates a magnetic field around the wire. *(Courtesy of Protek.)*

two iron vanes located within a coil. One vane is stationary and the other is movable. A pointer is attached to the movable vane and moves proportionally to the current passing through the coil. The magnetic field of the coil induces a North and South Pole into the iron vanes. Because like poles repel and both vanes have the same polarity, they move away from each other. The iron-vane movement requires more current for a full-scale deflection than does the moving coil meter movement. For this reason, iron-vane movements find little application in low-power circuits. The iron-vane movement is also inaccurate for frequencies above 100 hertz. It is used primarily for 60-hertz applications.

The clamp-on-meter is based on the principle that an AC current flowing in a conductor causes a magnetic field to expand and collapse as the current increases and decreases in value (Figure 13–5). Each time the AC current changes polarity, the magnetic field changes direction. The clamp-on AC meter uses a split-core transformer. This allows the core to be opened and placed around the conductor. At the end of the core is a coil, which is cut by the magnetic lines of force. This induces a volt-

age into the coil, creating a flow of alternating current. This AC current must be rectified before being sent to the meter movement, generally a moving coil. This type of meter is used for measuring high values of AC current. The current in the conductor must be large in order to produce a magnetic field strong enough to induce a current flow in the core of the meter.

The basic AC meter movement is a current-measuring device. However, this type of meter movement can also be used to measure AC voltage and power. Because alternating current changes direction periodically, the meter leads can be hooked into an AC circuit in either direction. However, the meter must be connected in series with the circuit when measuring current. When measuring voltage, the meter must always be connected in parallel.

Always be sure the current or voltage being measured is within the range of the meter. For additional protection, always start at the highest scale and work down to the appropriate scale.

13–1 Questions

1. How is a d'Arsonval meter movement used to measure an AC voltage?
2. What makes an iron-vane meter movement desirable for measuring an AC signal?
3. Explain the principle behind a clamp-on AC meter.
4. Draw a circuit showing how to connect an AC ammeter.
5. Describe the proper procedure for using an AC voltmeter in a circuit. (Include switch settings where possible).

13–2 Oscilloscopes

An **oscilloscope** is the most versatile piece of test equipment available for working on electronic equipment and circuits (Figure 13–6). It provides a visual display of what is occurring in the circuit.

An oscilloscope is capable of providing the following information about an electronic circuit:

1. The frequency of a signal
2. The duration of a signal
3. The phase relationship between signal waveforms
4. The shape of a signal's waveform
5. The amplitude of a signal

The basic parts of an oscilloscope are: a cathode-ray tube (CRT), a sweep generator, horizontal and vertical deflection amplifiers, and power supplies (Figure 13–7).

The sweep generator provides a sawtooth waveform as input to the horizontal deflection amplifier. The horizontal and vertical deflection amplifiers increase the amplitude of the input voltage to the proper level for deflection of the electron beam in the cathode-ray tube. The power supply provides DC voltages to operate the amplifiers and cathode-ray tube.

The cathode-ray tube consists of three parts: a phosphor screen, deflection plates, and an electron gun (Figure 13–8). The phosphor screen emits light when struck by electrons. The electron gun generates the electron beam that strikes the screen. As the electron beam approaches the screen, the deflection plates change the direction of the electron beam as it approaches the screen. The horizontal deflection plate is attached to the sweep generator, which moves the electron beam back and forth across the screen. The vertical amplifier is connected to the input signal and controls its amplitude.

Using a faceplate that has been marked in centimeters along both a vertical and a horizontal axis

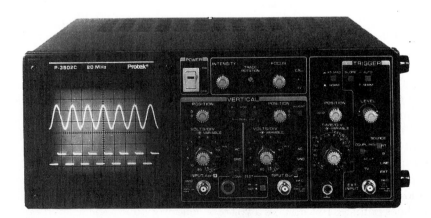

Figure 13–6 An oscilloscope is the most versatile piece of test equipment available to a technician. *(Courtesy of Protek.)*

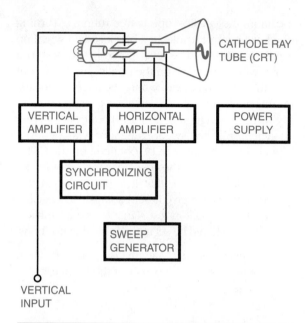

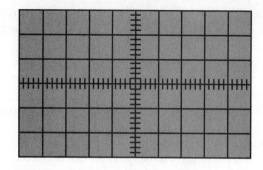

Figure 13–9 Oscilloscope faceplate.

The power switch of an oscilloscope is usually located on the front panel (Figure 13–10). It may be a toggle, push-button, or rotary switch. It may be mounted separately or mounted with another switch. Its purpose is to apply line voltage to operate the oscilloscope.

The intensity (also called brightness) switch is used to control the electron beam within the CRT. It is a rotary control and allows the operator to adjust the electron beam for proper viewing. Caution: Keep the brightness as low as possible. Too much intensity, too long, can burn a hole or etch a line in the phosphor screen, ruining the CRT.

The focus and astigmatism controls are connected to the electron gun and are used to adjust the electron beam size and shape before it reaches the deflection plates. Both of these controls are

Figure 13–7 Block diagram of the basic parts of an oscilloscope.

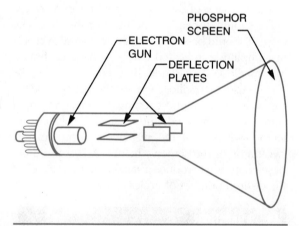

Figure 13–8 Basic parts of a cathode-ray tube (CRT).

(Figure 13–9), the oscilloscope can be calibrated with a known voltage prior to testing an unknown signal. Then, when the unknown signal is applied, its amplitude can be calculated. Rather than mark the actual faceplate of the CRT, oscilloscopes are provided with separate faceplates called *graticules* that are mounted in front of the CRT.

Figure 13–10 Oscilloscopes vary in the number of front panel controls. *(Courtesy of Protek.)*

rotary controls. To use them, the electron beam is slowly swept across the screen of the CRT and both controls are alternately adjusted until a perfectly round dot is obtained. On some newer oscilloscopes the astigmatism control may be recessed below the front panel.

The horizontal and vertical position controls are also rotary controls. They allow the electron beam to be positioned anywhere on the face of the CRT. They are initially set so that the electron beam sweeps across the center of the CRT. They are then adjusted to place the electron beam in a position to measure amplitude and time by its placement on a graticule.

The vertical block consists of a vertical input jack, and AC/DC switch, and a volts/cm rotary switch. The oscilloscope probe is connected to the input jack. The probe is then connected to the circuit to be tested. The AC/DC switch allows the signal to be sent directly to the vertical amplifier in the DC position or to a capacitor in the AC position. The capacitor in the AC position is used to block any DC component on the signal being checked. The volts/cm switch is used to adjust the amplitude of the input signal. If the signal is too large, the vertical amplifier distorts it. If the signal is too small, it is amplified. This control is calibrated to match the CRT faceplate. Where this control is set depends on the amplitude of the signal on the CRT.

The horizontal block, also referred to as the time base, consists of a time/cm rotary switch, a trigger-control switch, and a triggering level control. The time/cm switch sets the horizontal sweep rate that is to be represented by the horizontal graduations. The lower the setting, the fewer cycles per second are displayed. The trigger-control switch selects the source and polarity of the trigger signal. The trigger source can be line, internal, or external. The polarity can be positive or negative. When line is selected as the trigger source, the power-line frequency of 60 hertz is used as the trigger frequency. When internal (INT) is selected as the trigger frequency, the input-signal frequency is used as the trigger frequency. External (EXT) allows a trigger frequency from an external source to be used.

The level control sets the amplitude that the triggering signal must exceed before the sweep generator starts. If the level control is in the AUTO

position, the oscilloscope is free running. Turning the level control results in a blank screen when no signal is present. The level control is turned to a point where the trace on the oscilloscope goes out and then turned back to where the trace reappears. At this point, a stable presentation is on the screen. By using the triggering and level controls together to synchronize the sweep generator with the input signal, a stable trace can be obtained on the CRT.

Prior to use, the oscilloscope should be checked to determine that it is not faulty. Misadjusted controls can give false readings. Most oscilloscopes have a built-in test signal for setup. Initially the controls should be set in the following positions:

Intensity, Focus, Astigmatism, and Position controls (set to the center of their range):

Triggering: INT +
Level: AUTO
Time/cm: 1 msec
Volts/cm: 0.02
Power: ON

The oscilloscope probe should be connected to the test jack of the voltage calibrator. A square wave should appear on the trace of the CRT. The display should be stable and show several cycles at the amplitude indicated by the voltage calibrator. The oscilloscope can now be used.

To use the oscilloscope, set the volts/cm selector switch to its highest setting. Connect the input signal to be observed, and rotate the volts/cm selector switch until the display is approximately two-thirds of the screen's height. Adjust the time-base controls to produce a stable trace and to select the desired number of cycles.

13–2 Questions

1. What can be learned about a waveform by using an oscilloscope?
2. What are the basic parts of an oscilloscope?
3. Describe a basic procedure for setting up an oscilloscope prior to its initial use.
4. What are applications of an oscilloscope when working with a circuit?
5. What does the graticule marked on the screen of the oscilloscope represent?

13–3 Frequency Counters

A **frequency counter** (Figure 13–11) measures frequency by comparing a known frequency against an input frequency. All frequency counters consist of the same basic sections: a time base, an input-signal conditioner, a gate-control circuit, a main gate, a decade counter, and a display (Figure 13–12).

The signal conditioner converts the input signal to a waveshape and amplitude compatible with the circuitry in the counter. The main gate passes the conditioned input signal to the counter circuit if a signal from the time base is present at the same time. The time base drives the gate-control circuit, using a signal compatible with the signal being measured. The gate control acts as the synchronization center of the counter. It controls the opening and closing of the main gate and also provides a signal to latch (lock onto) the count at the end of the counting period and reset the circuitry for the next count. The decade counter keeps a running tally of all the pulses that pass through the main gate. One decade counter is required for each digit displayed. The display, which provides a visual readout of the frequency being measured, can be of several types. The more common displays are gas-discharge tubes, light-emitting diode displays, and liquid-crystal displays.

The electronic counter was once a piece of laboratory equipment but is now used in electronics repair shops, engineering departments, ham-radio shacks, and on industrial production lines. The wide use of the counter can be attributed to the integrated circuit, which has reduced the size and price of the counter while increasing its accuracy, stability, reliability, and frequency range. (Integrated circuits are discussed in Section 3.)

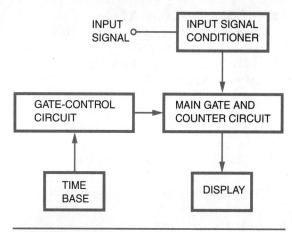

Figure 13–12 Block diagram of an electronic counter.

Figure 13–11 A frequency counter is used widely in repair shops and industry. *(Courtesy of Protek.)*

13–3 Questions

1. What is the function of an electronic counter?
2. What are the basic sections of a counter?
3. Draw and label a block diagram of an electronic counter.
4. What is the function of the signal conditioner in a counter?
5. Why has the counter increased in popularity?

SUMMARY

■ To measure AC current or voltage on a moving coil meter, the current or voltage must first be converted to DC.

■ Iron-vane meter movement does not require conversion to DC.

- A clamp-on AC meter is based on the principle that current flowing through a wire generates a magnetic field.
- An oscilloscope provides the following information about a circuit:
  —Frequency of the signal
  —Duration of the signal
  —Phase relationships between signal waveforms
  —Shape of the signal's waveform
  —Amplitude of the signal
- The basic parts of an oscilloscope are:
  —Cathode-ray tube
  —Sweep generator
  —Horizontal deflection amplifier
  —Vertical deflection amplifier
  —Power supply
- A counter measures frequency by comparing an unknown frequency to a known frequency.
- The basic parts of a frequency counter are:
  —Time base
  —Input signal conditioner
  —Gate-control circuit
  —Main gate
  —Decade counter
  —Display

Chapter 13 Self-Test

1. Describe how a DC meter can be adapted to measure an AC signal.
2. Explain how a clamp-on ammeter is used for measuring current.
3. What type of information does an oscilloscope provide?
4. Describe the process for checking out an oscilloscope to determine if it is operating properly.
5. Identify and describe the function of each of the major blocks of a counter.
6. What has been the major factor in bringing the counter to the work bench of repair shops?

14

RESISTIVE AC CIRCUITS

Objectives

After completing this chapter, the student will be able to:

- Describe the phase relationship between current and voltage in a resistive circuit.
- Apply Ohm's law to AC resistive circuits.
- Solve for unknown quantities in series AC resistive circuits.
- Solve for unknown quantities in parallel AC resistive circuits.
- Solve for power in AC resistive circuits.

The relationship of current, voltage, and resistance is similar in DC and AC circuits. The simple AC circuit must be understood before moving on to more complex circuits containing capacitance and inductance.

14–1 Basic AC Resistive Circuits

A basic AC circuit (Figure 14–1) consists of an AC source, conductors, and a resistive load. The AC source can be an AC generator or a circuit that generates an AC voltage. The resistive load can be a resistor, a heater, a lamp, or any similar device.

When an AC voltage is applied to the resistive load, the AC current's amplitude and direction vary in the same manner as those of the applied voltage. When the applied voltage changes polarity, the current also changes. They are said to be **in phase.** Figure 14–2 shows the in-phase relationship that exists between the current and the applied voltage in a pure resistive circuit. The current and voltage waveforms pass through zero and maximum values at the same time. However, the two waveforms do not have the same peak amplitudes because they represent different quantities, measured in different units.

The AC current flowing through the resistor varies with the voltage and the resistance in the circuit. The current at any instant can be determined by applying Ohm's law.

Effective values are used in most measurements. As stated previously, the effective value is the amount of AC voltage that produces the same degree of heat as a DC voltage of the same value. The effective value can be considered the DC equivalent value. Ohm's law can be used with effective AC values, just as with DC values, in a pure resistive circuit.

EXAMPLE If a circuit has an AC voltage source of 120 volts and a resistance of 1000 ohms, what is the effective current in the circuit? (Remember, an AC voltage or current is assumed to be effective unless otherwise specified.)

Figure 14–1 A basic AC circuit consists of an AC source, conductors, and a resistive load.

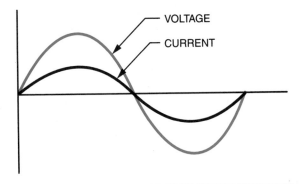

Figure 14–2 The voltage and current are in phase in a pure resistive circuit.

Given:

$I = ?$

$E = 120$ volts

$R = 1000$ ohms

Solution:

$$I = \frac{E}{R}$$

$$I = \frac{120}{1000}$$

$$I = 0.12 \text{ ampere}$$

EXAMPLE If an effective current of 1.7 amperes flows through a 68-ohm resistor, what is the effective value of the applied voltage?

Given:

$I = 1.7$ amps

$E = ?$

$R = 68$ ohms

Solution:

$$I = \frac{E}{R}$$

$$1.7 = \frac{E}{68}$$

$$\frac{1.7}{1} = \frac{E}{68}$$

$$(1)(E) = (1.7)(6.8)$$

$$E = 115.60 \text{ volts}$$

14–1 Questions

1. What is the phase relationship between current and voltage in a pure resistive circuit?
2. What AC value is used for most measurements?
3. What is the effective current in a circuit of 10,000 ohms with 12 volts applied?

4. What is the voltage (rms) of a circuit with 250 milliamperes flowing through 100 ohms?

5. What is the resistance of a circuit when 350 microamperes are produced with an AC voltage of 12 volts?

14–2 Series AC Circuits

The current in a resistive circuit depends on the applied voltage. The current is always in phase with the voltage regardless of the number of resistors in the circuit. At any point in the circuit, the current has the same value.

Figure 14–3 shows a simple series circuit. The current flow is the same through both resistors. Using Ohm's law, the voltage drop across each resistor can be determined. The voltage drops added together equal the applied voltage. Figure 14–4 shows the phase relationships of the voltage drops, the applied voltage, and the current in the circuit. All the voltages and the current are in phase with one another in a pure resistive circuit.

EXAMPLE If an AC circuit has an effective value of 120 volts applied across two resistors (R_1 = 470 ohms and R_2 = 1000 ohms), what is the voltage drop across each resistor?

Given:

$$R_1 = 470 \text{ ohms}$$
$$R_2 = 1000 \text{ ohms}$$
$$E_{R_1} = ?$$
$$E_{R_2} = ?$$
$$E_T = 120$$

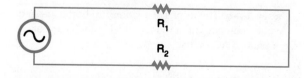

Figure 14–3 Simple series AC circuit.

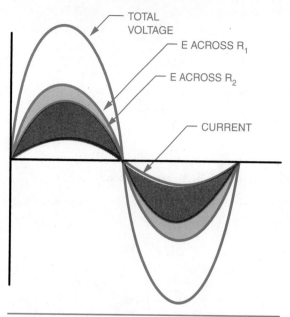

Figure 14–4 The in-phase relationship of the voltage drops, applied voltage, and current in a series AC circuit.

Solution:
First, find the total resistance (R_T).

$$R_T = R_1 + R_2$$
$$R_T = 470 + 1000$$
$$R_T = 1470 \text{ ohms}$$

Using R_T, now find the total current (I_T).

$$I_T = \frac{E_T}{R_T}$$
$$I_T = \frac{120}{1470}$$
$$I_T = 0.082 \text{ amp}$$

In a series circuit, $I_T = I_{R_1} = I_{R_2}$

$$\therefore I_{R_1} = 0.082 \text{ amp}$$
$$I_{R_2} = 0.082 \text{ amp}$$

Use I_1 to find the voltage drop across resistor R_1.

$$I_{R_1} = \frac{E_{R_1}}{R_1}$$

$$0.082 = \frac{E_{R_1}}{470}$$

$$\frac{0.082}{1} = \frac{E_{R_1}}{470}$$

$$(1)(E_{R_1}) = (0.082)(470)$$

$$E_{R_1} = 38.54 \text{ volts}$$

Use I_2 to find the voltage drop across resistor R_1.

$$I_{R_2} = \frac{E_{R_2}}{R_2}$$

$$0.082 = \frac{E_{R_2}}{1000}$$

$$\frac{0.082}{1} = \frac{E_{R_2}}{1000}$$

$$(1)(E_{R_2}) = (0.082)(1000)$$

$$E_{R_2} = 82 \text{ volts}$$

The voltage drop across each resistor is the effective voltage drop.

14–2 Questions

1. What is the voltage drop across two resistors, 22 kilohms and 47 kilohms, connected in series with 24 volts applied?
2. What is the voltage drop across the following series resistors:
   a. $E_T = 100$ V, $R_1 = 680$ ohms, $R_2 = 1200$ ohms
   b. $E_T = 24$ V, $R_1 = 22$ kilohms, $R_2 = 47$ kilohms
   c. $I_T = 250$ mA, $R_1 = 100$ ohms, $R_2 = 500$ ohms
   d. $R_T = 10$ kilohms, $I_{R_1} = 1$ mA, $R_2 = 4.7$ kilohms
   e. $E_T = 120$ V, $R_1 = 720$ ohms, $I_{R_2} = 125$ mA

14–3 Parallel AC Circuits

The voltage in a parallel circuit (Figure 14–5) remains constant across each individual branch. However, the total current divides among the individual branches. In a parallel AC circuit, the total current is in phase with the applied voltage (Figure 14–6). The individual currents are also in phase with the applied voltage.

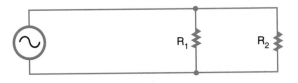

Figure 14–5 A simple parallel AC circuit.

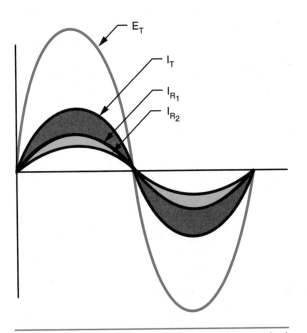

Figure 14–6 The in-phase relationship of the applied voltage, total current, and individual branch currents in a parallel AC circuit.

All current and voltage values are the effective values. These values are used the same way DC values are used.

EXAMPLE If an AC circuit has an effective voltage of 120 volts applied across two parallel resistors of 470 ohms and 1000 ohms respectively, what is the current flowing through each of the resistors?

Given:

$$I_{R_1} = ?$$

$$I_{R_2} = ?$$

$$E_T = 120 \text{ volts}$$

$$R_1 = 470 \text{ ohms}$$

$$R_2 = 1000 \text{ ohms}$$

Solution:

In a parallel circuit $E_T = E_{R_1} = E_{R_2}$

$$\therefore E_{R_1} = 120 \text{ volts}$$

$$E_{R_2} = 120 \text{ volts}$$

Use E_1 to find the current through resistor R_1.

$$I_{R_1} = \frac{E_{R_1}}{R_1}$$

$$I_{R_1} = \frac{120}{470}$$

$$I_{R_1} = 0.26 \text{ amp or } 260 \text{ mA}$$

Use E_2 to find the current through resistor R_2.

$$I_{R_2} = \frac{E_{R_2}}{R_2}$$

$$I_{R_2} = \frac{120}{1000}$$

$$I_{R_2} = 0.12 \text{ amp or } 120 \text{ mA}$$

14–3 Question

1. What is the current flow through the following parallel AC resistive circuits?
   a. E_T = 100 V, R_1 = 470 ohms, R_2 = 1000 ohms
   b. E_T = 24 V, R_1 = 22 kilohms, R_2 = 47 kilohms
   c. E_T = 150 V, R_1 = 100 ohms, R_2 = 500 ohms
   d. I_T = 0.0075 A, E_{R_1} = 10 V, R_2 = 4.7 kilohms
   e. R_T = 4700 ohms, I_{R_1} = 11 mA, E_{R_2} = 120 V

14–4 Power in AC Circuits

Power is dissipated in AC resistive circuits the same way as in DC resistive circuits. Power is measured in watts and is equal to the current times the voltage in the circuit.

The power consumed by the resistor in an AC circuit varies with the amount of current flowing through it and the voltage applied across it. Figure 14–7 shows the relationship of power, current, and voltage. The power curve does not fall below the reference line because the power is dissipated in the form of heat. It does not matter in which

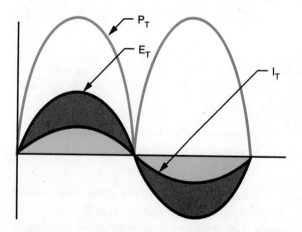

Figure 14–7 The relationship of power, current, and voltage in a resistive AC circuit.

direction the current is flowing, as power is assumed to have a positive value.

Power varies between the peak value and zero. Midway between peak value and zero is the average power consumed by the circuit. In an AC circuit, the average power is the power consumed. This can be determined by multiplying the effective voltage value by the effective current value. This is expressed as:

$$P = IE$$

EXAMPLE What is the power consumption in an AC circuit with 120 volts applied across 150 ohms of resistance? (Remember, when the voltage value is not specified for an AC source, it is assumed to be the effective value.)

Given:

$$I_T = ?$$
$$E_T = 120 \text{ volts}$$
$$R_T = 150 \text{ ohms}$$
$$P_T = ?$$

Solution:
First, find the total current (I_T).

$$I_T = \frac{E_T}{R_T}$$
$$I_T = \frac{120}{150}$$
$$I_T = 0.80 \text{ amp}$$

Now, find the total power (P_T).

$$P_T = I_T E_T$$
$$P_T = (0.80)(120)$$
$$P_T = 96 \text{ watts}$$

When the current is not given, the current must be determined prior to using the power formula. The resistance in this circuit consumes 96 watts of power.

14–4 Questions

1. What is the total power consumption of the following circuits:
   Series:
   a. $E_T = 100$ V, $R_1 = 680 \ \Omega$, $R_2 = 1200 \ \Omega$
   b. $I_T = 250$ mA, $R_1 = 100 \ \Omega$, $R_2 = 500 \ \Omega$
   Parallel:
   c. $E_T = 100$ V, $R_1 = 470 \ \Omega$, $R_2 = 1000 \ \Omega$
   d. $I_T = 7.5$ mA, $E_{R_1} = 10$ V, $R_2 = 4.7$ kΩ
2. Find the power consumption of each individual component in the following circuit.

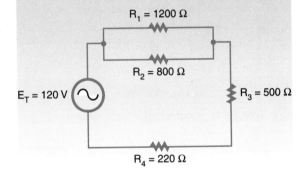

SUMMARY

- A basic AC resistive circuit consists of a voltage source, conductors, and a resistive load.
- The current is in phase with the applied voltage in a resistive circuit.
- The effective value of AC current or voltage produces the same results as the equivalent DC voltage or current.
- The effective values are the most widely used measurement values.
- Ohm's law can be used with all effective values.
- AC voltage or current values are assumed to be the effective values if not otherwise specified.

Chapter 14 Self-Test

1. Explain the phase relationship between current and voltage in a pure resistive circuit.

2. What is the effective voltage of an AC circuit with 25 mA flowing through 4.7 kilohms?

3. What is the voltage drop across two resistors of 4.7 kilohms and 3.9 kilohms in series with an AC voltage of 12 volts applied?

4. If two parallel resistors of 2.2 kilohms and 5.6 kilohms have an AC effective voltage of 120 V applied across their input, what is the current developed through each of the resistors?

5. What determines the power consumption in an AC circuit?

6. What is the power consumption in an AC circuit with 120 volts applied across a load of 1200 ohms?

15

CAPACITIVE AC CIRCUITS

Objectives

After completing this chapter, the student will be able to:

- Describe the phase relationship between current and voltage in a capacitive AC circuit.
- Determine the capacitive reactance in an AC capacitive circuit.
- Describe how resistor-capacitor networks can be used for filtering, coupling, and phase shifting.
- Explain how low-pass and high-pass RC filters operate.

Capacitors are key components of AC circuits. Capacitors combined with resistors and inductors form useful electronic networks.

15–1 Capacitors in AC Circuits

When an AC voltage is applied to a capacitor, it gives the appearance that electrons are flowing in the circuit. However, electrons do not pass through the dielectric of the capacitor. As the applied AC voltage increases and decreases in amplitude, the capacitor charges and discharges. The resulting movement of electrons from one plate of the capacitor to the other represents current flow.

The current and applied voltage in a capacitive AC circuit differ from those in a pure resistive circuit. In a pure resistive circuit the current flows in phase with the applied voltage. Current and voltage in a capacitive AC circuit do not flow in phase with each other (Figure 15–1). When the current is at maximum, the voltage is at zero. This relationship is described as 90 degrees out of phase. The current leads the applied voltage in a capacitive circuit.

In a capacitive AC circuit, the applied voltage is constantly changing, causing the capacitor to charge and discharge. After the capacitor is initially charged, the voltage stored on its plates opposes any change in the applied voltage. The opposition that the capacitor offers to the applied AC voltage is called **capacitive reactance.** Capacitive reactance is represented by X_C and is measured in ohms.

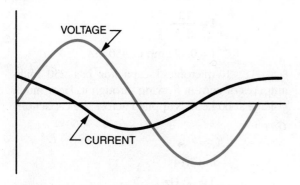

Figure 15–1 Note the out-of-phase relationship between the current and the voltage in a capacitive AC circuit. The current leads the applied voltage.

Capacitive reactance can be calculated by using the formula:

$$X_C = \frac{1}{2\pi fC}$$

where: π = pi, the constant 3.14
f = frequency in hertz
C = capacitance in farads

Capacitive reactance is a function of the frequency of the applied AC voltage and the capacitance. Increasing the frequency decreases the reactance, resulting in greater current flow. Decreasing the frequency increases the opposition and decreases current flow.

EXAMPLE What is the capacitive reactance of a 1-microfarad capacitor at 60 hertz?

Given:

$$X_C = ?$$
$$\pi = 3.14$$
$$f = 60 \text{ Hz}$$
$$C = 1 \ \mu F = 0.000001 \text{ F}$$

Solution:

$$X_C = \frac{1}{2\pi fC}$$

$$X_C = \frac{1}{(2)(3.14)(60)(0.000001)}$$

$$X_C = \frac{1}{0.000377}$$

$$X_C = 2653 \text{ ohms}$$

EXAMPLE What is the capacitive reactance of a 1-microfarad capacitor at 400 hertz?

Given:

$$X_C = ?$$
$$\pi = 3.14$$
$$f = 60 \text{ Hz}$$
$$C = 1 \ \mu F = 0.000001 \text{ F}$$

Solution:

$$X_C = \frac{1}{2\pi fC}$$

$$X_C = \frac{1}{(2)(3.14)(400)(0.000001)}$$

$$X_C = \frac{1}{0.00251}$$

$$X_C = 398 \text{ ohms}$$

EXAMPLE What is the capacitive reactance of a 0.1-microfarad capacitor at 60 hertz?

Given:

$$X_C = ?$$
$$\pi = 3.14$$
$$f = 60 \text{ Hz}$$
$$C = 0.1 \text{ }\mu\text{F} = 0.0000001 \text{ F}$$

Solution:

$$X_C = \frac{1}{2\pi fC}$$

$$X_C = \frac{1}{(2)(3.14)(60)(0.0000001)}$$

$$X_C = \frac{1}{0.0000377}$$

$$X_C = 26,525 \text{ ohms}$$

EXAMPLE What is the capacitive reactance of a 10-microfarad capacitor at 60 hertz?

Given:

$$X_C = ?$$
$$\pi = 3.14$$
$$f = 60 \text{ Hz}$$
$$C = 10 \text{ }\mu\text{F} = 0.00001 \text{ F}$$

Solution:

$$X_C = \frac{1}{2\pi fC}$$

$$X_C = \frac{1}{(2)(3.14)(60)(0.00001)}$$

$$X_C = \frac{1}{0.00377}$$

$$X_C = 265 \text{ ohms}$$

Capacitive reactance is the opposition to changes in the applied AC voltage by a capacitor. In an AC circuit, a capacitor is thus an effective way of controlling current. Using Ohm's law, the current is directly proportional to the applied voltage and inversely proportional to the capacitive reactance. This is expressed as:

$$I = \frac{E}{X_C}$$

Note: X_C (capacitive reactance) has been substituted for R (resistance) in Ohm's law.

It is important to keep in mind that capacitive reactance depends on the frequency of the applied voltage and the capacitance in the circuit.

EXAMPLE A 100-microfarad capacitor has 12 volts applied across it at 60 hertz. How much current is flowing through it?

Given:

$$I = ?$$
$$E = 12 \text{ VAC}$$
$$\pi = 3.14$$
$$f = 60 \text{ Hz}$$
$$C = 100 \text{ }\mu\text{F} = 0.0001 \text{ F}$$

Solution:
First, find the capacitive reactance (X_C).

$$X_C = \frac{1}{2\pi fC}$$

$$X_C = \frac{1}{(2)(3.14)(60)(0.0001)}$$

$$X_C = \frac{1}{0.0377}$$

$$X_C = 26.5 \text{ ohms}$$

Using X_C, now find the current flow:

$$I = \frac{E}{X_C}$$

$$I = \frac{12}{26.5}$$

$$I = 0.45 \text{ amp or 450 mA}$$

EXAMPLE 10-microfarad capacitor has 250 milli-amperes of current flowing through it. How much voltage at 60 hertz is applied across the capacitor?

Given:

$$X_C = ?$$
$$\pi = 3.14$$
$$f = 60 \text{ Hz}$$
$$C = 10 \text{ }\mu\text{F} = 0.00001 \text{ F}$$
$$I = 250 \text{ mA or 0.25 amp}$$
$$E = ?$$

Solution:

First, find the capacitive reactance (X_C).

$$X_C = \frac{1}{2\pi fC}$$

$$X_C = \frac{1}{(2)(3.14)(60)(0.00001)}$$

$$X_C = \frac{1}{0.00377}$$

$$X_C = 265 \text{ ohms}$$

Now, find the voltage drop (E).

$$I = \frac{E}{X_C}$$

$$0.25 = \frac{E}{265}$$

$$\frac{0.25}{1} = \frac{E}{265}$$

$$(I)(E) = (0.25)(265)$$

$$E = 66.25 \text{ volts}$$

When capacitors are connected in series, the capacitive reactance is equal to the sum of the individual capacitive reactance values.

$$X_{C_T} = X_{C_1} + X_{C_2} + X_{C_3} \ldots + X_{C_n}$$

When capacitors are connected in parallel, the reciprocal of the capacitive reactance is equal to the sum of the reciprocals of the individual capacitive reactance values.

$$\frac{1}{X_{C_T}} = \frac{1}{X_{C_1}} + \frac{1}{X_{C_2}} + \frac{1}{X_{C_3}} \ldots + \frac{1}{X_{C_n}}$$

15–1 Questions

1. Describe how an AC voltage appears to make current move through a capacitor.
2. What is the relationship between current and voltage in a capacitive circuit?
3. What is capacitive reactance?
4. What is the capacitive reactance of a 10-microfarad capacitor at 400 hertz?

15–2 Applications of Capacitive Circuits

Capacitors can be used alone or combined with resistors to form *RC (resistor-capacitor) networks*. RC networks are used for filtering, decoupling, DC blocking, or coupling phase-shift circuits.

A filter is a circuit that discriminates among frequencies, attenuating (weakening) some while allowing others to pass. It works by establishing a cut-off point between frequencies that are passed and frequencies that are attenuated. The two most common types of filters are low-pass filters and high-pass filters. A **low-pass filter** allows low frequencies to pass with little opposition while attenuating the high frequencies. A **high-pass filter** permits frequencies above the cut-off point to pass while attenuating frequencies below the cut-off point.

A low-pass filter (Figure 15–2) consists of a capacitor and a resistor in series. The input voltage is applied across both capacitor and resistor. The output is taken from across the capacitor. At low frequencies, the capacitive reactance is higher than the resistance, so most of the voltage is dropped across the capacitor. Therefore, most of the voltage appears across the output. As the input frequency increases, the capacitive reactance decreases, and less voltage is dropped across the capacitor. Therefore, more voltage is dropped across the resistor, with a decrease in the output voltage. The cut-off is gradual. Frequencies above the cut-off point pass with a gradual attenuation in the output voltage.

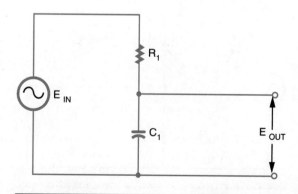

Figure 15–2 RC low-pass filter.

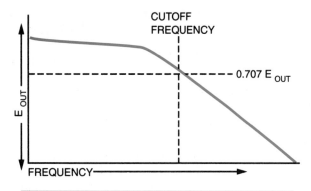

Figure 15–3 Frequency response of an RC low-pass filter.

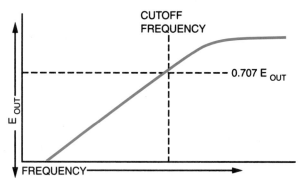

Figure 15–5 Frequency response of an RC high-pass filter.

Figure 15–3 shows a frequency response curve for an RC low-pass filter.

A high-pass filter also consists of a resistor and capacitor in series (Figure 15–4). However, the output is taken across the resistor. At high frequencies, the capacitive reactance is low, and most of the voltage is dropped across the resistor. As the frequency decreases, the capacitive reactance increases, and more voltage is dropped across the capacitor. This results in a decrease in the output voltage across the resistor. Again the drop in output voltage is gradual. Figure 15–5 shows a frequency response curve for an RC high-pass filter.

Most electronic circuits include both AC and DC voltages. This results in an AC signal superimposed on a DC signal. If the DC signal is used to operate equipment, it is advantageous to remove the AC signals. An RC low-pass filter can be used

for this purpose. A *decoupling network* (Figure 15–6) allows the DC signal to pass while attenuating or eliminating the AC signal. The AC signal may be in the form of oscillations, noises, or transient spikes. By adjusting the cut-off frequency, most of the AC signal can be filtered out, leaving only the DC voltage across the capacitor.

In another application, it may be desirable to pass the AC signal while blocking the DC voltage. This type of circuit is called a *coupling network* (Figure 15–7). An RC high-pass filter can be used. Initially the capacitor charges to the DC voltage level. Once the capacitor is charged, no more DC current can flow in the circuit. The AC source causes the capacitor to charge and discharge at the AC rate, creating current flow through the resistor. The component values are chosen so that the AC signal is passed with a minimum of attenuation.

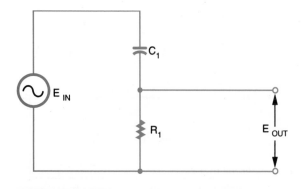

Figure 15–4 RC high-pass filter.

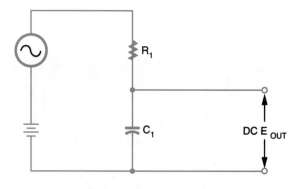

Figure 15–6 RC decoupling network.

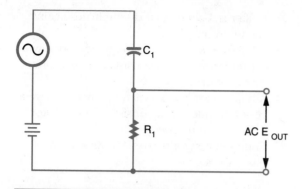

Figure 15–7 RC coupling network.

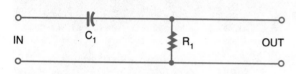

Figure 15–8 Leading output phase-shift network. The output voltage leads the input voltage.

At times it is necessary to shift the phase of the AC output signal with respect to the input signal. RC networks can be used for phase-shifting applications. The RC **phase-shift networks** are used only where small amounts of phase shift, less than 60 degrees, are desired.

Figure 15–8 shows a phase-shift network with the input applied across the resistor-capacitor combination and the output taken across the resistor. In the circuit, the current leads the voltage because of the capacitor. The voltage across the resistor is in phase with the resistor. This results in the output voltage leading the input voltage.

In Figure 15–9, the output is taken across the capacitor. The current in the circuit leads the applied voltage. However, the voltage across the capacitor lags the applied voltage.

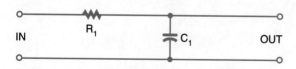

Figure 15–9 Lagging output phase-shift network. The voltage across the capacitor lags the applied voltage.

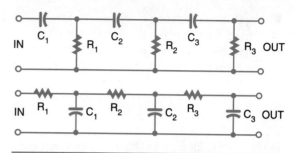

Figure 15–10 Cascaded RC phase-shift networks.

To achieve greater phase shifts, several RC networks may be cascaded (connected) together (Figure 15–10). Cascaded networks reduce the output voltage, however. An amplifier is needed to raise the output voltage to the proper operating level.

Phase-shift networks are valid for only one frequency, because the capacitive reactance varies with changes in frequency. Changing the reactance results in a different phase shift.

15–2 Questions

1. What are three uses of resistor-capacitor networks in electronic circuits?
2. Draw a diagram of a low-pass filter and describe how it works.
3. Draw a diagram of a high-pass filter and describe how it works.
4. What is the purpose of a decoupling network?
5. Where are RC phase-shift networks used?

SUMMARY

■ When an AC voltage is applied to a capacitor, it gives the appearance of current flow.
■ The capacitor charging and discharging represents current flow.
■ The current leads the applied voltage by 90 degrees in a capacitive circuit.
■ Capacitive reactance is the opposition a capacitor offers to the applied voltage.

- Capacitive reactance is represented by X_C.
- Capacitive reactance is measured in ohms.
- Capacitive reactance can be calculated by the formula:

$$X_C = \frac{1}{2\pi fC}$$

- RC networks are used for filtering, coupling, and phase shifting.

- A filter is a circuit that discriminates against certain frequencies.
- A low-pass filter passes frequencies below a cut-off frequency. It consists of a resistor and capacitor in series.
- A high-pass filter passes frequencies above a cut-off frequency. It consists of a capacitor and resistor in series.
- Coupling networks pass AC signals but block DC signals.

Chapter 15 Self-Test

1. What is the relationship between current and the applied voltage in a capacitive circuit?
2. What is the capacitive reactance of a 1000-microfarad capacitor at 60 hertz?
3. In question number 2, what is the current flow through the capacitor with 12 volts applied?
4. List three applications for capacitive circuits.
5. Why are capacitive coupling circuits important?

16

INDUCTIVE AC CIRCUITS

Objectives

After completing this chapter, the student will be able to:

- Describe the phase relationship between current and voltage in an inductive AC circuit.
- Determine the inductive reactance in an AC circuit.
- Explain impedance and its effect on inductive circuits.
- Describe how an inductor-resistor network can be used for filtering and phase shifting.
- Explain how low-pass and high-pass inductive circuits operate.

Inductors, like capacitors, oppose current flow in AC circuits. They may also introduce a phase shift between the voltage and the current in AC circuits. A large number of electronic circuits are composed of inductors and resistors.

16–1 Inductance in AC Circuits

Inductors in AC circuits offer opposition to current flow. When an AC voltage is placed across an inductor, it creates a magnetic field. As the AC voltage changes polarity, it causes the magnetic field to expand and collapse. It also induces a voltage in the inductor coil. This induced voltage is called a counter-electromotive force (CEMF), the greater the inductance, the greater the CEMF. The CEMF is out of phase with the applied voltage by 180 degrees (Figure 16–1) and opposes the applied voltage. This opposition is as effective in reducing current flow as a resistor.

The amount of voltage induced in the inductor depends on the rate of change of the magnetic field. The faster the magnetic field expands and collapses, the greater the induced voltage. The total effective voltage across the inductor is the difference between the applied voltage and the induced voltage. The induced voltage is always less than the applied voltage.

Figure 16–2 shows the relationship of the current to the applied voltage. Note that in a purely inductive circuit, the current lags behind the applied voltage by 90 degrees.

The opposition offered to current flow by an inductor in an AC circuit is called **inductive reactance.** Inductive reactance is measured in ohms. The amount of inductive reactance offered by an

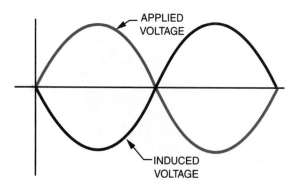

Figure 16–1 The applied voltage and the induced voltage are 180 degrees out of phase with each other in an inductive circuit.

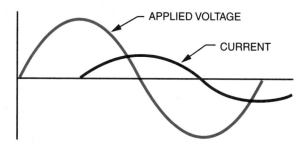

Figure 16–2 The current lags the applied voltage in an AC inductive circuit.

inductor depends on its inductance and the frequency of the applied voltage. The larger the inductance, the larger the magnetic field generated and the greater the opposition to current flow. Also, the higher the frequency, the greater the opposition to current flow.

Inductive reactance is expressed by the symbol X_L. Inductive reactance is determined by the formula:

$$X_L = 2\pi fL$$

where: π = pi or 3.14
 f = frequency in hertz
 L = inductance in henries

EXAMPLE What is the inductive reactance of a 0.15-henry coil at 60 hertz?

Given: *Solution:*
X_L = ? $X_L = 2\pi fL$
π = 3.14 $X_L = (2)(3.14)(60)(0.15)$
f = 60 Hz X_L = 56.52 ohms
L = 0.15 H

EXAMPLE What is the inductive reactance of a 0.15-henry coil at 400 hertz?

Given: *Solution:*
X_L = ? $X_L = 2\pi fL$
π = 3.14 $X_L = (2)(3.14)(400)(0.15)$
f = 400 Hz X_L = 376.80 ohms
L = 0.15 H

Notice that the inductive reactance increases with the increase in frequency.

Ohm's law applies to inductive reactance in AC circuits in the same manner that it applies to resistance. The inductive reactance in an AC circuit is directly proportional to the applied voltage and inversely proportional to the current. This relationship is expressed as:

$$I = \frac{E}{X_L}$$

The inductance increases with an increase in voltage or a decrease in current. Likewise, a decrease in inductive reactance results from a decrease in voltage or an increase in current.

EXAMPLE How much current flows through a 250-millihenry inductor when a signal of 12 volts and 60 hertz is applied across it?

Given:

$$X_L = ?$$
$$\pi = 3.14$$
$$f = 60 \text{ Hz}$$
$$L = 0.25 \text{ H}$$
$$I = ?$$
$$E = 12 \text{ V}$$

Solution:
First, find the inductive reactance (X_L).

$$X_L = 2\pi fL$$
$$X_L = (2)(3.14)(60)(0.25)$$
$$X_L = 94.20 \text{ ohms}$$

Using X_L, now find the current flow (I).

$$I = \frac{E}{X_L}$$
$$I = \frac{12}{94.2}$$
$$I = 0.127 \text{ amp or } 127 \text{ mA}$$

EXAMPLE How much voltage is necessary to cause 10 milliamperes to flow through a 15-millihenry choke at 400 Hz?

Given:

$$X_L = ?$$
$$\pi = 3.14$$

$$f = 400 \text{ Hz}$$
$$L = 0.015 \text{ H}$$
$$I = 0.01 \text{ A}$$
$$E = ?$$

Solution:
First, find the inductive reactance (X_L).

$$X_L = 2\pi fL$$
$$X_L = (2)(3.14)(400)(0.015)$$
$$X_L = 37.68 \text{ ohms}$$

Using X_L, now find the voltage (E).

$$I = \frac{E}{X_L}$$
$$0.01 = \frac{E}{37.68}$$
$$\frac{0.01}{1} = \frac{E}{37.68}$$
$$(I)(E) = (0.01)(37.68)$$
$$E = 0.38 \text{ volt}$$

EXAMPLE What is the inductive reactance of a coil that has 120 volts applied across it with 120 milliamperes of current flow?

Given:

$$I = 0.12 \text{ A}$$
$$E = 120 \text{ V}$$
$$X_L = ?$$

Solution:

$$I = \frac{E}{X_L}$$
$$0.12 = \frac{120}{X_L}$$
$$\frac{0.12}{1} = \frac{120}{X_L}$$
$$(1)(120) = (0.12)(X_L)$$
$$\frac{(1)(120)}{0.12} = \frac{(0.12)(X_L)}{0.12}$$
$$\frac{(1)(120)}{0.12} = 1 X_L$$
$$1000 \text{ ohms} = X_L$$

The impedance of a circuit containing both inductance and resistance is the total opposition to current flow by both the inductor and the resistor.

use of the phase shift caused by the inductor, inductive reactance and the resistance cannot e added directly. The impedance is the vector sum of the inductive reactance and the resistance in the circuit. The impedance is expressed in ohms and is designated by the letter Z. Impedance can be defined in terms of Ohm's law as:

$$I = \frac{E}{Z}$$

The most common inductive circuit consists of a resistor and an inductor connected in series. This is referred to as an *RL circuit*. The impedance of a series RL circuit is the square root of the sum of the squares of the inductive reactance and the resistance. This can be expressed as:

$$Z = \sqrt{R^2 + X_L^2}$$

EXAMPLE What is the impedance of a 100-millihenry choke in series with a 470-ohm resistor with 12 volts, 60 hertz applied across them?

Given:

$$X_L = ?$$
$$\pi = 3.14$$
$$f = 60 \text{ Hz}$$
$$100 \text{ mH} = 0.1 \text{ H}$$
$$Z = ?$$
$$R = 470 \text{ ohms}$$

Solution:
First, find the inductive reactance (X_L).

$$X_L = 2\pi fL$$
$$X_L = (2)(3.14)(60)(0.1)$$
$$X_L = 37.68 \text{ ohms}$$

Using X_L, now find the impedance (Z).

$$Z = \sqrt{R^2 + X_L^2}$$
$$Z = \sqrt{(470)^2 + (37.68^2)}$$
$$Z = \sqrt{220,900 + 1419.78}$$
$$Z = \sqrt{222,319.78}$$
$$Z = 471.51 \text{ ohms}$$

When inductors are connected in series, the inductive reactance is equal to the sum of the individual inductive reactance values.

$$X_{L_T} = X_{L_1} + X_{L_2} + X_{L_3} \ldots + X_{L_n}$$

When inductors are connected in parallel, the reciprocal of the inductive reactance is equal to the sum of the reciprocals of the individual inductive reactance values.

$$\frac{1}{X_{L_T}} = \frac{1}{X_{L_1}} + \frac{1}{X_{L_2}} + \frac{1}{X_{L_3}} \ldots + \frac{1}{X_{L_n}}$$

16–1 Questions

1. What does an AC voltage do when applied across an inductor?
2. What is the relationship between current and voltage in an inductive circuit?
3. What is inductive reactance?
4. What is the inductive reactance of a 200-millihenry inductor at 10,000 hertz?
5. How is the impedance determined for an inductor-resistor network?

16–2 Applications of Inductive Circuits

Inductive circuits are widely used in electronics. Inductors compete with capacitors for filtering and phase-shift applications. Because inductors are larger, heavier, and more expensive than capacitors, they have fewer applications than capacitors. However, inductors have the advantage of providing a reactive effect while still completing a DC circuit path. Capacitors can provide a reactive effect, but block the DC elements.

Inductors are sometimes combined with capacitors to improve the performance of a circuit. The reactive effect of the capacitor then opposes the reactive effect of the inductor. The end result is that they complement each other in a circuit.

Series RL networks are used as low- and high-pass filters. Figure 16–3 shows the two basic types

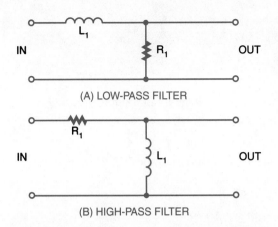

(A) LOW-PASS FILTER

(B) HIGH-PASS FILTER

Figure 16–3 RL filters.

of filters. The circuits are essentially resistor-inductor voltage dividers. Figure 16–3A is a low-pass filter. The input is applied across the inductor and resistor. The output is taken from across the resistor. At low frequencies, the reactance of the coil is low. Therefore it opposes little current and most of the voltage is dropped across the resistor. As the input frequency increases, the inductive reactance increases and offers more opposition to current flow, so that more voltage drops across the inductor. With more voltage dropped across the inductor, less voltage is dropped across the resistor. Increasing the input frequency decreases the output voltage. Low frequencies are passed with little reduction in amplitude while high frequencies are greatly reduced in amplitude.

Figure 16–3B is a high-pass filter. The input is applied across the inductor and resistor and the output is taken across the inductor. At high frequencies, the inductive reactance of the coil is high, causing most of the voltage to be dropped across the coil. As the frequency decreases, the inductive reactance decreases, offering less opposition to the current flow. This causes less voltage to be dropped across the coil and more voltage to be dropped across the resistor.

The frequency above or below the frequencies passed or attenuated is called the *cut-off frequency*. The symbol for the cut-off frequency is f_{CO}.

The cut-off frequency can be determined by the formula:

$$f_{CO} = \frac{R}{2\pi fL}$$

where: f_{CO} = cut-off frequency in hertz
R = resistance in ohms
π = 3.14
f = frequency in hertz
L = inductance in henries

16–2 Questions

1. What are the disadvantages of using inductors in circuits?
2. What are the advantages of using inductors in circuits?
3. Draw a diagram of an RL low-pass filter and explain how it operates.
4. Draw a diagram of an RL high-pass filter and explain how it operates.
5. How can the cut-off frequency of an RL circuit be determined?

SUMMARY

■ In a pure inductive circuit, the current lags the applied voltage by 90 degrees.
■ Inductive reactance is the opposition to current flow offered by an inductor in an AC circuit.
■ Inductive reactance is represented by X_L.
■ Inductive reactance is measured in ohms.
■ Inductive reactance can be calculated by the formula:

$$X_L = 2\pi fL$$

■ Impedance is the vector sum of the inductive reactance and the resistance in the circuit.
■ Series RL circuits are used for low- and high-pass filters.

Chapter 16 Self-Test

1. What is the relationship between current and the applied voltage in an inductive circuit?
2. What factor affects the inductive reactance of an inductive circuit?
3. What is the inductive reactance of a 100-millihenry coil at 60 hertz?
4. How much current would flow through the inductor in question number 3 if 24 volts were applied?
5. How are inductors used in circuits?
6. What is the cut-off frequency of an inductive circuit?

17

RESONANCE CIRCUITS

Objectives

After completing this chapter, the student will be able to:

- Identify the formulas for determining capacitive and inductive reactance.
- Identify how AC current and voltage react in capacitors and inductors.
- Determine the reactance of a series circuit, and identify whether it is capacitive or inductive.
- Define the term *impedance*.
- Solve problems for impedance that contain both resistance and capacitance or inductance.
- Discuss how Ohm's law must be modified prior to using it for AC circuits.
- Solve for X_C, X_L, X, Z, and I_T in RLC series circuits.
- Solve for I_C, I_L, I_X, I_R, and I_Z in RLC parallel circuits.

In previous chapters, resistance, capacitance, and inductance in AC circuits were looked at individually in a circuit. In this chapter, resistance, capacitance, and inductance will be observed as connected together in an AC circuit. The concepts covered in this chapter will not present any new material, but will apply all of the principles presented so far.

When the reactance of the inductor equals the reactance of the capacitor, it forms a resonant circuit. Resonant circuits are used in a variety of circuits in electronics.

17-1 Reactance in Series Circuits

A series circuit (Figure 17–1) DC voltage is generally supplied by a battery. The value of the amount of current flowing in a circuit may be determined by Ohm's law which uses the battery voltage and resistance of the circuit.

When an AC voltage source is applied to a circuit (Figure 17–2), the instantaneous value of current flowing through the resistor varies with the alternating output voltage applied. Ohm's law applies to the AC circuit just as it does with the DC circuit. Peak current may be calculated from the source's peak voltage, and rms current from the rms voltage.

Figure 17–3 shows a graph of one cycle of AC. It shows that when the voltage reaches a peak, the current also reaches a peak. Both the voltage and the current crosses the zero line together. The two waveforms are said to be *in phase*. This is the condition that occurs when a circuit contains pure resistance and no reactive components.

The effects of pure inductance or capacitance causes the voltage or current of a circuit to be 90 degrees *out of phase*. This situation becomes more complex when both reactive and resistance

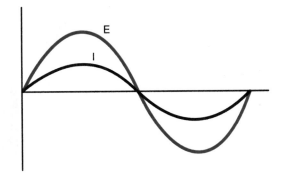

Figure 17–3 In a resistive AC circuit, current and voltage are in phase.

components are combined—a condition that is typical of an AC circuit.

In Figure 17–4, the components are shown connected in a series circuit, and the current flows equally through both components. Resistor R_1 has a value of 47 ohms, and the calculated value of inductive reactance for inductor L_1 is 25.25 ohms for the frequency of 60 hertz. The current flow in the circuit is 2.25 amperes, and the voltage drop across resistor R_1 is 105.75 volts and across inductor L_1, is 56.81 volts. These voltages appear to be incorrect because they total more than the supply voltage of 120 volts. This occurs because the voltage across the resistor is out of phase with the voltage across the inductor.

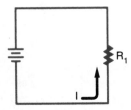

Figure 17–1 DC resistive circuit.

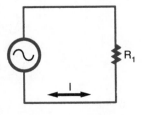

Figure 17–2 AC resistive circuit.

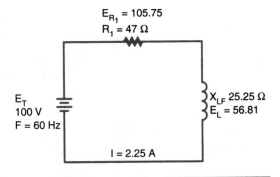

Figure 17–4 Voltage in either RL circuits such as this one or in RC circuits are not in phase and cannot be added directly.

Voltage and current are in phase in the resistive portion of the circuit. In the inductive portion of the circuit, voltage leads the current by 90 degrees. Since the current at all points in a series circuit must be in phase, then E_R and E_L are out of phase.

A way to represent voltage and current in circuits, such as the one in Figure 17–4, is to use vectors. Vectors are arrows that start at the origin of a coordinate system and point in a particular direction (Figure 17–5). The length of the arrow indicates the magnitude—the longer the arrow, the larger the value. The angle the arrow makes with the x-axis indicates its phase in degrees. The positive x-axis is zero degrees and the degrees increase as the arrow is moved in a counter-clockwise direction.

In Figure 17–4, the current flows through both components with the same phase. The current vector is used as the zero degree reference and will lie on the x-axis. Voltage E_R is in phase with the current and its vector is also placed on the x-axis.

The voltage across the inductor, E_L, is 90 degrees ahead of the current. Therefore, its vector points straight up, or 90 degrees from the x-axis. Each vector is drawn to scale. The source voltage vector (E) is started at the origin and ends at the maximum values of E_R and E_L as shown in Figure 17–5. The angle of the vector E (θ) is the phase between the source voltage and current, and its length indicates the voltage's magnitude.

The base voltage diagram may be rearranged to form a right triangle, with the hypotenuse representing the longest vector (Figure 17–6). Scale drawings are not needed to determine magnitude because the Pythagorean Theorem states:

$$E = \sqrt{E_R{}^2 + E_L{}^2}$$

This formula may be used to calculate any vector when the other two legs are known.

Vector representation also allows the use of trigonometric functions to determine voltage when only one voltage and the phase angle is known. It also determines the phase angle when the two voltages are known. These relationships are:

$$\text{Sin } \theta = E_L/E_T$$
$$\text{Cos } \theta = E_R/E_T$$
$$\text{Tan } \theta = E_L/E_R$$

Using the example in Figure 17–4 and any of the relationships shown, the phase difference may be determined between the supply voltage and current as 28.26 degrees.

Experimenting with different values of E_L and E_R will reveal several useful tips to remember:

■ When a circuit is purely resistive, the phase angle is zero because voltage and current are in phase.

■ As the inductive reactance increases, the phase angle becomes greater until it reaches 45 degrees when the resistance and the reactance are equal in value. As the inductive reactance increases further, the angle will approach 90 degrees.

■ When a circuit contains pure reactance with no resistance the phase angle will increase to 90 degrees.

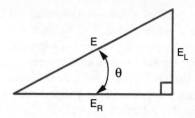

Figure 17–6 Vectors can be rearranged to form a right triangle.

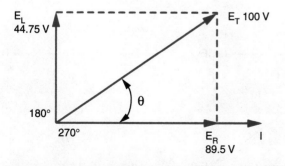

Figure 17–5 Vector can be used to show the relationship between voltages in a reactive circuit.

Current flow is the same through all components in a series circuit. The voltage drop across any component in the circuit is proportional to the resistance or reactance of that component. If vectors are drawn for the resistance and reactance of the circuit, they would be proportional to those drawn for the voltage (Figure 17–4). The resistance, R, would be drawn at 0 degrees and the inductive reactance, X_L, would be drawn at 90 degrees.

The combined effect of resistance and reactance is called impedance and is represented by the symbol Z. Impedance must be used to calculate the current in a reactive circuit when the supply voltage is known. Dividing the source voltage by the resistance added to the reactance will yield an incorrect answer because the voltages involved are not in phase with each other. Vectors may be used to describe the circuit impedance (Figure 17–7).

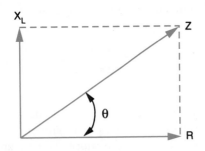

Figure 17–7 Vectors can also be used to describe impedance.

The series RC circuit in Figure 17–8A is described by the vector diagrams in Figure 17–8B and 17–8C. Again, the current is used as the zero-degree reference point, with E_R at 0 degrees since it is in phase with the current. Remember, in a capacitive circuit, the voltage lags 90 degrees behind the current, so its voltage vector (E_C) is drawn downward. The phase angle of such a circuit is sometimes given a negative value, although it is just as acceptable to specify "leading" or "lagging" instead. All the same trigonometric equations and the Pythagorean Theorem may be applied to the vectors.

17–1 Questions

1. Draw a graph showing the relationship of current and voltage in a purely resistive series circuit.
2. Show the relationship between current and voltage for the circuit shown in Figure 17–9.
3. For the circuit in Figure 17–9, draw a graph showing the vector relationship. Include all arrows and label each vector.
4. Using the Pythagorean Theorem, determine the impedance for the circuit in Figure 17–9.
5. Determine the phase angle for the circuit shown in Figure 17–9.
6. Verify the impedance for the circuit in Figure 17–9.
7. Verify the current flow for the circuit in Figure 17–9.

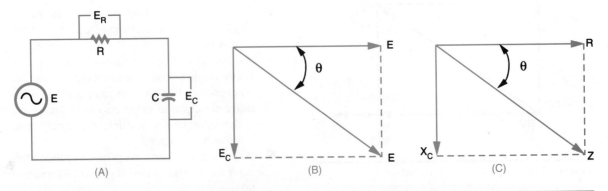

Figure 17–8 Vectors can be used to describe capacitive AC circuits, the same as inductive circuits.

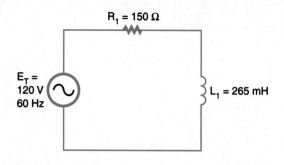

Figure 17-9

17-2 Reactance in Parallel Circuits

Parallel circuits containing inductors and capacitors may also be analyzed with vector diagrams. However, the vectors used are current vectors because the voltage across each component must be equal and in phase with each other.

Figure 17-10 shows a parallel RL circuit and its resulting vector diagram. The vector for the resistor current is placed at 0 degrees. It is a simple matter to plot the vector for the inductor's current, I_L. The current through the inductor lags behind the voltage by 90 degrees, so the vector is drawn downward.

Figure 17-11 shows the vector diagram for a parallel RC circuit. Notice that the capacitive and inductive vectors (Figure 17-10) for parallel circuits are drawn in the opposite direction to those of a series circuit. Remember, in parallel circuits, current is examined and not voltage.

Just as the resistance of a parallel circuit is always less than the value of the smallest resistor, the impedance of a parallel RL or RC circuit is smaller than both the individual resistance or reactance. Vectors may be used to determine parallel circuit impedance; however, the reciprocal values of R, X, and Z must be used. This makes impedance calculations a little more complex for parallel circuits. All the trigonometric functions mentioned previously are applicable to parallel circuits provided that the vectors are drawn correctly. Mathematically, for Figures 17-10 and 17-11:

Inductive circuits:

$$I = \sqrt{I_R^2 + I_L^2}$$
$$1/Z = 1/R + 1/X_L$$

Capacitive circuits:

$$I = \sqrt{I_R^2 + I_C^2}$$
$$1/Z = 1/R + 1/X_C$$

17-2 Questions

1. Using the Pythagorean Theorem, calculate the resultant current flow for the circuit shown in Figure B.
2. Calculate the impedance of the circuit in Figure 17-12.

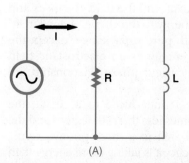

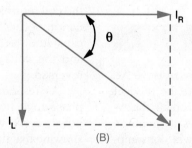

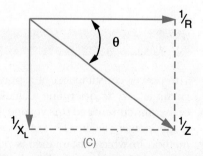

Figure 17-10 Vectors can be used to analyze parallel inductance circuits. Current flow, not voltage, is used because the voltage across each component is equal an in phase.

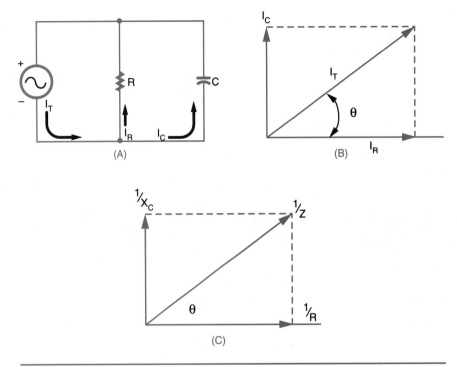

Figure 17–11 Vectors can be used to analyze parallel capacitance circuits.

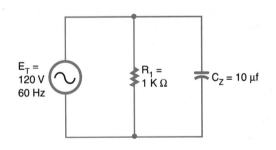

Figure 17–12

17–3 Power

The power consumption of a purely resistive AC circuit is easy to determine. Calculate the product of the rms current and rms voltage to obtain the average power. Figure 17–13A shows a graphical method in which instantaneous power consumption may be calculated by plotting current and voltage on the same axes and then performing successive multiplications to plot the power curve.

The same principle may be applied to the reactive circuit (Figure 17–13B). Remember that a circuit containing just pure inductance shows current lagging voltage by 90 degrees. Plotting the power curve for this circuit yields an alternating waveform that is centered on 0. The net power consumption of an inductive circuit is low.

During the positive portion of the waveform, the inductor takes energy and stores it in the form of a magnetic field. During the negative portion of the waveform, the field collapses and the coil returns energy to the circuit. A similar situation occurs with pure capacitance, except the capacitor stores energy as an electrostatic field and the voltage current phase relationship are reversed.

As resistance is introduced to a circuit, the phase angle becomes less than 90 degrees and the power curve will shift to a more positive value, showing that the circuit is taking more energy than it is returning. However, the capacitive or inductive part of the circuit still stores and releases energy and consumes no power. The power loss is due

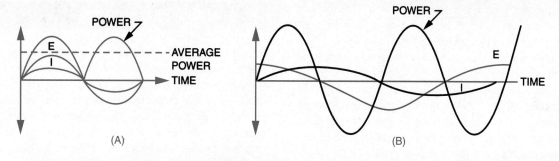

Figure 17–13 Power dissipation in a resistive circuit has a non-zero value (A). In a reactive circuit, there is no average or net power loss (B).

entirely to the resistance. Remember, only the resistive part of the circuit consumes power.

Figure 17–14 shows a capacitive circuit. Using the Pythagorean Theorem, the resistance of 100 ohms and the capacitance reactance of 50 ohms gives a combined impedance (Z) of approximately 112 ohms as follows:

$$Z = \sqrt{R^2 + X_{C^2}}$$
$$Z = \sqrt{100^2 + 50^2}$$
$$Z = \sqrt{10,000 + 2,500}$$
$$Z = \sqrt{12,500}$$
$$Z = 111.8 \ \Omega$$

Using Ohm's law, this would allow a current of approximately 1 amp to flow in the circuit. The voltage drop across the resistor would be 100 volts, so the true power dissipated by the resistor would be approximately 100 watts.

The capacitive part of the circuit consumes no power, yet multiplying the source voltage by circuit current yields an answer of 112 watts. The difference is accounted for by the difference in phases between the various voltages. As opposed to true power, the figure obtained by multiplying the source voltage and current is known as *apparent power*, and it is specified as 112 VA or 112 volt-amperes. The true power of the circuit can never

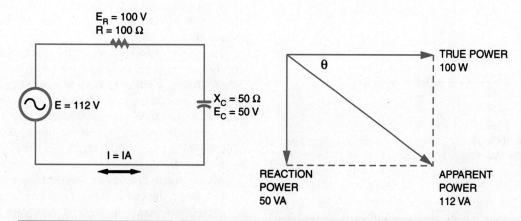

Figure 17–14 In a reactive circuit, the true power dissipated with resistance and the reactive power supplied to its reactance vectorly sum to produce an apparent power vector.

be greater than the apparent power. The ratio of true power in watts to apparent power in volt-amps is called the *power factor*.

Vector diagrams may be used to analyze the power factor. If the phase angle of the circuit is known, the power factor may be calculated by taking the cosine of the angle. In the circuit shown in Figure 17–14, it is the same as the ratio of E_R to source voltage. Several ways exist to obtain the power factor based on the data available. In a pure resistive circuit, the true power is the same as the apparent power, so the power factor is 1. In a pure inductive or capacitive circuit the true power is 0—no matter what the apparent power is—so the power factor is 0. Power factor is an important consideration in heavy industrial power distribution where the cables must be capable of handling the apparent power load.

17–3 Questions

1. What is the net power consumption of an inductive circuit? Why?
2. Calculate the apparent power for the circuit shown in Figure 17–15.
3. Define power factor.
4. What is the relationship between true power and apparent power?

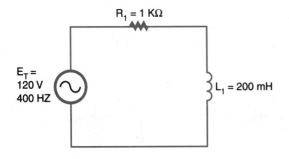

Figure 17–15

17–4 Introduction to Resonance

Resonance is a phenomenon that occurs in many fields, as well as in electronics. A resonance device produces a broadening and dampening effect. In electronics, resonant circuits pass desired frequencies and reject all others. The ability of a series or parallel combination of X_L or X_C to produce resonance provides a number of unique applications. Resonant circuits make it possible for a radio or television receiver to tune-in and receive a station at a particular frequency. The tuning circuit consists of a coil of wire in parallel with a capacitor. Parallel tuned circuits have maximum impedance at resonant frequencies. Tuned circuits are vital to a variety of types of communication equipment from radios to radar.

Resonance occurs when a circuit's inductive and capacitive reactance are balanced. Previously, it was mentioned that inductive reactance increases with frequency, and capacitive reactance increases with frequency. If both inductive and capacitive components are in an AC circuit, at one particular frequency their reactances will be equal but opposite. This condition is referred to as *resonance*. A circuit that contains this characteristics is called a *resonant circuit*. Both inductance and capacitance must be present for this condition to occur.

In any resonant circuit, some resistance is usually present. While the resistance does not have an effect on the resonant frequency, it does affect other resonant circuit parameters that will be explained later.

The value of inductance and capacitance determines the specific frequency of resonance of a circuit. Changing either or both will result in a different resonant frequency. Typically, the larger the value of inductance and capacitance, the lower the resonant frequency. Therefore, the smaller the value of inductance and capacitance, the higher the resonant frequency.

Above and below the resonant frequency of any LC circuit, the circuit behaves as any standard AC circuit. Resonance is desired with radio frequencies in tuning receivers and transmitters, certain industrial equipment, and test equipment. It is undesired in audio amplifiers and power supplies. Resonant circuits are not used in the audio bands of frequencies.

17—4 Questions

1. When does resonance occur?
2. Where are resonance circuits used?
3. What is the relationship between size of components and resonance frequencies?

SUMMARY

■ Ohm's law applies to AC circuits the same as it does to DC circuits.
■ The AC current lags the voltage by 90 degrees in an inductor (ELI).

■ The AC current leads the voltage by 90 degrees in a capacitor (ICE).
■ Vector representation allows the use of trignometric functions to determine voltage or current when the phase angle is known.
■ The combined effect of resistance and inductance or capacitance is called impedance.
■ Apparent power is obtained by multiplying the source voltage and current with units of volt-amperes.
■ The ratio of true power to apparent power in volts-amperes is called the power factor.
■ Power factor is very important in the consideration of heavy industrial power distribution.
■ Resonance circuits make it possible for a circuit to tune a station to a particular frequency.
■ Resonance is desired for radio frequency in tuning circuits.

Chapter 17 Self-Test

1. What are the values of X_C, X_L, X, Z, and I_T for the circuit shown in Figure 17–16?

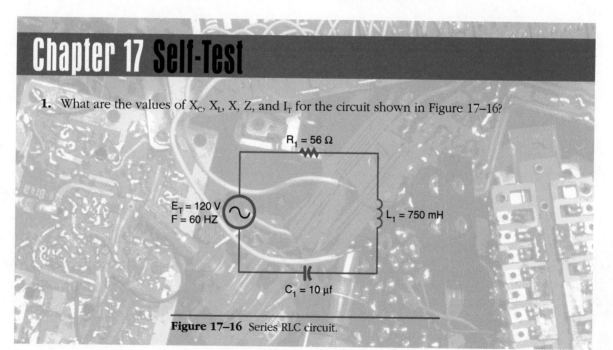

Figure 17–16 Series RLC circuit.

2. What are the values of I_C, I_L, I_X, I_R, and I_Z for the circuit shown in Figure 17–17?

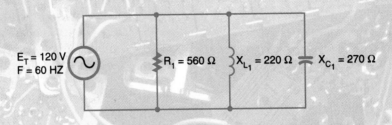

$E_T = 120 \text{ V}$
$F = 60 \text{ HZ}$

$R_1 = 560 \ \Omega$ $X_{L_1} = 220 \ \Omega$ $X_{C_1} = 270 \ \Omega$

Figure 17–17 Parallel RLC circuit.

18

TRANSFORMERS

Objectives

After completing this chapter, the student will be able to:

- Describe how a transformer operates.
- Explain how transformers are rated.
- Explain how transformers operate in a circuit.
- Describe the differences between step-up, step-down, and isolation transformers.
- Describe how the ratio of the voltage, current, and number of turns are related with a transformer.
- Describe applications of a transformer.
- Identify different types of transformers.

Transformers allow the transfer of an AC signal from one circuit to another. The transfer may involve stepping up the voltage, stepping down the voltage, or passing the voltage unchanged.

18-1 Electromagnetic Induction

If two electrically isolated coils are placed next to each other and an AC voltage is put across one coil, a changing magnetic field results. This changing magnetic field induces a voltage into the second coil. This action is referred to as **electromagnetic induction.** The device is called a *transformer*.

In a transformer, the coil containing the AC voltage is referred to as the *primary winding*. The other coil, in which the voltage is induced, is referred to as the *secondary winding*. The amount of voltage induced depends on the amount of mutual induction between the two coils. The amount of mutual induction is determined by the coefficient of coupling. The *coefficient of coupling* is a number from 0 to 1, with 1 indicating that all the primary flux lines cut the secondary windings and 0 indicating that none of the primary flux lines cut the windings.

The design of a **transformer** is determined by the frequency at which it will be used, the power it must handle, and the voltage it must handle. For example, the application of the transformer determines the type of core material that the coils are wound on. For low-frequency applications, iron cores are used. For high-frequency applications, air cores are used. Air cores are nonmetallic cores used to reduce losses associated with the higher frequencies.

Transformers are rated in **volt-amperes** (VA) rather than in power (watts). This is because of the loads that can be placed on the secondary winding. If the load is a pure capacitive load, the reactance could cause the current to be excessive. The power rating has little meaning where a volt-ampere rating can identify the maximum current the transformer can handle.

Figure 18–1 shows the schematic symbol of a transformer. The direction of the primary and secondary windings on the core determines the polarity of the induced voltage in the secondary winding. The AC voltage can either be in phase or 180 degrees out of phase with the induced voltage. Dots are used on the schematic symbol of the transformer to indicate polarity.

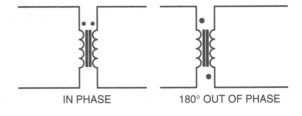

Figure 18–1 Transformer schematic symbol showing phase indication.

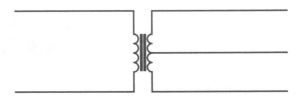

Figure 18–2 Transformer with a center-tapped secondary.

Transformers are wound with tapped secondaries (Figure 18–2). A center-tapped secondary is the equivalent of two secondary windings, each with half of the total voltage across them. The center tap is used for power supply to convert AC voltages to DC voltages. A transformer may have taps on the primary to compensate for line voltages that are too high or too low.

18-1 Questions

1. How does a transformer operate?
2. What determines the design of a transformer?
3. Give an example of how the application of a transformer determines its design.
4. How are transformers rated?
5. Draw and label the schematic symbol for a transformer.

18-2 Mutual Inductance

When a transformer is operated without a load (Figure 18–3) there is no secondary current flow.

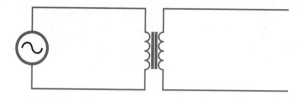

Figure 18–3 Transformer without a load on the secondary.

There is a primary current flow because the transformer is connected across a voltage source. The amount of primary current depends of the size of the primary windings. The primary windings act like an inductor. Exciting current is the small amount of primary current that flows. The exciting current overcomes the AC resistance of the primary winding and supports the magnetic field of the core. Because of inductive reactance, the exciting current lags behind the applied voltage. These conditions change when a load is applied across the secondary.

When a load is connected across the secondary winding (Figure 18–4) a current is induced into the secondary. Transformers are wound with the secondary on top of the primary. The magnetic field created by the primary current cuts the secondary windings. The current in the secondary establishes a magnetic field of its own. The expanding magnetic field in the secondary cuts the primary turns, inducing a voltage back into the primary. This magnetic field expands in the same direction as the current in the primary, aiding it and causing it to increase, with an effect called **mutual inductance.** The primary induces a voltage into the secondary and the secondary induces a voltage back into the primary.

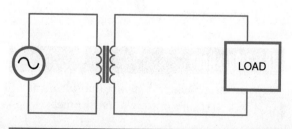

Figure 18–4 Transformer with a loaded secondary.

18–2 Questions

1. How does loading a transformer affect its operation?
2. Define *mutual inductance.*
3. Describe how a transformer induces a voltage into its secondary.

18–3 Turns Ratio

The turns ratio of a transformer determines whether the transformer is used to step up, step down, or pass voltage unchanged. The **turns ratio** is the number of turns in the secondary winding divided by the number of turns in the primary winding. This can be expressed as:

$$\text{turns ratio} = \frac{N_S}{N_P}$$

where: N = number of turns

A transformer with secondary voltage greater than its primary voltage is called a **step-up transformer.** The amount the voltage is stepped up depends on the turns ratio. The ratio of secondary to primary voltage is equal to the ratio of secondary to primary turns. This is expressed as:

$$\frac{E_S}{E_P} = \frac{N_S}{N_P}$$

Thus the turns ratio of a step-up transformer is always greater than one.

EXAMPLE A transformer has 400 turns on the primary and 1200 turns on the secondary. If 120 volts of AC current are applied across the primary, what voltage is induced into the secondary?

Given:

$E_S = ?$

$E_P = 120$ volts

$N_S = 1200$ turns

$N_P = 400$ turns

Solution:

$$\frac{E_S}{E_P} = \frac{N_S}{N_P}$$

$$\frac{E_S}{120} = \frac{1200}{400}$$

$$(120)(1200) = (E_S)(400)$$

$$\frac{(120)(1200)}{400} = E_S$$

$$360 \text{ volts} = E_S$$

A transformer that produces a secondary voltage less than its primary voltage is called a **step-down transformer.** The amount the voltage is stepped down is determined by the turns ratio. In a step-down transformer the turns ratio is always less than one.

EXAMPLE A transformer has 500 turns on the primary and 100 turns on the secondary. If 120 volts AC are applied across the primary, what is the voltage induced in the secondary?

Given:
$E_S = ?$
$N_P = 500$ turns
$N_S = 100$ turns
$E_P = 120$ volts

Solution:
$$\frac{E_S}{E_P} = \frac{N_S}{N_P}$$
$$\frac{E_S}{120} = \frac{100}{500}$$
$$(120)(100) = (E_S)(500)$$
$$\frac{(120)(100)}{500} = E_S$$
$$24 \text{ volts} = E_S$$

Assuming no transformer losses, the power in the secondary must equal the power in the primary. Although the transformer can step up voltage, it cannot step up power. The power removed from the secondary can never be more than the power supplied to the primary. Therefore, when a transformer steps up the voltage, it steps down the current so the output power remains the same. This can be expressed as:

$$P_P = P_S$$
$$(I_P)(E_P) = (I_S)(E_S)$$

The current is inversely proportional to the turns ratio. This can be expressed as:

$$\frac{I_P}{I_S} = \frac{N_S}{N_P}$$

EXAMPLE A transformer has a 10:1 turns ratio. If the primary has a current of 100 milliamperes, how much current flows in the secondary? (Note: The first number in the ratio refers to the primary, the second number to the secondary.)

Given:
$I_S = ?$
$N_P = 10$
$N_S = 1$
$I_P = 100 \text{ mA} = 0.1 \text{ A}$

Solution:
$$\frac{I_P}{I_S} = \frac{N_S}{N_P}$$
$$\frac{0.1}{I_S} = \frac{1}{10}$$
$$(1)(I_S) = (0.1)(10)$$
$$I_S = 1 \text{ amp}$$

An important application of transformers is in impedance matching. Maximum power is transferred when the impedance of the load matches the impedance of the source. When the impedance does not match, power is wasted.

For example, if a transistor amplifier can efficiently drive a 100-ohm amplifier, it will not efficiently drive a 4-ohm speaker. A transformer used between the transistor amplifier and speaker can make the impedance of the speaker appear to be in proportion. This is accomplished by choosing the proper turns ratio.

The *impedance ratio* is equal to the turns ratio squared. This is expressed as:

$$\frac{Z_P}{Z_S} = \left(\frac{N_P}{N_S}\right)^2$$

EXAMPLE What must the turns ratio of a transformer be to match a 4-ohm speaker to a 100-ohm source?

Given:
$N_P = ?$
$N_S = ?$
$Z_P = 100$
$Z_S = 4$

Solution:
$$\frac{Z_P}{Z_S} = \left(\frac{N_P}{N_S}\right)^2$$
$$\frac{100}{4} = \left(\frac{N_P}{N_S}\right)^2$$
$$\sqrt{25} = \frac{N_P}{N_S}$$
$$\frac{5}{1} = \frac{N_P}{N_S}$$

The turns ratio is 5:1.

18—3 Questions

1. What determines whether a transformer is a step-up or a step-down transformer?

2. Write the formula for determining the turns ratio of a transformer.
3. Write the formula for determining voltage based on the turns ratio of a transformer.
4. What is the secondary output of a transformer with 100 turns on the primary and 1800 turns on the secondary and 120 volts applied?

18—4 Applications

Transformers have many applications. Among them are: stepping up and stepping down voltage and current, impedance matching, phase shifting, isolation, blocking DC while passing AC, and producing several signals at various voltage levels.

Transmitting electrical power to homes and industry requires the use of transformers. Power stations are located next to sources of energy, and electrical power must often be transmitted over great distances. The wires used to carry the power have resistance, which causes power loss during the transmission. The power is equal to the current times the voltage:

$$P = IE$$

Ohm's law states that current is directly proportional to voltage and inversely proportional to resistance:

$$I = \frac{E}{R}$$

The amount of power lost, then, is proportional to the amount of resistance in the line. The easiest way to reduce power losses is to keep the current low.

EXAMPLE A power station produces 8500 volts at 10 amperes. The power lines have 100 ohms of resistance. What is the power loss of the lines?

Given:

$$P = ?$$
$$I = 10 \text{ amps}$$
$$E = ?$$
$$R = 100 \text{ ohms}$$

Solution:
First, find the amount of voltage drop.

$$I = \frac{E}{R}$$
$$10 = \frac{E}{100}$$
$$\frac{10}{1} = \frac{E}{100}$$
$$(1)(E) = (10)(100)$$
$$E = 1000 \text{ volts}$$

Using E, find the power loss.

$$P = IE$$
$$P = (10)(1000)$$
$$P = 10,000 \text{ watts}$$

Using a transformer to step the voltage up to 85,000 volts at 1 ampere, what is the power loss?

Given:

$$I = 1 \text{ amp}$$
$$E = ?$$
$$R = 100 \text{ ohms}$$

Solution:
First, find the amount of voltage drop.

$$I = \frac{E}{R}$$
$$1 = \frac{E}{100}$$
$$\frac{1}{1} = \frac{E}{100}$$
$$(1)(E) = (1)(100)$$
$$E = 100 \text{ volts}$$

Using E, find the power loss.

$$P = IE$$
$$P = (1)(100)$$
$$P = 100 \text{ watts}$$

How the transformer is wound determines whether it produces a phase shift or not. The application determines how important the phase shift

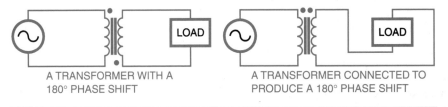

A TRANSFORMER WITH A
180° PHASE SHIFT

A TRANSFORMER CONNECTED TO
PRODUCE A 180° PHASE SHIFT

Figure 18–5 A transformer can be used to generate a phase shift.

is. Note: The phase can be shifted by simply reversing the leads to the load (Figure 18–5).

If DC voltage is applied to a transformer, nothing occurs in the secondary once the magnetic field is established. A changing current is necessary to induce a voltage in the secondary. A transformer can be used to isolate the secondary from any DC voltage in the primary (Figure 18–6).

Transformers are used to isolate electronic equipment from 120-volts AC, 60-hertz power while it is being tested (Figure 18–7). The reason for using a transformer is to prevent shocks. Without the transformer, one side of the power source is connected to the chassis. When the chassis is removed from the cabinet, the "hot" chassis presents

a shock hazard. This condition is more likely to occur if the power cord can be plugged in either way. A transformer prevents connecting either side of the equipment to ground. An **isolation transformer** does not step up or step down the voltage.

An **autotransformer** is a device used to step up or step down applied voltage. It is a special type of transformer in which the primary and the secondary windings are both part of the same core. Figure 18–8A shows an autotransformer stepping down a voltage. Because the secondary consists of fewer turns, the voltage is stepped down. Figure 18–8B shows an autotransformer stepping up a voltage. Because the secondary has more turns than the primary, the voltage is stepped up. A disadvantage of the autotransformer is that the secondary is not isolated from the primary. The advantage is that the autotransformer is cheaper and easier to construct than a transformer.

A special type of autotransformer is a variable autotransformer, in which the load is connected to a movable arm and one side of the autotransformer. Moving the arm varies the turns ratio,

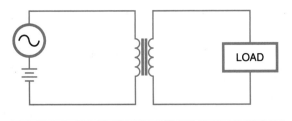

Figure 18–6 A transformer can be used to block DC voltage.

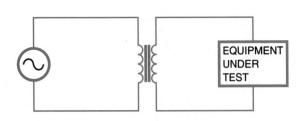

Figure 18–7 An isolation transformer prevents electrical shock by isolating the equipment from ground.

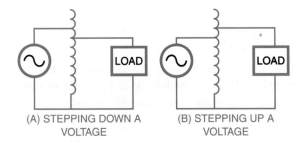

(A) STEPPING DOWN A
VOLTAGE

(B) STEPPING UP A
VOLTAGE

Figure 18–8 An autotransformer is a special type of transformer used to step up or step down the voltage.

producing a change in the voltage across the load. The output voltage can be varied from 0 to 130 volts AC.

18—4 Questions

1. What are the applications of transformers?
2. How are transformers used in transmitting electrical power to the home?
3. How can a transformer produce a phase shift of the input signal?
4. Why are isolation transformers important when working on electronic equipment?
5. What is an autotransformer used for?

SUMMARY

■ A transformer consists of two coils, a primary winding, and a secondary winding.
■ An AC voltage is put across the primary winding, inducing a voltage in the secondary winding.
■ Transformers allow an AC signal to be transferred from one circuit to another.
■ Transformers allow stepping up, stepping down, or passing the signal unchanged.

■ Transformers are designed to operate at certain frequencies.
■ Transformers are rated in volt-amperes (VA).
■ The schematic symbol used for iron core transformers is:

■ The turns ratio determines whether a transformer is used to step up, step down, or pass voltage unchanged.

$$\text{turns ratio} = \frac{N_S}{N_P}$$

■ The ratio of secondary to primary voltage is equal to the ratio of secondary to primary turns.

$$\frac{E_S}{E_P} = \frac{N_S}{N_P}$$

■ A transformer that produces a secondary voltage greater than its primary voltage is called a step-up transformer.
■ The turns ratio of a step-up transformer is always greater than 1.

- A transformer that produces a secondary voltage less than its primary voltage is called a step-down transformer.
- The turns ratio of a step-down transformer is always less than 1.
- The amount the voltage is stepped up or down is determined by the turns ratio.
- Transformer applications include: impedance matching, phase shifting, isolation, blocking DC while passing AC, and producing several signals at different voltage levels.
- An isolation transformer passes the signal unchanged.
- An isolation transformer is used to prevent electric shocks.
- An autotransformer is used to step up or step down voltage.
- An autotransformer is a special transformer that does not provide isolation.

Chapter 18 Self-Test

1. Explain how electromagnetic induction induces a voltage into the secondary of a transformer.

2. Why are transformers rated in volt-amperes rather than watts?

3. What is the difference between two transformers, one that has voltage applied to the primary without a load on the secondary and one that has a load on the secondary?

4. What turns ratio is required on the secondary of a transformer if the primary has 400 turns? The applied voltage is 120 volts AC and the secondary voltage is 12 volts.

5. What turns ratio is required for an impedance-matching transformer to match a 4-ohm speaker to a 16-ohm source?

6. Explain why transformers are important for transmitting electrical power to residential and industrial needs.

7. How does an isolation transformer prevent electrical shock?

Semiconductor Devices

19

SEMICONDUCTOR FUNDAMENTALS

Objectives

After completing this chapter, the student will be able to:

- Identify materials that act as semiconductors.
- Define *covalent bonding*.
- Describe the doping process for creating N- and P-type semiconductor materials.
- Explain how doping supports current flow in a semiconductor material.

Semiconductors are the basic components of electronic equipment. The more commonly used semiconductors are the *diode* (used to rectify), the *transistor* (used to amplify), and the *integrated circuit* (used to switch or amplify). The primary function of semiconductor devices is to control voltage or current for some desired result.

Advantages of semiconductors include the following:

- Small size and weight
- Low power consumption at low voltages
- High efficiency
- Great reliability
- Ability to operate in hazardous environments
- Instant operation when power is applied
- Economic mass production

Disadvantages of semiconductors include:

- Great susceptibility to changes in temperature
- Extra components required for stabilization
- Easily damaged (by exceeding power limits, by reversing polarity of operating voltage, by excess heat when soldering into circuit)

19—1 Semiconduction in Germanium and Silicon

Semiconductor materials have characteristics that fall between those of insulators and conductors. Three pure semiconductor elements are carbon (C), germanium (Ge), and silicon (Si). Those suitable for electronic applications are germanium and silicon.

Germanium is a brittle, grayish-white element discovered in 1886. A powder, germanium dioxide, is recovered from the ashes of certain types of coal. The powder is then reduced to the solid form of pure germanium.

Silicon was discovered in 1823. It is found extensively in the earth's crust as a white or sometimes colorless compound, silicon dioxide. Silicon dioxide (silica) can be found abundantly in sand, quartz, agate, and flint. It is then chemically reduced to pure silicon in a solid form. Silicon is the most commonly used semiconductor material.

Once the pure or **intrinsic material** is available, it must be modified to produce the qualities necessary for semiconductor devices.

As described in Chapter 1, the center of the atom is the nucleus, which contains protons and neutrons. The protons have a positive charge and the neutrons have no charge. Electrons orbit around the nucleus and have a negative charge. Figure 19–1 shows the structure of the silicon atom. The first orbit contains two electrons, the second orbit contains eight electrons, and the outer orbit, or valence shell, contains four electrons. **Valence** is an indication of the atom's ability to gain or lose electrons and determines the electrical and chemical properties of the atom. Figure 19–2 shows a simplified drawing of the silicon atom, with only the four electrons in the valence shell.

Materials that need electrons to complete their valence shell are not stable and are referred to as *active materials*. To gain stability, an active material must acquire electrons in its valence shell. Silicon atoms are able to share their valence electrons with other silicon atoms in a process called covalent bonding (Figure 19–3). **Covalent bonding** is the process of sharing valence electrons, resulting in the formation of crystals.

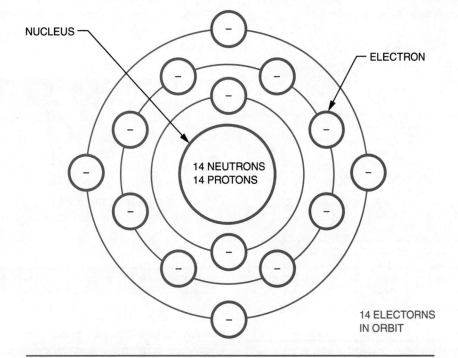

Figure 19–1 Atomic structure of silicon.

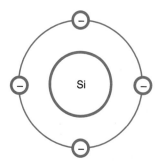

Figure 19–2 Simplified silicon atom shown with only valence electrons.

Each atom in such a crystalline structure has four of its own electrons and four shared electrons from four other atoms, a total of eight valence electrons. This covalent bond cannot support electrical activity because of its stability.

At room temperature, pure silicon crystals are poor conductors. They behave like insulators. If heat energy is applied to the crystals, however, some of the electrons absorb the energy and move to a higher orbit, breaking the covalent bond. This allows the crystals to support current flow.

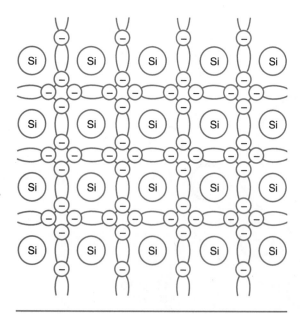

Figure 19–3 Crystalline structure of silicon with covalent bonding.

Silicon, like other semiconductor materials, is said to have a **negative temperature coefficient** because, as the temperature increases, its resistance decreases. The resistance is cut in half for every 6 degrees celsius of rise in temperature.

Like silicon, germanium has four electrons in its valence shell and can form a crystalline structure. Germanium's resistance is cut in half for every 10 degrees celsius of temperature rise. Thus germanium appears to be more stable with respect to temperature change than silicon. However, germanium requires less heat energy to dislodge its electrons than does silicon. Silicon has a thousand times more resistance than germanium at room temperature.

Heat is a potential source of trouble for semiconductors that is not easy to control. Good circuit design minimizes heat changes. Its resistance is what makes silicon preferable to germanium in most circuits. In some applications, heat-sensitive devices are necessary. In these the germanium temperature coefficient can be an advantage, therefore, germanium is used.

All early transistors were made of germanium. The first silicon transistor was not made until 1954. Today, silicon is used for most solid-state applications.

19–1 Questions

1. What is a semiconductor material?
2. Define the following terms:
   a. Covalent bonding
   b. Negative temperature coefficient
3. Why are silicon and germanium considered semiconductor materials?
4. Why is silicon preferred over germanium?

19–2 Conduction in Pure Germanium and Silicon

The electrical activity in semiconductor material is highly dependent on temperature. At extremely low temperatures, valence electrons are held tightly to the parent atom through the covalent bond. Because these valence electrons cannot drift, the material cannot support current flow. Germanium and silicon crystals function as insulators at low temperatures.

As the temperature increases, the valence electrons become agitated. Some of the electrons break the covalent bonds and drift randomly from one atom to the next. These free electrons are able to carry a small amount of electrical current if an electrical voltage is applied. At room temperature, enough heat energy is available to produce a small number of free electrons and to support a small amount of current. As the temperature increases, the material begins to acquire the characteristics of a conductor. Only at extremely high temperatures does silicon conduct current as ordinary conductors do. Typically, such high temperatures are not encountered under normal usage.

When an electron breaks away from its covalent bond, the space previously occupied by the electron is referred to as a hole (Figure 19–4). As described in Chapter 2, a **hole** simply represents the absence of an electron. Because an electron

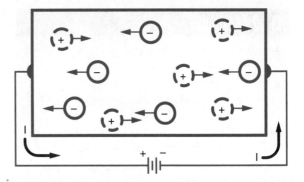

Figure 19–5 Current flow in pure semiconductor material.

has a negative charge, its absence represents the loss of a negative charge. A hole then has the characteristic of a positively charged particle. As an electron jumps from one valence shell to another valence shell with a hole, it leaves a hole behind it. If this action continues, the hole appears to move in the opposite direction to the electron.

Each corresponding electron and hole are referred to as an *electron-hole pair*. The number of electron-hole pairs increases with an increase in temperature. At room temperature a small number of electron-hole pairs exist.

When pure semiconductor material is subjected to a voltage, the free electrons are attracted to the positive terminal of the voltage source (Figure 19–5). The holes created by movement of the free electrons drift toward the negative terminal. As the free electrons flow into the positive terminal, an equal number leave the negative terminal. As the holes and electrons recombine, both holes and free electrons cease to exist.

In review, holes constantly drift toward the negative terminal of the voltage source. Electrons always flow toward the positive terminal. Current flow in a semiconductor consists of the movement of both electrons and holes. The amount of current flow is determined by the number of electron-hole pairs in the material. The ability to support current flow increases with the temperature of the material.

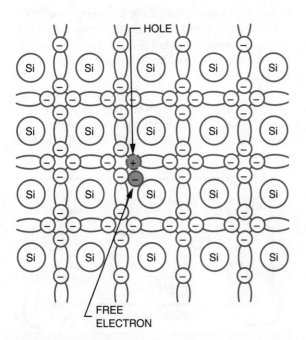

Figure 19–4 A hole is created when an electron breaks its covalent bond.

19–2 Questions

1. How can pure germanium support a current flow?
2. When a potential is applied to pure germanium, in what direction do the electrons and holes move?
3. What determines the amount of current flow in a pure semiconductor material?

19–3 Conduction in Doped Germanium and Silicon

Pure semiconductors are mainly of theoretical interest. Development and research are concerned with the effects of adding impurities to pure materials. If it were not for these impurities, most semiconductors would not exist.

Pure semiconductor materials, such as germanium and silicon, support only a small number of electron-hole pairs at room temperature. This allows for conduction of very little current. To increase their conductivity, a process called doping is used.

Doping is the process of adding impurities to a semiconductor material. Two types of impurities are used. The first, called **pentavalent,** is made of atoms with five valence electrons. Examples are arsenic and antimony. The other, called **trivalent,** is made of atoms with three valence electrons. Examples are indium and gallium.

When pure semiconductor material is doped with a pentavalent material such as arsenic (As), some of the existing atoms are displaced with arsenic atoms (Figure 19–6). The arsenic atom shares four of its valence electrons with adjacent silicon atoms in a covalent bond. Its fifth electron is loosely attached to the nucleus and is easily set free.

The arsenic atom is referred to as a *donor atom* because it gives its extra electron away. There are many donor atoms in a semiconductor material that has been doped. This means that many free electrons are available to support current flow.

At room temperature the number of donated free electrons exceeds the number of electron-hole

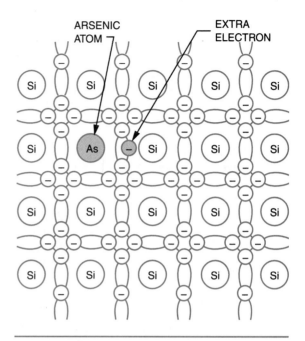

Figure 19–6 Silicon semiconductor material doped with an arsenic atom.

pairs. This means that there are more electrons than holes. The electrons are therefore called the *majority carrier.* The holes are *minority carriers.* Because the negative charge is the majority carrier, the material is called *N-type.*

If voltage is applied to **N-type material** (Figure 19–7) the free electrons contributed by the donor atoms flow toward the positive terminal. Additional electrons break away from their covalent

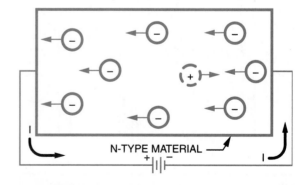

Figure 19–7 Current flow in N-type material.

bonds and flow toward the positive terminal. These free electrons, in breaking their covalent bonds, create electron-hole pairs. The corresponding holes move toward the negative terminal.

When semiconductor materials are doped with trivalent materials such as indium (I), the indium atom shares its three valence electrons with three adjacent atoms (Figure 19–8). This creates a hole in the covalent bond.

The presence of additional holes allows the electrons to drift easily from one covalent bond to the next. Because holes easily accept electrons, atoms that contribute extra holes are called *acceptor atoms.*

Under normal conditions, the number of holes greatly exceeds the number of electrons in such material. Therefore, the holes are the majority carrier and the electrons are the minority carrier. Because the positive charge is the majority carrier, the material is called **P-type material.**

If voltage is applied to P-type material, it causes the holes to move toward the negative terminal and the electrons to move toward the posi-

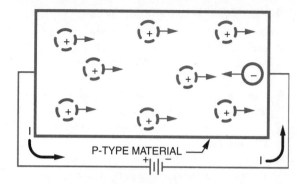

Figure 19–9 Current flow in P-type material.

tive terminal (Figure 19–9). In addition to the holes provided by the acceptor atom, holes are produced as electrons break away from their covalent bonds creating electron-hole pairs.

N- and P-type semiconductor materials have much higher conductivity than pure semiconductor materials. This conductivity can be increased or decreased by the addition or deletion of impurities. The more heavily a semiconductor material is doped, the lower its electrical resistance.

19–3 Questions

1. Describe the process of doping a semiconductor material.
2. What are the two types of impurities used for doping?
3. What determines whether a material, when doped, is an N-type or P-type semiconductor material?
4. How does doping support current flow in a semiconductor material?
5. What determines the conductivity of a semiconductor material?

SUMMARY

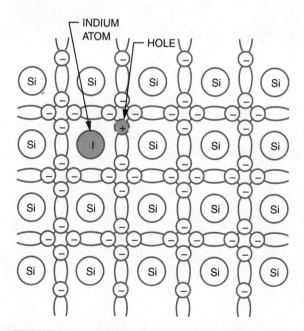

Figure 19–8 Silicon semiconductor material doped with an indium atom.

■ Semiconductor materials are any materials with characteristics that fall between those of insulators and conductors.

■ Pure semiconductor materials are germanium (Ge), silicon (Si), and carbon (C).

■ Silicon is used for most semiconductor devices.

■ Valence is an indication of an atom's ability to gain or lose electrons.

■ Semiconductor materials have valence shells that are half full.

■ Crystals are formed by atoms sharing their valance electrons through covalent bonding.

■ Semiconductor materials have a negative temperature coefficient: As the temperature rises, their resistance decreases.

■ Heat creates problems with semiconductor materials by allowing electrons to break their covalent bonds.

■ As the temperature increases in a semiconductor material, electrons drift from one atom to another.

■ A hole represents the absence of an electron in the valence shell.

■ A difference of potential, applied to pure semiconductor material, creates a current flow toward the positive terminal and a hole flow toward the negative terminal.

■ Current flow in semiconductor materials consists of both electron flow and hole movement.

■ Doping is the process of adding impurities to a semiconductor material.

■ Trivalent materials have atoms with three valence electrons and are used to make P-type material.

■ Pentavalent materials have atoms with five valence electrons and are used to make N-type material.

■ In N-type material, electrons are the majority carrier and holes are the minority carrier.

■ In P-type material, holes are the majority carrier and electrons are the minority carrier.

■ N- and P-type semiconductor materials have a higher conductivity than pure semiconductor materials.

Chapter 19 Self-Test

1. What makes silicon more desirable to use than germanium?

2. Why is covalent bonding important in the formation of semiconductor materials?

3. Describe how an electron travels through a block of pure silicon at room temperature.

4. Describe the process of how to convert a block of pure silicon to N-type material.

5. Describe what happens to a block of N-type material when a voltage is applied.

20

PN JUNCTION DIODES

Objectives

After completing this chapter, the student will be able to:

- Describe what a junction diode is and how it is made.
- Define *depletion region* and *barrier voltage*.
- Explain the difference between forward bias and reverse bias of a diode.
- Draw and label the schematic symbol for a diode.
- Describe three diode construction techniques.
- Identify the most common diode packages.
- Test diodes using an ohmmeter.

Diodes are the simplest type of semiconductor. They allow current to flow in only one direction. The knowledge of semiconductors that is acquired by studying diodes is also applicable to other types of semiconductor devices.

20–1 PN Junctions

When pure or intrinsic semiconductor material is doped with a pentavalent or trivalent material, the doped material is called N- or P-type based on the majority carrier. The electrical charge of each type is neutral because each atom contributes an equal number of protons and electrons.

Independent electrical charges exist in each type of semiconductor material, because electrons are free to drift. The electrons and holes that drift are referred to as *mobile charges.* In addition to the mobile charges, each atom that loses an electron has more protons than electrons and therefore assumes a positive charge. Similarly, each atom that gains an electron has more electrons than protons and assumes a negative charge. As described in Chapter 1, these individual charged atoms are called positive and negative ions. There is always an equal number of mobile and ionic charges within N-type and P-type semiconductor materials.

A **diode** is created by joining N- and P-type materials together (Figure 20–1). Where the materials come in contact with each other, a junction is formed. This device is referred to as a *junction diode.*

When the junction is formed, the mobile charges in the vicinity of the junction are strongly attracted to their opposites and drift toward the junction. As the charges accumulate, the action increases. Some electrons move across the junction and fill some of the holes near the junction in the P-type material. In the N-type material, the electrons become depleted near the junction. This region near the junction where the electrons and holes are depleted is called the **depletion region.** It extends only a short distance on either side of the junction.

There are no majority carriers in the depletion region, and the N-type and P-type materials are no longer electrically neutral. The N-type material takes on a positive charge near the junction, and the P-type material takes on a negative charge.

The depletion region does not get larger. The combining action tapers off quickly and the region remains small. The size is limited by the opposite charges that build up on each side of the junction. As the negative charge builds up, it repels further electrons and keeps them from crossing the junction. The positive charge absorbs free electrons and aids in holding them back.

These opposite charges that build up on each side of the junction create a voltage, referred to as the **barrier voltage.** It can be represented as an external voltage source, even though it exists across the PN junction (Figure 20–2).

The barrier voltage is quite small, measuring only several tenths of a volt. Typically, the barrier voltage is 0.3 volt for a germanium PN junction and 0.7 volt for a silicon PN junction. This voltage becomes apparent when an external voltage source is applied.

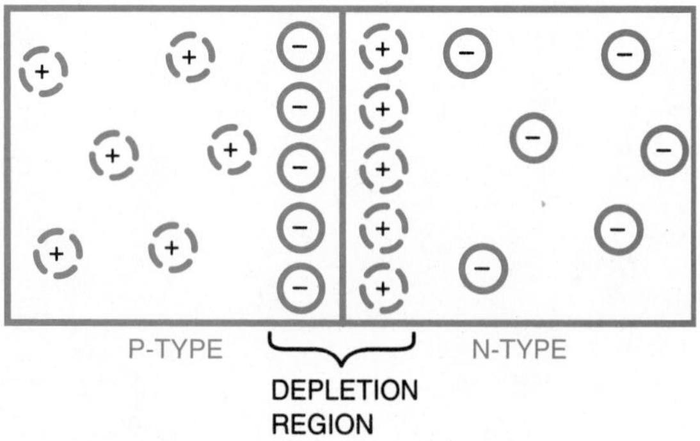

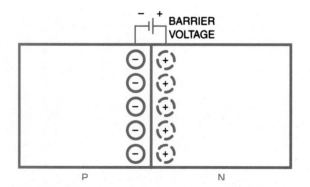

Figure 20–1 Diode formed by joining P- and N-type material to form a PN junction.

Figure 20–2 Barrier voltage as it exists across a PN junction.

1. Define the following terms:
   a. Donor atom
   b. Acceptor atom
   c. Diode
2. What occurs when an N-type material is joined with a P-type material?
3. How is the depletion region formed?
4. What is the barrier voltage?
5. What are typical barrier voltages for a germanium and a silicon diode?

20–2 Diode Biasing

When a voltage is applied to a diode, it is referred to as a **bias voltage.** Figure 20–3 shows a PN junction diode connected to a voltage source. A resistor is added for limiting current to a safe value.

In the circuit shown, the negative terminal of the voltage source is connected to the N-type material. This forces electrons away from the terminal, toward the PN junction. The free electrons that accumulate on the P side of the junction are attracted by the positive terminal. This action cancels the negative charge on the P side, the barrier voltage is eliminated, and a current is able to flow. Current flow occurs only if the external voltage is greater than the barrier voltage.

The voltage source supplies a constant flow of electrons, which drift through the N-type material along with the free electrons contained in it. Holes in the P material also drift toward the junction. The

holes and electrons combine at the junction and appear to cancel each other. However, as the electrons and holes combine, new electrons and holes appear at the terminals of the voltage source. The majority carriers continue to move toward the PN junction as long as the voltage source is applied.

Electrons flow through the P side of the diode, attracted by the positive terminal of the voltage source. As electrons leave the P material, holes are created that drift toward the PN junction where they combine with other electrons. When the current flows from the N-type toward the P-type material, the diode is said to have a **forward bias.**

The current that flows when a diode is forward biased is limited by the resistance of the P and N materials and the external resistance of the circuit. The diode resistance is small. Therefore, connecting a voltage directly to a forward-biased diode creates a large current flow. This can generate enough heat to destroy the diode. To limit the forward current flow, an external resistor must be connected in series with the diode.

A diode conducts current in the forward direction only if the external voltage is larger than the barrier voltage and is connected properly. A germanium diode requires a minimum forward bias of 0.3 volt; a silicon diode requires a minimum forward bias of 0.7 volt.

Once a diode starts conducting, a voltage drop occurs. This voltage drop is equal to the barrier voltage and is referred to as the *forward voltage drop* (E_F). The voltage drop is 0.3 volt for a germanium diode and 0.7 volt for a silicon diode. The amount of forward current (I_F) is a function of the external voltage (E), the forward voltage drop (E_F), and the external resistance (R). The relationship can be shown using Ohm's law:

$$I = \frac{E}{R}$$

$$I_F = \frac{E - E_F}{R}$$

EXAMPLE A silicon diode has an external bias voltage of 12 volts with an external resistor of 150 ohms. What is the total forward current?

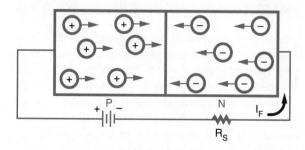

Figure 20–3 PN junction diode with forward bias.

Given:

$I_F = ?$

$E = 12$ volts

$R = 150$ ohms

$E_F = 0.7$ volt

Solution:

$$I_F = \frac{E - E_F}{R}$$

$$I_F = \frac{12 - 0.7}{150}$$

$$I_F = \frac{11.3}{150}$$

$$I_F = 0.075 \text{ amp}$$

or 75 mA

In a diode that is forward biased, the negative terminal of the external voltage source is connected to the N material, and the positive terminal is connected to the P material. If these terminals are reversed, the diode does not conduct and is said to be connected in **reverse bias** (Figure 20–4). In this configuration, the free electrons in the N material are attracted toward the positive terminal of the external voltage source. This increases the number of positive ions in the area of the PN junction, which increases the width of the depletion region on the N side of the junction. Electrons also leave the negative terminal of the voltage source and enter the P material. These electrons fill holes near the PN junction, causing the holes to move toward the negative terminal, which increases the width of the depletion region on the P side of the junction. The overall effect is that the depletion region is wider than in an unbiased or forward-biased diode.

The reverse-biased voltage increases the barrier voltage. If the barrier voltage is equal to the external voltage source, holes and electrons cannot support current flow. A small current flows with a reverse bias applied. This leakage current is referred to as *reverse current* (I_R) and exists because of minority carriers. At room temperature the minority carriers are few in number. As the temperature increases, more electron-hole pairs are created. This increases the number of majority carriers and the leakage current.

All PN junction diodes produce a small leakage current. In germanium diodes it is measured in microamperes; in silicon diodes it is measured in nanoamperes. Germanium has more leakage current because it is more sensitive to temperature. This disadvantage of germanium is offset by its smaller barrier voltage.

In summary, a PN junction diode is a one-directional device. When it is forward biased, a current flows. When it is reverse biased, only a small leakage current flows. It is this characteristic that allows the diode to be used as a rectifier. A rectifier converts an AC voltage to a DC voltage.

20—2 Questions

1. What is a bias voltage?
2. What is the minimum amount of voltage needed to produce current flow across a PN junction diode?
3. What is the difference between forward and reverse biasing?
4. What is leakage current in a PN junction diode?

20—3 Diode Characteristics

Both germanium and silicon diodes can be damaged by excessive heat and excessive reverse voltage. Manufacturers specify the maximum forward current (I_F *max*) that can be handled safely. They also specify the maximum-safe reverse voltage **(peak inverse voltage, or PIV).** If the PIV is exceeded, a large reverse current flows, creating excess heat and damaging the diode.

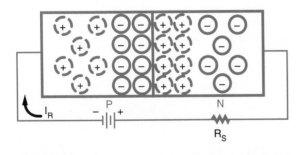

Figure 20–4 PN junction diode with reverse bias.

ANODE   CATHODE

Figure 20–5 Diode schematic symbol.

At room temperature, the reverse current is small. As the temperature increases, the reverse current increases, interfering with proper operation of the diode. In germanium diodes, the reverse current is higher than in silicon diodes, doubling with approximately every 10 degrees celsius of increased temperature.

The diode symbol is shown in Figure 20–5. The P section is represented by an arrow, and the N section by a bar. Forward current flows from the N section to the P section (against the arrow). The N section is called the *cathode,* and the P section is called the *anode.* The cathode supplies and the anode collects the electrons.

Figure 20–6 shows a properly connected forward-biased diode. The negative terminal is connected to the cathode. The positive terminal is connected to the anode. This setup conducts a forward current. A resistor (R_S) is added in series to limit the forward current to a safe value.

Figure 20–7 shows a diode connected in reverse bias. The negative terminal is connected to

the anode. The positive terminal is connected to the cathode. In reverse bias, a small reverse current (I_R) flows.

20–3 Questions

1. What problem can a reverse current create in a germanium or silicon diode?
2. Draw and label the schematic symbol for a diode.
3. Draw a circuit that includes a forward-biased diode.
4. Draw a circuit that includes a reverse-biased diode.
5. Why should a resistor be connected in series with a forward-biased diode?

20–4 Diode Construction Techniques

The PN junction of a diode may be one of three types: a grown junction, an alloyed junction, or a diffused junction. Each involves a different construction technique.

In grown junction method (the earliest technique used) intrinsic semiconductor material and P-type impurities are placed in a quartz container and heated until they melt. A small semiconductor crystal, called a seed, is then lowered into the molten mixture. The seed crystal is slowly rotated and withdrawn from the molten mixture slowly enough to allow the molten mixture to cling to the seed. The molten mixture, clinging to the seed crystal, cools and rehardens, assuming the same crystalline characteristics as the seed. As the seed crystal is withdrawn, it is alternately doped with N- and P-type impurities. *Doping* is the process of adding impurities to pure semiconductor crystals to increase the number of free electrons or the number of holes. This creates N and P layers in the crystal as it is grown. The resulting crystal is then sliced into many PN sections.

The alloyed junction method of forming a semiconductor is extremely simple. A small pellet

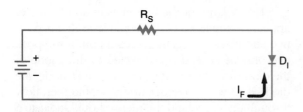

Figure 20–6 Diode connected with forward bias.

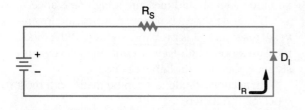

Figure 20–7 Diode connected with reverse bias.

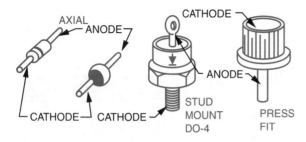

Figure 20–8 Common diode packages.

of trivalent material, such as indium, is placed on an N-type semiconductor crystal. The pellet and crystal are heated until the pellet melts and partially fuses with the semiconductor crystal. The area where the two materials combine forms the P-type material. When the heat is removed, the material recrystallizes and a solid PN junction is formed.

The diffused junction method is the method most in use today. A mask with openings is placed on a thin section of N- or P-type semiconductor material called a wafer. The wafer is then placed in an oven and exposed to an impurity in a gaseous state. At an extremely high temperature, the impure atoms penetrate or diffuse through the exposed surfaces of the water. The depth of diffusion is controlled by the length of the exposure and the temperature.

Once the PN junction is formed, the diode must be packaged to protect it from both environmental and mechanical stresses. The package must also provide a means of connecting the diode to a circuit. Package style is determined by the purpose or application of the diode (Figure 20–8). If large currents are to flow through the diode, the package must be designed to keep the junction from overheating. Figure 20–9 shows the package for diodes rated at 3 amperes or less. The cathode is identified by a white or silver band on the end.

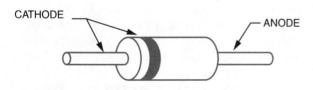

Figure 20–9 Package for diode rated at 3 amperes or less.

20–4 Questions

1. Describe three methods of diode fabrication.
2. Which method of diode fabrication is favored over the others?
3. Draw four common diode packages.
4. How is the cathode identified on diode packages rated at less than 3 amperes?

20–5 Testing PN Junction Diodes

A diode can be tested by checking the forward-to-reverse-resistance ratio with an ohmmeter. The resistance ratio indicates the ability of the diode to pass current in one direction and block current in the other direction.

A germanium diode has a low forward resistance of several hundred ohms. The reverse resistance is high, greater than 100,000 ohms. Silicon diodes have a higher forward and reverse resistance than germanium. An ohmmeter test of a diode should show a low forward resistance and a high reverse resistance.

Caution: Some ohmmeters use high-voltage batteries which can destroy a PN junction.

The polarity of the terminals in an ohmmeter appears at the leads of the ohmmeter: Red is positive and black is negative or common. If the positive ohmmeter lead is connected to the anode of the diode and the negative lead to the cathode, the diode is forward biased. Current should then flow through the diode and the meter should indicate a low resistance. If the meter leads are reversed, the diode is reverse biased. Little current should flow and the meter should measure a high resistance.

If a diode shows both low forward and low reverse resistance, it is probably shorted. If the diode measures both high forward and high reverse resistance, then it is probably opened.

An accurate diode test can be made with most types of ohmmeters.

Caution: Some ohmmeters used for troubleshooting have an open-circuit lead voltage of less than

0.3 volt. This type of meter cannot be used to measure the forward resistance of a diode.

The forward resistance voltage must be larger than the barrier voltage of the diode (0.7 volt for silicon and 0.3 volt for germanium) for conduction to take place.

An ohmmeter can also be used to determine the cathode and anode of an unmarked diode. When there is a low reading, the positive lead is connected to the anode and the negative lead is connected to the cathode.

20—5 Questions

1. How is a diode tested with an ohmmeter?
2. What precaution(s) must be taken when testing diodes with an ohmmeter?
3. How does the ohmmeter indicate that a diode is shorted?
4. How does the ohmmeter indicate that a diode is opened?
5. How can an ohmmeter be used to determine the cathode end of an unmarked diode?

SUMMARY

■ A junction diode is created by joining N-type and P- type materials together.
■ The region near the junction is referred to as the depletion region. Electrons cross the junction from the N-type to the P-type material and thus both the holes and the electrons near the junction are depleted.
■ The size of the depletion region is limited by the charge on each side of the junction.

■ The charge at the junction creates a voltage called the barrier voltage.
■ The barrier voltage is 0.3 volt for germanium and 0.7 volt for silicon.
■ A current flows through a diode only when the external voltage is greater than the barrier voltage.
■ A diode that is forward biased conducts current. The P-type material is connected to the positive terminal, and the N-type material is connected to the negative terminal.
■ A diode that is reverse biased conducts only a small leakage current.
■ A diode is a one-directional device.
■ A diode's maximum forward current and reverse voltage are specified by the manufacturer.
■ The schematic symbol for a diode is:

■ In a diode, the cathode is the N-type material, and the anode is the P-type material.
■ Diodes can be constructed by the grown junction, alloyed junction, or diffused junction method.
■ The diffused junction method is the one most often used.
■ Packages for diodes of less than 3 amperes identify the cathode end of the diode with a white or silver band.
■ A diode is tested by comparing the forward to the reverse resistance with an ohmmeter.
■ When a diode is forward biased, the resistance is low.
■ When a diode is reverse biased, the resistance is high.

Chapter 20 Self-Test

1. What does a PN junction diode accomplish?
2. Under what conditions will a silicon PN junction diode turn on?
3. Draw examples of a PN junction diode in forward and reverse bias. (Use schematic symbols.)

21

ZENER DIODES

Objectives

After completing this chapter, the student will be able to:

- Describe the function and characteristics of a zener diode. *reference point*
- Draw and label the schematic symbol for a zener diode.
- Explain how a zener diode operates as a voltage regulator.
- Describe the procedure for testing zener diodes.

Zener diodes are closely related to PN junction diodes. They are constructed to take advantage of reverse current. Zener diodes find wide application for controlling voltage in all types of circuits.

21–1 Zener Diode Characteristics

As previously stated, a high reverse-biased voltage applied to a diode may create a high reverse current, which can generate excessive heat and cause a diode to break down. The applied reverse voltage at which the breakdown occurs is called the *breakdown voltage,* or *peak reverse voltage.* A special diode called a **zener diode** is connected to operate in the reverse-biased mode. It is designed to operate at those voltages that exceed the breakdown voltage. This breakdown region is called the **zener region.**

When a reverse-biased voltage is applied that is high enough to cause breakdown in a zener diode, a high reverse current (I_Z) flows. The reverse current is low until breakdown occurs. After breakdown, the reverse current increases rapidly. This occurs because the resistance of the zener diode decreases as the reverse voltage increases.

The breakdown voltage of a zener diode (E_Z) is determined by the resistivity of the diode. This, in turn, is controlled by the doping technique used during manufacture. The rated breakdown voltage represents the reverse voltage at the zener test current (I_{ZT}). The *zener test current* is somewhat less than the maximum reverse current the diode can handle. The breakdown voltage is typically rated with 1 to 20% tolerance.

The ability of a zener diode to dissipate power decreases as the temperature increases. Therefore *power dissipation ratings* are given for specific temperatures. Power ratings are also based on lead lengths: shorter leads dissipate more power. A *derating factor* is given by the diode manufacturer to determine the power rating at different temperatures from the ones specified in their tables. For example, a derating factor of 6 milliwatts per degree celsius means that the diode power rating decreases 6 milliwatts for each degree of change in temperature.

Zener diodes are packaged like PN junction diodes (Figure 21–1). Low-power zener diodes are mounted in either glass or epoxy. High-power zener diodes are stud mounted with a metal case. The schematic symbol for the zener diode is simi-

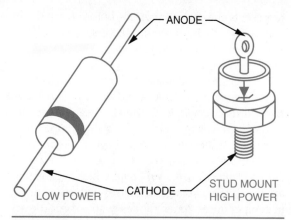

Figure 21–1 Zener diode packages.

Figure 21–2 Schematic symbol for a zener diode.

lar to the PN junction diode except for the diagonal lines on the cathode bar (Figure 21–2).

21–1 Questions

1. What is the unique feature of a zener diode?
2. How is a zener diode connected into a circuit?
3. What determines the voltage at which a zener diode breaks down?
4. What considerations go into determining the power dissipation rating of a zener diode?
5. Draw and label the schematic symbol used to represent a zener diode.

21–2 Zener Diode Ratings

The *maximum zener current* (I_{ZM}) is the maximum reverse current that can flow in a zener diode without exceeding the power dissipation rating specified by the manufacturer. The reverse current (I_R)

represents the leakage current before breakdown and is specified at a certain reverse voltage (E_R). The reverse voltage is approximately 80% of the zener voltage (E_Z).

Zener diodes that have a breakdown voltage of 5 volts or more have a *positive zener voltage-temperature coefficient,* which means that the breakdown voltage increases as the temperature increases. Zener diodes that have a breakdown voltage of less than 4 volts have a *negative zener voltage-temperature coefficient,* which means that the breakdown voltage decreases with an increase in temperature. Zener diodes with a breakdown voltage between 4 and 5 volts may have a positive or negative voltage-temperature coefficient.

A *temperature-compensated zener diode* is formed by connecting a zener diode in series with a PN junction diode, with the PN junction diode forward biased and the zener diode reverse biased. By careful selection of the diodes, temperature co-efficients can be selected that are equal and opposite. More than one PN junction diode may be needed for proper compensation.

21–2 Questions

1. What determines the maximum zener current of a zener diode?
2. What is the difference between the maximum zener current and the reverse current for a zener diode?
3. What does a positive zener voltage-temperature coefficient signify?
4. What does a negative zener voltage-temperature coefficient signify?
5. How can a zener diode be temperature compensated?

21–3 Voltage Regulation with Zener Diodes

A zener diode can be used to stabilize or regulate voltage. For example, it can be used to compensate for power-line voltage changes or load-resistance changes while maintaining a constant DC output.

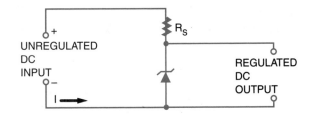

Figure 21–3 Typical zener diode regulator circuit.

Figure 21–3 shows a typical zener diode regulator circuit. The zener diode is connected in series with resistor R_S. The resistor allows enough current to flow for the zener diode to operate in the zener breakdown region. The DC input voltage must be higher than the zener diode breakdown voltage. The voltage drop across the zener diode is equal to the zener diode's voltage rating. Zener diodes are manufactured to have a specific breakdown voltage rating that is often referred to as the diode's *zener voltage rating* (V_Z). The voltage drop across the resistor is equal to the difference between the zener (breakdown) voltage and the input voltage.

The input voltage may increase or decrease. This causes the current through the zener diode to increase or decrease accordingly. When the zener diode is operating in the zener voltage, or breakdown region, a large current will flow through the zener with an increase in input voltage. However, the zener voltage remains the same. The zener diode opposes an increase in input voltage, because when the current increases the resistance drops. This allows the zener diode's output voltage to remain constant as the input voltage changes. The change in the input voltage appears across the series resistor. The resistor is in series with the zener diode, and the sum of the voltage drop must equal the input voltage. The output voltage is taken across the zener diode. The output voltage can be increased or decreased by changing the zener diode and the series resistor.

The circuit just described supplies a constant voltage. When a circuit is designed, the current in the circuit must be considered as well as the

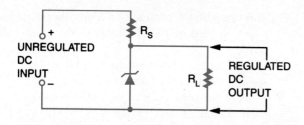

Figure 21–4 Zener diode voltage regulator with load.

voltage. The external load requires a specific load current (I_L) determined by the load resistance and output voltage (Figure 21–4). The load current and the zener current flow through the series resistor. The series resistor must be chosen so that the zener current is adequate to keep the zener diode in the breakdown zone and allow the current to flow.

When the load resistance increases, the load current decreases, which should increase the voltage across the load resistance. But the zener diode opposes any change by conducting more current. The sum of the zener current and the load current through the series resistor remains constant. This action maintains the same voltage across the series resistor.

Similarly, when the load current increases, the zener current decreases, maintaining a constant voltage. This action allows the circuit to regulate for change in output current as well as input voltage.

21–3 Questions

1. What is a practical function for a zener diode?
2. Draw a schematic diagram of a zener diode regulator circuit.
3. How can the voltage of a zener diode voltage regulator circuit be changed?
4. What must be considered in designing a zener diode voltage regulator?
5. Describe how a zener diode voltage regulator maintains a constant output voltage.

21–4 Testing Zener Diodes

Zener diodes can be quickly tested for opens, shorts, or leakage with an ohmmeter. The ohmmeter is connected in forward and reverse bias in the same manner as with PN junction diodes. However, these tests do not provide information on whether the zener diode is regulating at the rated value. For that, a regulation test must be performed with a metered power supply that can indicate both voltage and current.

Figure 21–5 shows the proper setup for a zener diode regulation test. The output of the power supply is connected with a limiting resistor in series with a zener diode to be tested. A voltmeter is connected across the zener diode under test to

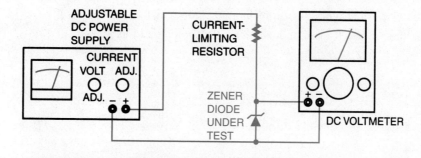

Figure 21–5 Setup for testing zener diode regulation.

monitor the zener voltage. The output voltage is slowly increased until the specified current is flowing through the zener diode. The current is then varied on either side of the specified zener current (I_Z). If the voltage remains constant, the zener diode is operating properly.

21—4 Questions

1. Describe the process for testing a zener diode with an ohmmeter.
2. What parameters are not tested when using an ohmmeter to test a zener diode?
3. Draw a schematic diagram showing how to connect a zener diode to check its breakdown voltage.
4. Describe how the circuit in question 3 operates to determine whether the zener diode is operating properly.
5. How can the cathode end of a zener diode be determined using an ohmmeter?

SUMMARY

■ Zener diodes are designed to operate at voltages greater than the breakdown voltage (peak reverse voltage).

■ The breakdown voltage of a zener diode is determined by the resistivity of the diode.
■ Zener diodes are manufactured with a specific breakdown (zener) voltage.
■ Power dissipation of a zener diode is based on temperature and lead lengths.
■ The schematic symbol for a zener diode is:

■ Zener diodes are packaged the same way as PN junction diodes.
■ Zener diodes with a breakdown voltage greater than 5 volts have a positive zener voltage-temperature coefficient.
■ Zener diodes with a breakdown voltage less than 4 volts have a negative zener voltage-temperature coefficient.
■ Zener diodes are used to stabilize or regulate voltage.
■ Zener diode regulators provide a constant output voltage despite changes in the input voltage or output current.
■ Zener diodes can be tested for opens, shorts, or leakage with an ohmmeter.
■ To determine whether a zener diode is regulating at the proper voltage, a regulation test must be performed.

Chapter 21 Self-Test

1. Explain how a zener diode functions in a voltage regulator circuit.
2. Describe the process for testing a zener diode's voltage rating.

22

BIPOLAR TRANSISTORS

Objectives

After completing this chapter, the student will be able to:

- Describe how a transistor is constructed and its two different configurations.
- Draw and label the schematic symbol for an NPN and a PNP transistor.
- Identify the ways of classifying transistors.
- Identify the function of a transistor using a reference manual and the identification number (2NXXXX).
- Identify commonly used transistor packages.
- Describe how to bias a transistor for operation.
- Explain how to test a transistor with both a transistor tester and an ohmmeter.
- Describe the process used for substituting a transistor.

In 1948, Bell Laboratories developed the first working junction transistor. A transistor is a three-element, two junction device used to control electron flow. By varying the amount of voltage applied to the three elements, the amount of current can be controlled for purposes of amplification, oscillation, and switching. These applications are covered in Chapters 28, 29, and 30.

22–1 Transistor Construction

When a third layer is added to a semiconductor diode, a device is produced that can amplify power, current, or voltage. The device is called a **bipolar transistor,** also referred to as a **junction transistor** or **transistor.** The term *transistor* will be used here.

A transistor, like a junction diode, can be constructed of germanium or silicon, but silicon is more popular. A transistor consists of three alternately doped regions (as compared to two in a diode). The three regions are arranged in one of two ways.

In the first method, the P-type material is sandwiched between two N-type materials, forming an NPN transistor (Figure 22–1). In the second method, a layer of N-type material is sandwiched between two layers of P-type material, forming a PNP transistor (Figure 22–2).

In both types of transistor, the middle region is called the *base* and the outer regions are called the *emitter* and *collector.* The emitter, base, and collector are identified by the letters *E, B,* and *C,* respectively.

(A) BLOCK DIAGRAM OF AN NPN TRANSISTOR

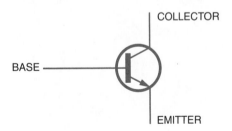

(B) SCHEMATIC SYMBOL FOR A NPN TRANSISTOR

Figure 22–1 NPN transistor.

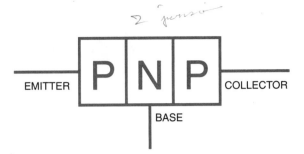

(A) BLOCK DIAGRAM OF A PNP TRANSISTOR

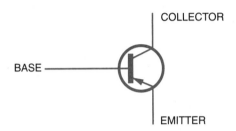

(B) SCHEMATIC SYMBOL FOR A PNP TRANSISTOR

Figure 22–2 PNP transistor.

22–1 Questions

1. How does the construction of a transistor differ from the construction of a PN junction diode?
2. What are the two types of transistor?
3. What are the three parts of a transistor called?
4. Draw and label the schematic symbols for an NPN and a PNP transistor.
5. What are transistors used for?

22–2 Transistor Types and Packaging

Transistors are classified by the following methods:

1. According to type (either NPN or PNP)
2. According to the material used (germanium or silicon)
3. According to major use (high or low power, switching, or high frequency)

Most transistors are identified by a number. This number begins with a 2 and the letter *N* and

Figure 22–3 Various transistor packages. *(Photo courtesy of International Rectifier.)*

has up to four more digits. These symbols identify the device as a transistor and indicate that it has two junctions.

A package serves as protection for the transistor and provides a means of making electrical connections to the emitter, base, and collector regions.

The package also serves as a heat sink, or an area from which heat can be radiated, removing excess heat from the transistor and preventing heat damage. Many different packages are available, covering a wide range of applications (Figure 22–3).

Transistor packages are designated by size and configuration. The most common package identifier consists of the letters *TO* (*transistor outline*) followed by a number. Some common transistor packages are shown in Figure 22–4.

Because of the large assortment of transistor packages available, it is difficult to develop rules for identifying the emitter, base, and collector leads of each device. It is best to refer to the manufacturer's specification sheet to identify the leads of each device.

22–2 Questions

1. How are transistors classified?
2. What symbols are used to identify transistors?

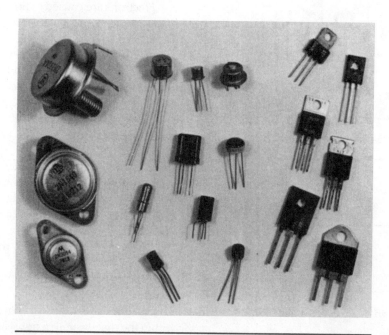

Figure 22–4 Typical transistor packages.

3. What purposes do the packaging of a transistor serve?

4. How are transistor packages labeled?

5. What is the best method for determining which leads of a transistor are the base, emitter, and collector?

22—3 Basic Transistor Operation

A diode is a rectifier and a transistor is an amplifier. A transistor may be used in a variety of ways, but its basic functions are to provide current amplification of a signal or to switch the signal.

A transistor must be properly biased by external voltages so that the emitter, base, and collector regions interact in the desired manner. In a properly biased transistor, the emitter junction is forward biased and the collector junction is reverse biased. A properly biased NPN transistor is shown in Figure 22–5.

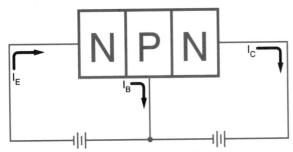

(A) BLOCK DIAGRAM OF A BIASED NPN TRANSISTOR

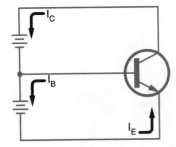

(B) SCHEMATIC DIAGRAM OF A BIASED NPN TRANSISTOR

Figure 22–5 Properly biased NPN transistor.

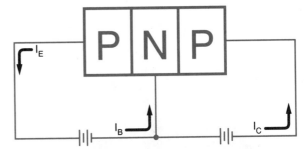

(A) BLOCK DIAGRAM OF A BIASED PNP TRANSISTOR

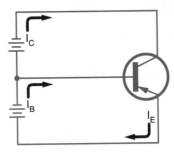

(B) SCHEMATIC DIAGRAM OF A BIASED PNP TRANSISTOR

Figure 22–6 Properly biased PNP transistor.

Electrons are caused to flow from an NPN transistor emitter by a forward bias. Forward bias is a positive voltage on the base terminal with respect to the emitter terminal. A positive potential attracts electrons, creating an electron flow from the emitter. The electrons that are attracted into the base are now influenced by the positive potential applied to the collector. The majority of electrons are attracted to the collector and into the positive side of the reverse-biased voltage source. A few electrons are absorbed into the base region and support a small electron flow from it. For this action to occur, the base region must be extremely thin. In a properly biased PNP transistor, the batteries must be reversed (Figure 22–6).

The difference between the NPN and PNP transistors is twofold: The batteries have opposite polarities, and the direction of the electron flow is reversed.

As with the diode, a barrier voltage exists within the transistor. In a transistor, the barrier voltage is produced across the emitter-base junction.

This voltage must be exceeded before electrons can flow through the junction. The internal barrier voltage is determined by the type of semiconductor material used. As in diodes, the internal barrier voltage is 0.3 volt for germanium transistors and 0.7 for silicon transistors.

The collector-base junction of a transistor must also be subjected to a positive potential that is high enough to attract most of the electrons supplied by the emitter. The reverse-bias voltage applied to the collector-base junction is usually much higher than the forward-bias voltage across the emitter-base junction, supplying this higher voltage.

22—3 Questions

1. What are the basic functions of a transistor?
2. What is the proper method for biasing a transistor?
3. What is the difference between biasing an NPN and a PNP transistor?
4. What is the barrier voltage for a germanium and a silicon transistor?
5. What is the difference between the collector-base junction and the emitter-base junction bias voltages?

22—4 Transistor Testing

Transistors are semiconductor devices that usually operate for long periods of time without failure. If a transistor does fail, the failure is generally caused by excessively high temperature, current, or voltage. Failure can also be caused by extreme mechanical stress. As a result of this electrical or mechanical abuse, a transistor may open or short internally, or its characteristics may alter enough to affect its operation. There are two methods for checking a transistor to determine if it is functioning properly: with an ohmmeter and with a transistor tester.

A conventional ohmmeter can help detect a defective transistor in an out-of-circuit test. Resistance tests are made between the two junctions of

a transistor in the following way: emitter to base, collector to base, and collector to emitter. In testing the transistor, the resistance is measured between any two terminals with the meter leads connected one way. The meter leads are then reversed. In one meter connection, the resistance should be high, 10,000 ohms or more. In the other meter connection, the resistance should be lower, less than 10,000 ohms.

Each junction of a transistor exhibits a low resistance when it is forward biased and a high resistance when reverse biased. The battery in the ohmmeter is the source of the forward- and reverse-bias voltage. The exact resistance measured varies with different types of transistors, but there is always a change when the ohmmeter leads are reversed. This method of checking works for either NPN or PNP transistors (Figure 22–7).

If a transistor fails this test, it is defective. If it passes, it may still be defective. A more reliable means of testing a transistor is by using a transistor tester.

Caution: As with diodes, the ohmmeter terminal voltage must never exceed the maximum voltage

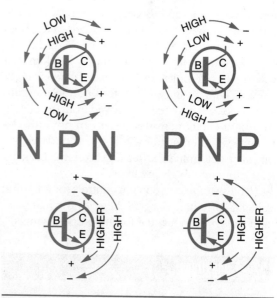

Figure 22–7 Resistance measurements of transistor junctions.

Figure 22–8 Transistor tester. *(Courtesy of Huntron Instruments, Inc.)*

rating between the junctions of a transistor. The lower scales of some ohmmeters can supply a damaging current to the transistor under test. As a precautionary measure, it is best to start out at the safest range and then change to a scale that gives an adequate reading.

Transistor testers are designed specifically for testing transistors and diodes. There are two types: an in-circuit tester and an out-of-circuit tester. Both may be housed in the same package (Figure 22–8).

The transistor's ability to amplify is taken as a rough measure of its performance. There is an advantage with an in-circuit tester because the transistor does not have to be removed from the circuit. An out-of-circuit transistor tester can not only determine whether the transistor is good or defective, but also determine the leakage current. Leakage tests cannot be made in-circuit.

Transistor testers contain controls for adjusting voltage, current, and signal. Refer to the manufacturer's instruction manual for the proper settings.

22–4 Questions

1. What may cause a transistor to fail?
2. What are two methods of testing a transistor?

3. When using an ohmmeter, what should the results be for an NPN transistor?
4. What are the two types of commercial transistor testers?

22–5 Transistor Substitution

Numerous guides have been prepared by manufacturers to provide cross references for transistor substitution. Most substitutions can be made with confidence.

If the transistor is unlisted or the number of the transistor is missing, the following procedure can be used to make an accurate replacement selection.

1. NPN or PNP? The first source of information is the symbol on the schematic diagram. If a schematic is not available, the polarity of the voltage source between the emitter and collector must be determined. If the collector voltage is positive with respect to the emitter voltage, then it is an NPN device. If the collector voltage is negative with respect to the emitter voltage, then it is a PNP device. An easy way to remember the polarity of the collector voltage for each type of transistor is shown in Figure 22–9.
2. Germanium or silicon? Measure the voltage from the emitter to the base. If the voltage is approximately 0.3 volt, the transistor is germanium. If the voltage is approximately 0.7 volt, the transistor is silicon.
3. Operating frequency range? Identify the type of circuit and determine whether it is working in the audio range, the kilohertz range, or the megahertz range.

NP OSITIVE N

PN EGATIVE P

Figure 22–9 How to remember the polarity of the collector voltage.

4. Operating voltage? Voltages from collector to emitter, collector to base, and emitter to base should be noted either from the schematic diagram or by actual voltage measurement. The transistor selected for replacement should have voltage ratings that are at least three to four times the actual operating voltage. This helps to protect against voltage spikes, transients, and surges that are inherent in most circuits.

5. Collector current requirements? The easiest way to determine the actual current is to measure the current in the collector circuit with an ammeter. This measurement should be taken under maximum power conditions. Again, a safety factor of three to four times the measured current should be allowed.

6. Maximum power dissipation? Use maximum voltage and collector current requirements to determine maximum power requirements (P = IE). The transistor is a major factor in determining power dissipation in the following types of circuits:
   • Input stages, AF, or RF (50 to 200 milliwatts)
   • IF stages and driver stages (200 mW to 1 watt)
   • High power output stages (1 watt and higher)

7. Current gain? The common emitter small signal DC current gain referred to as h_{fe} or Beta (β) should be considered. Some typical gain categories are:
   • RF mixers, IF and AF (80 to 150)
   • RF and AF drivers (25 to 80)
   • RF and AF output (4 to 40)
   • High gain preamps and sync separators (150 to 500)

8. Case style? Frequently, there is no difference between the case styles of original parts and recommended replacements. Case types and sizes need only be considered where an exact mechanical fit is required. Silicon grease should always be used with power devices to promote heat transfer.

9. Lead configuration? This is not a prime consideration for replacement transistors al-

though it may be desirable for ease of insertion and appearance.

22–5 Questions

1. Where can suggestions for transistor replacements be found?
2. Why does it matter whether a transistor is germanium or silicon?
3. Why are the operating frequency, voltage, current, and power ratings important when replacing a transistor?
4. What does the transistor's Beta refer to?
5. Are the transistor's case and lead configuration important when substituting a transistor?

SUMMARY

■ A transistor is a three-layer device used to amplify and switch power and voltage.
■ A bipolar transistor is also called a junction transistor or simply a transistor.
■ Transistors can be configured as NPN or PNP.
■ The middle region of the transistor is called the base, and the two outer regions are called the emitter and collector.
■ The schematic symbols used for NPN and PNP transistors are:

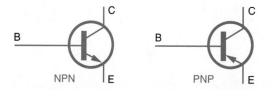

■ A transistor is classified according to whether it is NPN or PNP, silicon or germanium, high or low power, and switching or high frequency.
■ Transistors are identified with a prefix of 2N followed by up to four digits.
■ The transistor package provides protection, a heat sink, and a support for the leads.

- Transistor packages are identified with the letters *TO* (*transistor outline*).
- In a properly biased transistor, the emitter-base junction is forward biased and the collector-base junction is reverse biased.
- PNP transistor bias sources are the reverse of NPN bias sources.
- The internal barrier voltage for germanium transistors is 0.3 volt and for silicon transistors is 0.7 volt.

- The reverse-bias voltage applied to the collector-base junction is higher than the forward-bias voltage applied to the emitter-base junction.
- When a transistor is tested with an ohmmeter, each junction exhibits a low resistance when it is forward biased and a high resistance when it is reverse biased.
- Transistor testers are available for testing transistors in and out of circuit.

Chapter 22 Self-Test

1. The junction of a transistor can be forward-biased, reverse-biased, or unbiased. What are the normal conditions of bias across the emitter-base and collector-base junctions of a transistor?

2. When checking a good transistor with an ohmmeter, what kind of resistance should exist across each junction?

3. Using an ohmmeter, what difficulty, if any, would be experienced in identifying the transistor material type and emitter, base, and collector leads of an unknown transistor?

4. When connecting a transistor in a circuit, why must the technician know whether a transistor is NPN or PNP?

5. How does the testing of a transistor with an ohmmeter compare to testing a transistor with a transistor tester?

Chapter

23

FIELD EFFECT TRANSISTORS (FETs)

Objectives

After completing this chapter, the student will be able to:

- Describe the difference between transistors, JFETs, and MOSFETs.
- Draw schematic symbols for both P-channel and N-channel JFETs, depletion MOSFETs, and enhancement MOSFETs.
- Describe how a JFET, depletion MOSFET, and enhancement MOSFET operate.
- Identify the parts of JFETs and MOSFETs.
- Describe the safety precautions that must be observed when handling MOSFETs.
- Describe the procedure for testing JFETs and MOSFETs with an ohmmeter.

The history of field effect transistors (FETs) dates back to 1925, when Julius Lillenfield invented both the junction FET and the insulated gate FET. Both of these devices currently dominate electronics technology. This chapter is an introduction to the theory of junction and insulated gate FETs.

23–1 Junction FETs

The **junction field effect transistor (JFET)** is a unipolar transistor that functions using only majority carriers. The JFET is a voltage-operated device. JFETs are constructed from N-type and P-type semiconductor materials and are capable of amplifying electronic signals, but they are constructed differently from bipolar transistors and operate on different principles. Knowing how a JFET is constructed helps to understand how it operates.

Construction of a JFET begins with a substrate, or base, of lightly doped semiconductor material. The **substrate** can be either P- or N-type material. The PN junction in the substrate is made using both the diffusion and growth methods (see Chapter 20). The shape of the PN junction is important. Figure 23–1 shows a cross section of the embedded region within the substrate. The U-shaped region is called the **channel** and is flush with the upper surface of the substrate. When the channel is made of N-type material in a P-type substrate, an *N-channel JFET* is formed. When the channel is made of P-type material in an N-type substrate, a *P-channel JFET* is formed.

Three electrical connections are made to a JFET (Figure 23–2). One lead is connected to the substrate to form the *gate* (G). One lead is connected to each end of the channel to form the *source* (S) and the *drain* (D). It does not matter which lead is attached to the source or drain, because the channel is symmetrical.

The operation of a JFET requires two external bias voltages. One of the voltage sources (E_{DS}) is

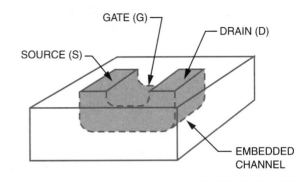

Figure 23–2 Lead connections for an N-channel JFET.

connected between the source and the drain, forcing the current to flow through the channel. The other voltage source (E_{GS}) is connected between the gate and the source. It controls the amount of current flowing through the channel. Figure 23–3 shows a properly biased N-channel JFET.

Voltage source E_{DS} is connected so that the source is made negative with respect to the drain. This causes a current to flow, because the majority carriers are electrons in the N-type material. The source-to-drain current is called the FET's *drain current* (I_D). The channel serves as resistance to the supply voltage (E_{DS}).

The gate-to-source voltage (E_{GS}) is connected so that the gate is negative with respect to the source. This causes the PN junction formed by the gate and channel to be reverse biased. This creates a depletion region in the vicinity of the PN junction, which spreads inward along the length of the channel. The depletion region is wider at the drain

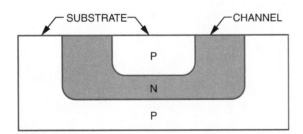

Figure 23–1 Cross section of an N-channel JFET.

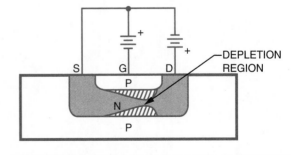

Figure 23–3 Properly biased N-channel JFET.

end because the E_{DS} voltage adds to the E_{GS} voltage creating a higher reverse bias voltage than that appearing across the source end.

The size of the depletion region is controlled by E_{GS}. As E_{GS} increases, so does the depletion region. A decrease in E_{GS} causes a decrease in the depletion region. When the depletion region increases, it effectively reduces the size of the channel. This reduces the amount of current that is able to flow through it. E_{GS} can thus be used to control the drain current (I_D) that flows through the channel. An increase in E_{GS} causes a decrease in I_D.

In normal operation, the input voltage is applied between the gate and the source. The resulting output current is the drain current (I_D). In a JFET, the input voltage is used to control the output current. In a transistor, it is the input current, not the voltage, that is used to control the output current.

Because the gate-to-source voltage is reverse biased, the JFET has an extremely high input resistance. If the gate-to-source voltage were forward biased, a large current would flow through the channel causing the input resistance to drop and reducing the gain of the device. The amount of gate-to-source voltage required to reduce I_D to zero is called the *gate-to-source cut-off voltage* ($E_{GS(off)}$). This value is specified by the manufacturer of the device.

The drain-to-source voltage (E_{DS}) has control over the depletion region within the JFET. As E_{DS} increases, I_D also increases. A point is then reached where I_D levels off, increasing only slightly as E_{DS} continues to rise. This occurs because the size of the depletion region has increased also, to the point where the channel is depleted of minority carriers and cannot allow I_D to increase proportionally with E_{DS}. The resistance of the channel also increases with an increase of E_{DS} with the result that I_D increases at a slower rate. However, I_D levels off because the depletion region expands and reduces the channel's width. When this occurs, I_D is said to *pinch off*. The value of E_{DS} required to pinch off or limit I_D is called the *pinch-off voltage* (E_P). E_P is usually given by the manufacturer for an E_{GS} of zero. E_P is always close to $E_{GS(off)}$ when E_{GS} is equal to zero. When E_P is equal to E_{GS} the drain current is pinched off.

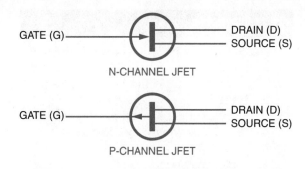

Figure 23–4 Schematic symbols for JFETs.

P-channel and N-channel JFETs have the same characteristics. The main difference between them is the direction of the drain current (I_D) through the channel. In a P-channel JFET, the polarity of the bias voltages (E_{GS}, E_{DS}) is opposite to that in an N-channel JFET.

The schematic symbols used for N-channel and P-channel JFETs are shown in Figure 23–4. The polarities required to bias an N-channel JFET are shown in Figure 23–5 and for a P-channel JFET in Figure 23–6.

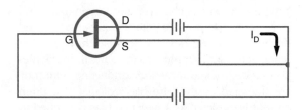

Figure 23–5 The polarities required to bias an N-channel JFET.

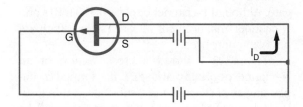

Figure 23–6 The polarities required to bias a P-channel JFET.

23–1 Questions

1. Describe how a JFET differs in construction from a bipolar transistor.
2. Identify the three electrical connections to the JFET.
3. How is the current shut off in a JFET?
4. Define the following with reference to JFETs:
   a. Depletion region
   b. Pinch-off voltage
   c. Source
   d. Drain
5. Draw and label a schematic diagram of a P-channel and an N-channel JFET.

23–2 Depletion Insulated Gate FETs (MOSFETs)

Insulated gate FETs do not use a PN junction. Instead they use a metal gate, which is electrically isolated from the semiconductor channel by a thin layer of oxide. This device is known as a **metal oxide semiconductor field effect transistor** (**MOSFET** or MOST).

There are two important types of MOSFETs: N-type units with N channels and P-type units with P channels. N-type units with N channels are called **depletion mode** devices because they conduct when zero bias is applied to the gate. In the depletion mode, the electrons are conducting until they are depleted by the gate bias voltage. The drain current is depleted as a negative bias is applied to the gate. P-type units with P channels are **enhancement mode** devices. In the enhancement mode, the electron flow is normally cut off until it is aided or enhanced by the bias voltage on the gate. Although P-channel depletion MOSFETs and N-channel enhancement MOSFETs exist, they are not commonly used.

Figure 23–7 shows a cross section of an N-channel **depletion MOSFET.** It is formed by implanting an N channel in a P substrate. A thin insulating layer of silicon dioxide is then deposited on the channel, leaving the ends of the channel exposed to be attached to wires and act as the source

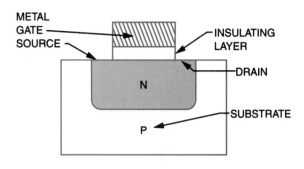

Figure 23–7 N-channel depletion MOSFET.

and the drain. A thin metallic layer is attached to the insulating layer over the N channel. The metallic layer serves as the gate. An additional lead is attached to the substrate. The metal gate is insulated from the semiconductor channel so that the gate and the channel do not form a PN junction. The metal gate is used to control the conductivity of the channel as in the JFET.

The MOSFET shown in Figure 23–8 has an N channel. The drain is always made positive with respect to the source, as in the JFET. The majority carriers are electrons in the N channel, which allow drain current (I_D) to flow from the source to the drain. The drain current is controlled by the gate-to-source bias voltage (E_{GS}), as in the JFET. When the source voltage is zero, a substantial drain current flows through the device, because a large number of majority carriers (electrons) are present

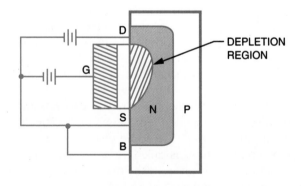

Figure 23–8 N-channel depletion MOSFET with bias supply.

in the channel. When the gate is made negative with respect to the source, the drain current decreases as the majority carriers are depleted. If the negative gate voltage is increased sufficiently, the drain current drops to zero. One difference between MOSFETs and JFETs is that the gate of the N-channel depletion MOSFET can also be made positive with respect to the source. This cannot be done with a JFET because it would cause the gate and channel PN junction to be forward biased.

When the gate voltage of a depletion MOSFET is made positive, the silicon dioxide insulating layer prevents any current from flowing through the gate lead. The input resistance remains high, and more majority carriers (electrons) are drawn into the channel, enhancing the conductivity of the channel. A positive gate voltage can be used to increase the MOSFET's drain current, and a negative gate voltage can be used to decrease the drain current. Because a negative gate voltage is required to deplete the N-channel MOSFET, it is called a depletion mode device. A large amount of drain current flows when the gate voltage is equal to zero. All depletion mode devices are considered to be normally turned on when the gate voltage is equal to zero.

An N-channel depletion MOSFET is represented by the schematic symbol shown in Figure 23–9. Note that the gate lead is separated from the source and drain leads. The arrow on the substrate lead points inward to represent an N-channel device. Some MOSFETs are constructed with the substrate connected internally to the source lead and the separate substrate lead is not used.

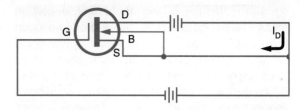

Figure 23–10 Properly biased N-channel depletion MOSFET.

A properly biased N-channel depletion MOSFET is shown in Figure 23–10. Note that it is biased the same way as an N-channel JFET. The drain-to-source voltage (E_{DS}) must always be applied so that the drain is positive with respect to the source. The gate-to-source voltage (E_{DS}) can be applied with the polarity reversed. The substrate is usually connected to the source, either internally or externally. In special applications, the substrate may be connected to the gate or another point within the FET circuit.

A depletion MOSFET may also be constructed as a P-channel device. P-channel devices operate in the same manner as N-channel devices. The difference is that the majority carriers are holes. The drain lead is made negative with respect to the source, and the drain current flows in the opposite direction. The gate may be positive or negative with respect to the source.

The schematic symbol for a P-channel depletion MOSFET is shown in Figure 23–11. The only difference between the N-channel and P-channel symbols is the direction of the arrow on the substrate lead.

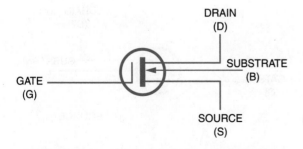

Figure 23–9 Schematic symbol for an N-channel depletion MOSFET.

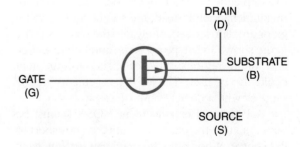

Figure 23–11 Schematic symbol for a P-channel depletion MOSFET.

Both N-channel and P-channel depletion MOSFETs are symmetrical. The source and drain leads may be interchanged. In special applications, the gate may be offset from the drain region to reduce capacitance between the gate and drain. When the gate is offset, the source and drain leads cannot be interchanged.

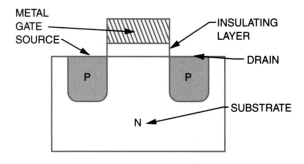

Figure 23–12 P-channel enhancement MOSFET.

23—2 Questions

1. How does a MOSFET differ in construction from a JFET?
2. Describe how a MOSFET conducts a current.
3. What is the major difference in operation between a MOSFET and a JFET?
4. Draw and label a schematic diagram of a P-channel and an N-channel MOSFET.
5. What leads can be interchanged on JFETs and MOSFETs?

23—3 Enhancement Insulated Gate FETs (MOSFETs)

Depletion MOSFETs are devices that are normally on. That is, they conduct a substantial amount of drain current when the gate-to-source voltage is zero. This is useful in many applications. It is also useful to have a device that is normally off; that is, a device that conducts only when a suitable value of E_{GS} is applied. Figure 23–12 shows a MOSFET that functions as a normally off device. It is similar to a depletion MOSFET, but it does not have a conducting channel. Instead, the source and drain regions are diffused separately into the substrate. The figure shows an N-type substrate and P-type source and drain regions. The opposite arrangement could also be used. The lead arrangements are the same as with a depletion MOSFET.

A P-channel enhancement MOSFET must be biased so that the drain is negative with respect to the source. When only the drain-to-source voltage (E_{DS}) is applied, a drain current does not flow. This is because there is no conducting channel between the source and drain. When the gate is made negative with respect to the source, holes are drawn toward the gate where they gather to create a P-type channel that allows current to flow from the drain to the source. When the negative gate voltage is increased, the size of the channel increases, allowing even more current to flow. An increase in gate voltage tends to enhance the drain current.

The gate of a P-channel enhancement MOSFET can be made positive with respect to the source without affecting the operation. The MOSFET's drain current is zero and cannot be reduced with the application of a positive gate voltage.

The schematic symbol for a P-channel enhancement MOSFET is shown in Figure 23–13. It is the same as that for a P-channel depletion MOSFET except that a broken line is used to interconnect the source, drain, and substrate region. This indicates the normally off condition. The arrow points outward to indicate a P channel.

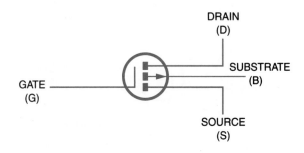

Figure 23–13 Schematic symbol for a P-channel enhancement MOSFET.

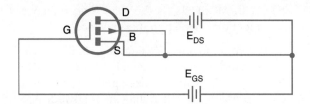

Figure 23–14 Properly biased P-channel enhancement MOSFET.

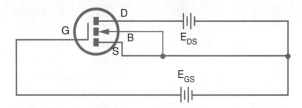

Figure 23–16 Properly biased N-channel enhancement MOSFET.

A properly biased P-channel enhancement MOSFET is shown in Figure 23–14. Notice that E_{DS} makes the MOSFET's drain negative with respect to the source. E_{GS} also makes the gate negative with respect to the source. Only when E_{GS} increases from zero volts and applies a negative voltage to the gate does a substantial amount of drain current flow. The substrate is normally connected to the source, but in special applications the substrate and source may be at different potentials.

N-channel enhancement MOSFETs may also be constructed. These devices operate with a positive gate voltage so that electrons are attracted toward the gate to form an N-type channel. Otherwise, these devices function like P-channel devices.

Figure 23–15 shows the schematic symbol for an N-channel enhancement MOSFET. It is similar to the P-channel device except that the arrow points inward to identify the N channel. Figure 23–16 shows a properly biased N-channel enhancement MOSFET.

MOSFETs are usually symmetrical, like JFETs. Therefore the source and drain can usually be reversed or interchanged.

23–3 Questions

1. How do depletion and enhancement MOSFETs differ from each other?
2. Describe how an enhancement insulated gate FET operates.
3. Draw and label the schematic symbol for a P-channel and an N-channel enhancement MOSFET.
4. Why are there four leads on a MOSFET?
5. What leads on an enhancement MOSFET can be reversed?

23–4 MOSFET Safety Precautions

Certain safety precautions must be observed when handling and using MOSFETs. It is important to check the manufacturer's specification sheet for maximum rating of E_{GS}. If E_{GS} is increased too much, the thin insulating layer ruptures, ruining the device. The insulating layer is so sensitive that it can be damaged by a static charge that has built up on the leads of the device. Electrostatic charges on fingers can be transferred to the MOSFET's leads when handling or mounting the device.

To avoid damage to the device, MOSFETs are usually shipped with the leads shorted together. Shorting techniques include wrapping leads with a shorting wire, inserting the device into a shorting ring, pressing the device into conductive foam,

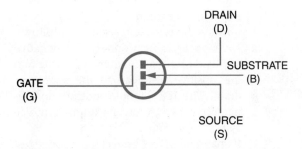

Figure 23–15 Schematic symbol for an N-channel enhancement MOSFET.

taping several devices together, shipping in antistatic tubes, and wrapping the device in metal foil.

Newer MOSFETs are protected with zener diodes electrically connected between the gate and source internally. The diodes protect against static discharges and in-circuit transients and eliminate the need for external shorting devices. In electronics, a *transient* is a temporary component of current existing in a circuit during adjustment to a load change, voltage source difference or line impulse.

If the following procedures are followed, unprotected MOSFETs can be handled safely:

1. Prior to installation into a circuit, the leads should be kept shorted together.
2. The hand used to handle the device should be grounded with a metallic wrist band.
3. The soldering iron tip should be grounded.
4. A MOSFET should never be inserted or removed from its circuit when the power is on.

23–4 Questions

1. What is the reason that MOSFETs have to be handled very carefully?
2. What voltage source, if exceeded, can ruin a MOSFET?
3. What methods are used to protect MOSFETs during shipping?
4. What precautions have been taken to protect newer MOSFETs?
5. Describe the procedures that must be observed when handling unprotected MOSFETs.

23–5 Testing FETs

Testing a field effect transistor is more complicated than testing a normal transistor. The following points must be considered prior to the actual testing of a FET.

1. Is the device a JFET or a MOSFET?
2. Is the FET an N-channel or a P-channel device?

3. With MOSFETs, is the device an enhancement or a depletion mode device?

Before removing a FET from a circuit or handling it, check to see whether it is a JFET or a MOSFET. MOSFETs can be damaged easily unless certain precautions in handling are followed.

1. Keep all of the leads of the MOSFET shorted until ready to use.
2. Make sure the hand used to handle the MOSFET is grounded.
3. Ensure that the power to the circuit is removed prior to insertion or removal of the MOSFET.

Both JFETs and MOSFETs can be tested using commercial transistor test equipment or an ohmmeter.

In using commercial transistor test equipment, refer to the operations manual for proper switch settings.

Testing JFETs with an Ohmmeter

1. Use a low-voltage ohmmeter in the R × 100 range.
2. Determine the polarity of the test leads. Red is positive and black is negative.
3. Determine the forward resistance as follows:
   a. N-channel JFETs: Connect the positive lead to the gate and the negative lead to the source or drain. Because the source and drain are connected by a channel, only one side needs to be tested. The forward resistance should be a low reading.
   b. P-channel JFETs: Connect the negative lead to the gate and the positive lead to the source or drain.
4. Determine the reverse resistance as follows:
   a. N-channel JFETs: Connect the negative test lead of the ohmmeter to the gate and the positive test lead to the source or drain. The JFET should indicate an infinite resistance. A lower reading indicates a short or leakage.
   b. P-channel JFETs: Connect the positive test lead of the ohmmeter to the gate and the negative test lead to the source or drain.

Testing MOSFETs with an Ohmmeter

The forward and reverse resistance should be checked with a low-voltage ohmmeter on its highest scale. MOSFETs have extremely high input resistance because of the insulated gate. The meter should register an infinite resistance in both the forward and reverse resistance test between the gate and source or drain. A lower reading indicates a breakdown of the insulation between the gate and source or drain.

23—5 Questions

1. What questions must be answered prior to the actual testing of a FET?
2. Why is it important to know whether a device is a JFET or a MOSFET before removing it from a circuit?
3. Describe how to test a JFET using an ohmmeter.
4. Describe how to test a MOSFET using an ohmmeter.
5. What procedure is used to test either a JFET or MOSFET with a commercial transistor tester?

SUMMARY

■ A JFET uses a channel instead of junctions (as in transistors) for controlling a signal.
■ The three leads of a JFET are attached to the gate, source, and drain.
■ The input signal is applied between the gate and the source for controlling a JFET.
■ JFETs have extremely high input resistance.
■ The schematic symbols for JFETs are:

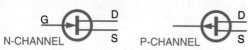

■ MOSFETs (insulated gate FETs) isolate the metal gate from the channel with a thin oxide layer.
■ Depletion mode MOSFETs are usually N-channel device and are classified as normally on.
■ Enhancement mode MOSFETs are usually P-channel devices and are normally off.
■ One difference between JFETs and MOSFETs is that the gate can be made positive or negative.
■ The schematic symbol for an enhancement MOSFET is:

■ The source and drain leads can be interchanged on most JFETs and MOSFETs because the devices are symmetrical.
■ The schematic symbol for an enhancement MOSFET is:

■ MOSFETs must be handled carefully to avoid rupture of the thin oxide layer separating the metal gate from the channel.
■ Electrostatic charges from fingers can damage a MOSFET.
■ Prior to use, keep the leads of a MOSFET shorted together.
■ Wear a grounded metallic wrist strap when handling MOSFETs.
■ Use a grounded soldering iron when soldering MOSFETs into a circuit and make sure the power to the circuit is off.
■ JFETs and MOSFETs can be tested using a commercial transistor tester or an ohmmeter.

Chapter 23 Self-Test

1. Explain what is meant by pinch-off voltage for a FET.
2. How is the pinch-off voltage for a JFET determined?
3. Describe what is meant by depletion-type MOSFET.
4. In what mode of operation is an enhancement MOSFET likely to be cut off?
5. Develop a list of safety precautions that must be observed when handling MOSFETs.

24

THYRISTORS

Objectives

After completing this chapter, the student will be able to:

- Identify common types of thyristors.
- Describe how an SCR, TRIAC, or DIAC operates in a circuit.
- Draw and label schematic symbols for an SCR, TRIAC, and DIAC.
- Identify circuit applications of the different types of thyristors.
- Identify the packaging used with the different types of thyristors.
- Test thyristors using an ohmmeter.

Thyristors are a broad range of semiconductor components used for electronically controlled switches. They are semiconductor devices with a bistable action that depends on a PNPN regenerative feedback to stay turned on or off. *Bistable action* refers to locking onto one of two stable states. *Regenerative feedback* is a method of obtaining an increased output by feeding part of the output back to the input.

Thyristors are widely used for applications where DC and AC power must be controlled. They are used to apply power to a load or remove power from a load. In addition, they can also be used to regulate power or adjust the amount of power applied to a load; for example, a dimmer control for a light or a motor speed control.

24–1 Silicon-Controlled Rectifiers

Silicon-controlled rectifiers are the best known of the thyristors and are generally referred to as SCRs. They have three terminals (anode, cathode, and gate) and are used primarily as switches. An SCR is basically a rectifier because it controls current in only one direction. The advantage of the SCR over a power transistor is that it can control a large current, dependent on an external circuit, with a small trigger signal. An SCR requires a current flowing to stay turned on after the gate signal is removed. If the current flow drops to zero, the SCR shuts off and a gate signal must be reapplied to turn the SCR back on. A power transistor would require ten times the trigger signal of an SCR to control the same amount of current.

An SCR is a solid-state device consisting of four alternately doped semiconductor layers. It is made from silicon by the diffusion or diffusion-alloy method (see Chapter 20). Figure 24–1 shows a simplified diagram of an SCR. The four layers are sandwiched together to form three junctions. Leads are attached to only three of the layers to form anode, cathode, and gate.

Figure 24–2 shows the four layers divided into two three-layer devices. They are a PNP and an NPN transistor interconnected to form a regenerative feedback pair. Figure 24–3 shows the schematic symbols for these transistors. The figure shows that the anode is positive with respect to the cathode, and the gate is open. The NPN transistor does not conduct because its emitter junction is not subject to a forward bias voltage (provided by the PNP transistor's collector or gate signal). Because the NPN transistor's collector is not conducting, the PNP transistor is not conducting (the NPN transis-

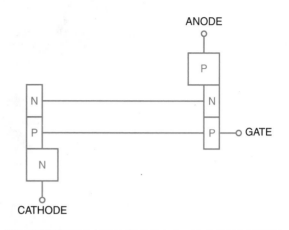

Figure 24–2 Equivalent SCR.

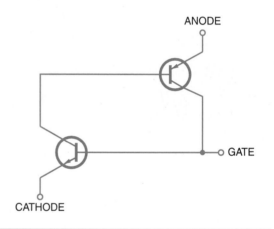

Figure 24–3 Schematic representation of an equivalent SCR.

tor's collector provides the base drive for the PNP transistor). The circuit does not allow current to flow from the cathode to the anode under these conditions.

If the gate is made positive with respect to the cathode, the emitter junction of the NPN transistor becomes forward biased and the NPN transistor conducts. This causes base current to flow through the PNP transistor, which, in turn, allows the PNP transistor to conduct. The collector current flowing through the PNP transistor causes the base current to flow through the NPN transistor. The two transistors hold each other in the conducting state,

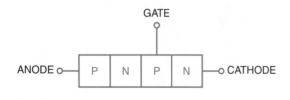

Figure 24–1 Simplified SCR.

allowing current to flow continuously from the cathode to the anode. This action takes place even though the gate voltage is applied only momentarily. The momentary gate voltage causes the circuit to switch to the conducting state and the circuit to continue conducting even though the gate voltage is removed. The anode current is limited only by the external circuit. To switch the SCR off, it is necessary to reduce the anode-to-cathode voltage to zero. This causes both transistors to turn off and remain off until a gate voltage is again applied.

The SCR is turned on by a positive input gate voltage and turned off by reducing the anode-to-cathode voltage to zero. When the SCR is turned on and is conducting a high cathode-to-anode current, it is conducting in the forward direction. If the polarity of the cathode-to-anode bias voltage is reversed, only a small leakage current flows in the reverse direction.

Figure 24–4 shows the schematic symbol for an SCR. This is a diode symbol with a gate lead attached. The leads are typically identified by the letters K (cathode), A (anode), and G (gate). Figure 24–5 shows several SCR packages.

A properly biased SCR is shown in Figure 24–6. The switch is used to apply and remove gate voltage. The resistor R_G is used to limit current to the specified gate current. The anode-to-cathode voltage is provided by the AC voltage source. The series resistor (R_L) is used to limit the anode-to-cathode current to the specified gate current when the device is turned on. Without resistor R_L the SCR would conduct an anode-to-cathode current high enough to damage the SCR.

SCRs are used primarily to control the application of DC and AC power to various types of loads. They can be used as switches to open or close circuits. They can also be used to vary the amount of power applied to a load. In using an SCR, a small gate current can control a large load current.

When an SCR is used in a DC circuit, there is no inexpensive method of turning off the SCR without removing power from the load. This problem can be solved by connecting a switch across

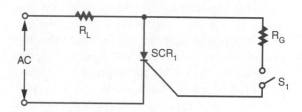

Figure 24–6 Properly biased SCR.

Figure 24–4 Schematic symbol for an SCR.

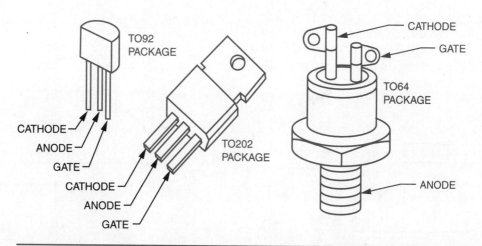

Figure 24–5 Common SCR packages.

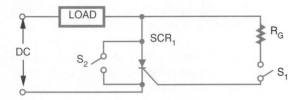

Figure 24–7 Removing power in a DC circuit.

the SCR (Figure 24–7). When switch S_2 is closed, it shorts out the SCR. This reduces the anode-to-cathode voltage to zero, reducing the forward current to below the holding value, and turning off the SCR.

When an SCR is used in an AC circuit it is capable of conducting only one of the alternations of each AC input cycle, the alternation that makes the anode positive with respect to the cathode. When the gate current is applied continuously, the SCR conducts continuously. When the gate current is removed, the SCR turns off within one-half of the AC signal and remains off until the gate current is reapplied. It should be noted that this means only half of the available power is applied to the load. It is possible to use the SCR to control current during both alternations of each cycle. This is accomplished by rectifying the AC signal so that both alternations of each cycle are made to flow in the same direction before being applied to the SCR.

Figure 24–8 shows a simple variable half-wave circuit. The circuit provides a phase shift from 0 to 90 electrical degrees of the anode voltage signal. Diode D_1 blocks the reverse gate voltage on the negative half-cycle of the anode supply voltage.

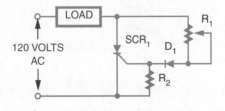

Figure 24–8 Variable half-wave circuit.

1. Why is an SCR considered better for switching than a transistor?
2. Describe how an SCR is constructed.
3. Explain how an SCR operates.
4. Draw and label the schematic symbol for an SCR.
5. What are some applications of an SCR?

24–2 TRIACs

TRIAC is an acronym for *triode AC* semiconductor. TRIACs have the same switching characteristics as SCRs but conduct both directions of AC current flow. A TRIAC is equivalent to two SCRs connected in parallel, back to back (Figure 24–9).

Because a TRIAC can control current flowing in either direction, it is widely used to control application of AC power to various types of loads. It can be turned on by applying a gate current and turned off by reducing the operating current to below the holding level. It is designed to conduct forward and reverse current through the terminals.

Figure 24–10 shows a simplified diagram of a TRIAC. A TRIAC is a four-layer NPNP device in parallel with a PNPN device. It is designed to respond to a gating current through a single gate. The input

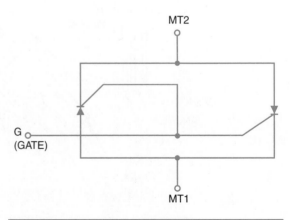

Figure 24–9 Equivalent TRIAC.

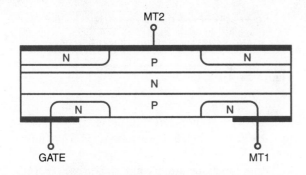

Figure 24–10 Simplified TRIAC.

and output terminals are identified as *main terminal 1 (MT1)* and *main terminal 2 (MT2)*. The terminals are connected across the PN junctions at the opposite ends of the device. Terminal MT1 is the reference point for measurement of voltage and current at the gate terminal. The gate (G) is connected to the PN junction at the same end as MT1. From MT1 to MT2, the signal must pass through an NPNP series of layers or a PNPN series of layers.

The schematic symbol for a TRIAC is shown in Figure 24–11. It consists of two parallel diodes connected in opposite directions with a single gate lead. The terminals are identified as MT1, MT2, and G (gate). Several TRIAC packages are shown in Figure 24–12.

A TRIAC can be used as an AC switch (Figure 24–13). It can also be used to control the amount of AC power applied to a load (Figure 24–14). TRIACs apply all available power to the load. When a TRIAC is used to vary the amount of AC power applied to the load, a special triggering device is

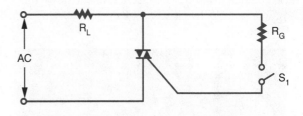

Figure 24–13 TRIAC AC switch circuit.

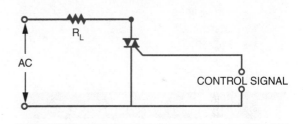

Figure 24–14 TRIAC AC control circuit.

Figure 24–11 Schematic symbol for a TRIAC.

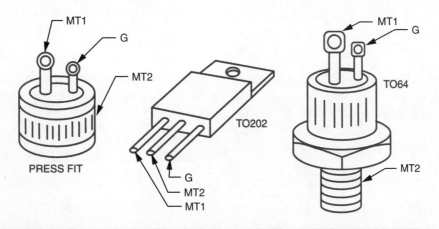

Figure 24–12 Common TRIAC packages.

needed to ensure that the TRIAC functions at the proper time. The triggering device is necessary because the TRIAC is not equally sensitive to the gate current flowing in opposite directions.

TRIACs have disadvantages when compared to SCRs. TRIACs have current ratings as high as 25 amperes, but SCRs are available with current ratings as high as 1400 amperes. A TRIACs maximum voltage rating is 500 volts, compared to an SCR's 2600 volts. TRIACs are designed for low frequency (50 to 400 hertz) where SCRs can handle up to 30,000 hertz. TRIACs also have difficulty switching power to inductive loads.

24—2 Questions

1. What is the difference between a TRIAC and an SCR?
2. Describe how a TRIAC is constructed.
3. Draw and label the schematic symbol for a TRIAC.
4. What are some applications of a TRIAC?
5. Compare the advantages and disadvantages of TRIACs and SCRs.

24—3 Bidirectional Trigger Diodes

Bidirectional (two-directional) trigger diodes are used in TRIAC circuits because TRIACs have nonsymmetrical triggering characteristics; that is, they are not equally sensitive to gate current flowing in opposite directions. The most frequently used triggering device is the **DIAC** (*diode AC*).

The DIAC is constructed in the same manner as the transistor. It has three alternately doped layers (Figure 24–15). The only difference in the construction is that the doping concentration around both junctions in the DIAC is equal. Leads are only

(CAN BE INSTALLED IN EITHER DIRECTION)

Figure 24–16 Equivalent DIAC.

attached to the outer layers. Because there are only two leads, the device is packaged like a PN junction diode.

Both junctions are equally doped, so a DIAC has the same effect on current regardless of the direction of flow. One of the junctions is forward biased and the other is reverse biased. The reverse-biased junction controls the current flowing through the DIAC. The DIAC performs as if it contained two PN junction diodes connected in series back-to-back (Figure 24–16). The DIAC remains in the off state until an applied voltage in either direction is high enough to cause its reverse-biased junction to break over and start conducting. *Break over* is the point at which conduction starts to occur. This causes the DIAC to turn on and the current to rise to a value limited by a series resistor in the circuit.

The schematic symbol for a DIAC is shown in Figure 24–17. It is similar to the symbol for a TRIAC. The difference is that a DIAC does not have a gate lead.

DIACs are most commonly used as a triggering device for TRIACs. Each time the DIAC turns on, it allows current to flow through the TRIAC gate, thus turning the TRIAC on. The DIAC is used in conjunction with the TRIAC to provide full-wave control of AC signals.

Figure 24–18 shows a variable full-wave phase-control circuit. Variable resistor R_1 and capacitor C_1 form a phase-shift network. When the voltage across C_1 reaches the breakover voltage of the DIAC, C_1 partially discharges through the DIAC into the gate of the TRIAC. This discharge creates a pulse that triggers the TRIAC into conduction. This

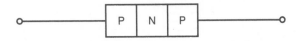

Figure 24–15 Simplified DIAC.

Figure 24–17 Schematic symbol for a DIAC.

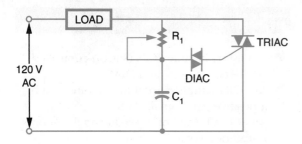

Figure 24–18 Variable full-wave phase-control circuit.

circuit is useful for controlling lamps, heaters, and speeds of small electrical motors.

24—3 Questions

1. Where are DIACs used in circuits?
2. Describe how a DIAC is constructed.
3. Explain how a DIAC works in a circuit.
4. Draw the schematic symbol for a DIAC.
5. Draw the schematic for a full-wave phase-control circuit using DIACs and TRIACs.

24—4 Testing Thyristors

Like other semiconductor devices, thyristors may fail. They can be tested with commercial test equipment or with an ohmmeter.

To use commercial test equipment for testing thyristors, refer to the operator's manual for proper switch settings and readings.

An ohmmeter can detect the majority of defective thyristors. It cannot detect marginal or voltage-sensitive devices. It can, however, give a good indication of the condition of the thyristor.

Testing SCRs with an Ohmmeter

1. Determine the polarity of the ohmmeter leads. The red lead is positive and the black lead is negative.
2. Connect the ohmmeter leads, positive to the cathode and negative to the anode. The resistance should exceed 1 megohm.
3. Reverse the leads, negative to the cathode and positive to the anode. The resistance should again exceed 1 megohm.
4. With the ohmmeter leads connected as in step 3, short the gate to the anode (touch the gate lead to the anode lead). The resistance should drop to less than 1 megohm.
5. Remove the short between the gate and the anode. If a low-resistance range of the ohmmeter is used, the resistance should stay low. If a high-resistance range is used, the resistance should return to above 1 megohm. In the higher resistance ranges, the ohmmeter does not supply enough current to keep the gate latched (turned on) when the short is removed.
6. Remove the ohmmeter leads from the SCR and repeat the test. Because some ohmmeters do not give significant results on step 5, step 4 is sufficient.

Testing TRIACs with an Ohmmeter

1. Determine the polarity of the ohmmeter leads.
2. Connect the positive lead to MT1 and the negative lead to MT2. The resistance should be high.
3. With the leads still connected as in step 2, short the gate to MT1. The resistance should drop.
4. Remove the short. The low resistance should remain. The ohmmeter may not supply enough current to keep the TRIAC latched if a large gate current is required.
5. Remove the leads and reconnect as specified in step 2. The resistance should again be high.
6. Short the gate to MT2. The resistance should drop.
7. Remove the short. The low resistance should remain.
8. Remove and reverse the leads, the negative lead to MT1 and the positive lead to MT2. The resistance should read high.
9. Short the gate to MT1. The resistance should drop.
10. Remove the short. The low resistance should remain.

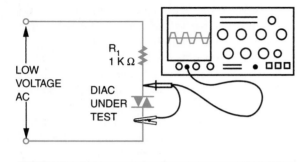

Figure 24–19 Dynamic DIAC test.

11. Remove the leads and reconnect in the same configuration. The resistance should again be high.
12. Short the gate to MT2. The resistance should drop.
13. Remove the short. The low resistance should remain.
14. Remove and reconnect the leads. The resistance should be high.

Testing DIACs with an Ohmmeter

In testing DIACs with an ohmmeter, a low resistance in either direction indicates that the device is not opened (defective). This does not indicate a shorted device. Further testing of the DIAC requires a special circuit setup to check the voltage at the terminals (Figure 24–19).

24–4 Questions

1. Describe the switch settings and indications for using a transistor tester for testing an SCR. (Refer to your manual.)
2. Describe the switch settings and indications for using a transistor tester for testing a TRIAC. (Refer to your manual.)
3. Describe the procedure for testing an SCR with an ohmmeter.
4. Describe the procedure for testing a TRIAC with an ohmmeter.
5. Describe the procedure for testing a DIAC with an ohmmeter.

SUMMARY

- Thyristors include SCRs (silicon-controlled rectifiers), TRIACs, and DIACs.
- An SCR controls current in one direction by a positive gate signal.
- An SCR is turned off by reducing the anode-to-cathode voltage to zero.
- SCRs can be used to control current in both AC and DC circuits.
- The schematic symbol for an SCR is:

- TRIACs are bidirectional triode thyristors.
- TRIACs control current in either direction by either a positive or negative gate signal.
- The schematic symbol for a TRIAC is:

- SCRs can handle up to 1400 amperes compared to 25 amperes for TRIACs.
- SCRs have voltage ratings up to 2600 volts compared to 500 volts for TRIACs.
- SCRs can handle frequencies of up to 30,000 hertz compared to 400 hertz for TRIACs.
- Because TRIACs have nonsymmetrical triggering characteristics, they require the use of a DIAC.
- DIACs are bidirectional trigger diodes.
- The schematic symbol for a DIAC is:

- DIACs are mostly used as triggering devices for TRIACs.
- Thyristors can be tested using commercial transistor testers or ohmmeters.

Chapter 24 Self-Test

1. What is the difference between a PN junction diode and an SCR?
2. After the SCR has been turned on, what effect does the anode supply voltage have on the anode current that flows through the SCR?
3. What effect does the load resistance have on the current flowing through an SCR circuit?
4. Describe the process for testing an SCR.
5. Why is a DIAC used in the gate circuit of a TRIAC?

25

INTEGRATED CIRCUITS

Objectives

After completing this chapter, the student will be able to:

- Explain the importance of integrated circuits.
- Identify advantages and disadvantages of integrated circuits.
- Identify the major components of an integrated circuit.
- Describe the four processes used to construct integrated circuits.
- Identify the major integrated circuit packages.
- List the families of integrated circuits.

Transistors and other semiconductor devices have made it possible to reduce the size of electronic circuits because of their small size and low power consumption. It is now possible to extend the principles behind semiconductors to complete circuits as well as individual components. The goal of the integrated circuit is to develop a single device to perform a specific function, such as amplification or switching, eliminating the separation between components and circuits.

Several factors have made the integrated circuit popular:

- It is reliable with complex circuits.
- It meets the need for low power consumption.
- It offers small size and weight.
- It is economical to produce.
- It offers new and better solutions to system problems.

25-1 Introduction to Integrated Circuits

An **integrated circuit (IC)** is a complete electronic circuit in a package no larger than that of a conventional low-power transistor (Figure 25–1). The circuit consists of diodes, transistors, resistors, and capacitors. Integrated circuits are produced with the same technology and materials used in making transistors and other semiconductor devices.

The most obvious advantage of the integrated circuit is its small size. An integrated circuit is constructed of a chip of semiconductor material approximately one-eighth of an inch square. It is because of the integrated circuit's small size that it finds extensive use in military and aerospace programs. The integrated circuit has also transformed the calculator from a desktop to a handheld instrument. Computer systems, once the size of rooms, are now available in portable models because of integrated circuits.

This small, integrated circuit consumes less power and operates at higher speeds than a conventional transistor circuit. The electron travel time is reduced by direct connection of the internal components.

Integrated circuits are more reliable than directly connected transistor circuits. In the integrated circuit, internal components are connected permanently. All the components are formed at the same time, reducing the chance for error. After the integrated circuit is formed, it is pretested before final assembly.

Many integrated circuits are produced at the same time. This results in substantial cost savings. Manufacturers offer a complete and standard line of integrated circuits. Special-purpose integrated circuits can still be produced to specification, but this results in higher costs if quantities are small.

Integrated circuits reduce the number of parts needed to construct electronic equipment. This reduces inventory, resulting in less overhead for the manufacturer and further reducing the cost of electronic equipment.

Integrated circuits do have some disadvantages. They cannot handle large amounts of current or voltage. High current generates excessive heat, damaging the device. High voltage breaks down the insulation between the various internal components. Most integrated circuits are low-power devices, consuming from 5 to 15 volts in the milliamp range. This results in power consumption of less than 1 watt.

Figure 25–1 Integrated circuit (IC) packages. (*Courtesy of Motorola Semiconductor Products, Inc.*)

Only four types of components are included in integrated circuits: diodes, transistors, resistors, and capacitors. Diodes and transistors are the easiest to construct. Resistors increase in size as they increase in resistance. Capacitors require more space than resistors and also increase in size as their capacity increases.

Integrated circuits cannot be repaired. This is because the internal components cannot be separated. Therefore, problems are identified by individual circuit instead of by individual component. The advantage of this disadvantage is that it greatly simplifies maintaining highly complex systems. It also reduces the amount of time required for maintenance personnel to service equipment.

When all factors are considered, the advantages outweigh the disadvantages. Integrated circuits reduce the size, weight, and cost of electronic equipment while increasing its reliability. As integrated circuits become more sophisticated, they become capable of performing a wider range of tasks.

25–1 Questions

1. Define *integrated circuit*.
2. What are the advantages of integrated circuits?
3. What are the disadvantages of integrated circuits?
4. What components can be included in integrated circuits?
5. What is the procedure for repairing a faulty integrated circuit?

25–2 Integrated Circuit Construction Techniques

Integrated circuits are classified according to their construction techniques. The more common construction techniques are monolithic, thin film, thick film, and hybrid.

Monolithic integrated circuits are constructed in the same manner as transistors but include a few additional steps (Figure 25–2). The integrated circuit begins with a circular silicon wafer 3 to 4 inches in diameter and about 0.010 inch thick. This serves as a substrate (base) on which the integrated circuits are formed. Many integrated circuits are formed at the same time on this wafer, up to several hundred, depending on the size of the wafer. The integrated circuits on the wafer are generally all the same size and type and contain the same number and type of component.

After fabrication, the integrated circuits are tested on the wafer. After testing, the wafer is sliced into individual chips. Each chip represents one complete integrated circuit containing all the components and connections associated with that circuit. Each chip that passes quality-control tests is then mounted into a package. Even though a large number of integrated circuits are fabricated at the same time, only a certain number are usable. This is referred to as the yield. The *yield* is the maximum number of usable integrated circuits compared with the rejects.

Thin-film integrated circuits are formed on the surface of an insulating substrate of glass or ceramic, usually less than an inch square. The components (resistors and capacitors) are formed by an extremely thin film of metal and oxides deposited on the substrate. Thin strips of metal are then deposited to connect the components. Diodes and transistors are formed as separate semiconductor devices and attached in the proper location. Resistors are formed by depositing tantalum or nichrome on the surface of the substrate in a thin film 0.0001 inch thick. The value of the resistor is determined by the length, width, and thickness of each strip. Conductors are formed of low-resistance metal such as gold, platinum, or aluminum. This process can produce a resistor accurate to within ±0.1%. It can also obtain a ratio between resistors accurate to ±0.01%. Accurate ratios are important for the proper operation of a particular circuit.

Thin-film capacitors consist of two thin layers of metal separated by an extremely thin dielectric. A metal layer is deposited on the substrate. An oxide coating is then formed over the metal for a dielectric. The dielectric is formed from an insulating material such as tantalum oxide, silicon oxide, or

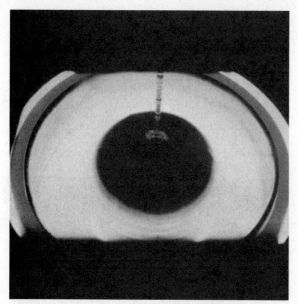

(1) CRYSTAL GROWING

(3) LAPPING AND POLISHING

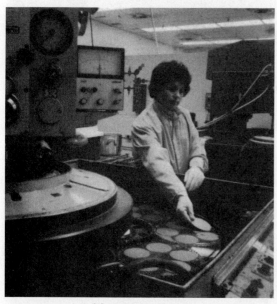

(2) INGOT SAWING

(4) AN EPITAXIAL FURNACE

Figure 25–2 Monolithic construction technique. (*Courtesy of Motorola Semiconductor Products, Inc.*)

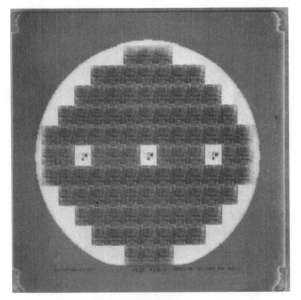

(5) MASKING

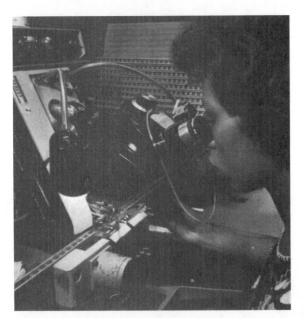

(6) INSPECTION OF A CIRCUIT

(7) INSPECTION OF COMPLETED DEVICE

Figure 25–2 Continued.

aluminum oxide. The top plate, formed from gold, tantalum, or platinum, is deposited on the dielectric. The capacitor value is obtained by adjusting the area of the plate and by varying the thickness and type of dielectric.

Diode and transistor chips are formed using the monolithic technique. The diode and transistor chips are permanently mounted on the substrate. They are then electrically connected to the thin-film circuit using extremely thin wires.

The materials used for components and conductors are deposited on the substrate using an evaporation or sputtering process. The *evaporation process* requires that the material deposited on the substrate be placed in a vacuum and heated until it evaporates. The vapors are then allowed to condense on the substrate forming a thin film.

The *sputtering process* subjects a gas-filled chamber to a high voltage. The high voltage ionizes the gas, and the material to be deposited is bombarded with ions. This dislodges the atoms within the material, which drift toward the substrate where they are deposited as a thin film. A mask is used to ensure proper location of the film deposit. Another method is to coat the substrate completely and cut or etch away the undesired portions.

In thick-film construction, the resistors, capacitors, and conductors are formed on the substrate using a screen printing process: A fine wire screen is placed over the substrate and metallized ink is forced through the screen with a squeegee. The screen acts as a mask. The substrate and ink are then heated to over 600 degrees celsius to harden the ink.

Thick-film capacitors have low values (in picofarads). When a higher value is required, discrete capacitors are used. The thick-film components are 0.001 inch thick; they have no discernable thickness. The thick-film components resemble conventional discrete components.

Hybrid integrated circuits are formed using monolithic, thin-film, thick-film, and discrete components. This allows for a high degree of circuit complexity by using monolithic circuits while at the same time taking advantage of extremely accurate component values and tolerances with the film techniques. Discrete components are used because they can handle relatively large amounts of power.

If only a few circuits are to be made, it is cheaper to use hybrid integrated circuits. The major expense in hybrid construction is the wiring and assembly of components and the final packaging of the device. Because hybrid circuits use discrete components, they are larger and heavier than monolithic integrated circuits. The use of discrete components tends to make hybrid circuits less reliable than monolithic circuits.

25–2 Questions

1. What methods are used to construct integrated circuits?
2. Describe the monolithic process of integrated circuit fabrication.
3. What are the differences between thin-film and thick-film fabrication processes?
4. How is a hybrid integrated circuit constructed?
5. What determines which process will be used when fabricating an integrated circuit?

25–3 Integrated Circuit Packaging

Integrated circuits are mounted in packages designed to protect them from moisture, dust, and other contaminants. The most popular package design is the **dual-inline package (DIP).** It is manufactured in several sizes to accommodate the various sizes of integrated circuits: small-scale integration (SSI), medium-scale integration (MSI), large-scale integration (LSI), and very large-scale integration (VLSI) (Figure 25–3). The package is made in either ceramic or plastic. Plastic is less expensive and suitable for most applications where the operating temperature falls between 0 to 70 degrees celsius. Ceramic devices are more expensive but offer better protection against moisture and contamination. They also operate in a wider range of temperatures (–55 to +125 degrees celsius).

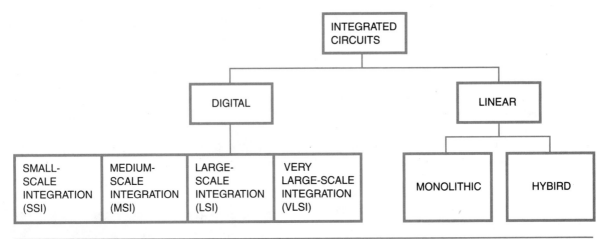

Figure 25-3 Families of integrated circuits.

Ceramic devices are recommended for military, aerospace, and severe industrial applications.

A small 8-pin DIP, called a mini-DIP, is used for devices with a minimum number of inputs and outputs. It is used mostly with monolithic integrated circuits.

A flat-pack is smaller and thinner than a DIP and is used where space is limited. It is made from metal or ceramic and operates in a temperature range of −55 to +125 degrees celsius.

After the integrated circuit is packaged, it is tested to ensure that it meets all electrical specifications. It is tested over a wide range of temperatures.

25—3 Questions

1. What are the functions of an integrated circuit package?
2. What package is most often used for integrated circuits?
3. What materials are used for integrated circuit packages?
4. What are the advantages of ceramic packages?
5. What is the advantage of the flat-pack integrated circuit package?

SUMMARY

■ Integrated circuits are popular because they:
   —Are reliable with complex circuits
   —Consume little power
   —Are small and light
   —Are economical to produce
   —Provide new and better solutions to problems
■ Integrated circuits cannot handle large amounts of current or voltage.
■ Only diodes, transistors, resistors, and capacitors are available as integrated circuits.
■ Integrated circuits cannot be repaired, only replaced.
■ Integrated circuits are constructed by monolithic, thin-film, thick-film, or hybrid techniques.
■ The most popular integrated circuit package is the DIP (dual-inline package).
■ Integrated circuit packages are made of ceramic or plastic, with plastic most often used.

Chapter 25 Self-Test

1. What components does a hybrid integrated circuit contain?
2. What is identified by the word "chip"?
3. In fabricating an integrated circuit using the monolithic construction method, what is the problem with resistors and capacitors?

26

OPTOELECTRIC DEVICES

Objectives

After completing this chapter, the student will be able to:

- Identify the three categories of semiconductor devices that react to light.
- Classify the major frequency ranges of light.
- Identify major light-sensitive devices and describe their operation and applications.
- Identify major light-emitting devices and describe their operation and applications.
- Draw and label the schematic symbols associated with optoelectric devices.
- Identify packages used for optoelectric devices.

Semiconductors in general, and semiconductor diodes in particular, have important uses in opto-electronics. Here, a device is designed to interact with electromagnetic radiation (light energy) in the visible, infrared, and ultraviolet ranges.

Three types of devices interact with light:

- Light-detection devices
- Light-conversion devices
- Light-emitting devices

The semiconductor material and the doping techniques used determine the relevant light wavelength of a particular device.

26–1 Basic Principles of Light

Light is electromagnetic radiation that is visible to the human eye. Light is thought to travel in a form similar to radio waves. Like radio waves, light is measured in wavelengths.

Light travels at 186,000 miles per second or 30,000,000,000 centimeters per second through a vacuum. The velocity is reduced as it passes through various types of mediums. The frequency range of light is 300 to 300,000,000 gigahertz (giga = 1,000,000,000). Of this frequency range only a small band is visible to the human eye. The visible region spans from approximately 400,000 to 750,000 gigahertz. Infrared light falls below 400,000 gigahertz and ultraviolet light lies above 750,000 gigahertz. Light waves at the upper end of the frequency range have more energy than light waves at the lower end.

26–1 Questions

1. What is light?
2. What frequency range of light is visible to the human eye?
3. What is infrared light?
4. What is ultraviolet light?
5. What type of light wave has the most energy?

26–2 Light-Sensitive Devices

The **photoconductive cell (photo cell)** is the oldest optoelectric device. It is a light-sensitive device in which the internal resistance changes with a change in light intensity. The resistance change is not proportional to the light striking it. The photo cell is made from light-sensitive material such as cadmium sulfide (CdS) or cadmium selenide (CdSe).

Figure 26–1 shows a typical photo cell. The light-sensitive material is deposited on an insulating substrate of glass or ceramic in an S-shape to allow greater contact length. The photo cell is more

Figure 26–1 Photo cell.

sensitive to light than any other device. The resistance can vary from several hundred million ohms to several hundred ohms. The photo cell is useful for low-light applications. It can stand high operating voltages of 200 to 300 volts with a low power consumption of up to 300 milliwatts. A disadvantage of the photo cell is its slow response to light change.

Figure 26–2 shows the schematic symbols used to represent a photo cell. The arrows indicate a light-sensitive device. Sometimes the Greek letter lambda (λ) is used to represent a light-sensitive device.

Photo cells are used in light meters for photographic equipment, instruction detectors, automatic door openers, and various kinds of test equipment to measure light intensity.

The **photovoltaic cell (solar cell)** converts light energy directly into electrical energy. The solar cell finds most of its applications in converting solar energy into electrical energy.

The solar cell is basically a PN junction device made from semiconductor materials. It is most commonly made from silicon. Figure 26–3 shows the construction technique. The P and N layers form a PN junction. The metal support and the

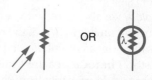

Figure 26–2 Schematic symbols for a photo cell.

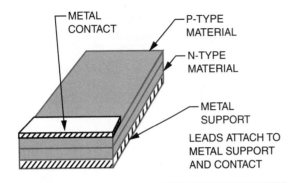

Figure 26-3 Construction of a solar cell.

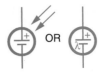

Figure 26-4 Schematic symbols for a solar cell.

metal contact act as the electrical contacts. It is designed with a large surface area.

The light striking the surface of the solar cell imparts much of its energy to the atoms in the semiconductor material. The light energy knocks valence electrons from their orbits, creating free electrons. The electrons near the depletion region are drawn to the N-type material, producing a small voltage across the PN junction. The voltage increases with an increase in light intensity. All the light energy striking the solar cell does not create free electrons, however. In fact, the solar cell is a highly inefficient device, with a top efficiency of 15 to 20%. The cell is inefficient when expressed in terms of electrical power output compared to the total power contained in the input light energy.

Solar cells have a low voltage output of about 0.45 volt at 50 milliamperes. They must be connected in a series and parallel network to obtain the desired voltage and current output.

Applications include light meters for photographic equipment, motion-picture projector soundtrack decoders, and battery chargers on satellites.

Schematic symbols for the solar cell are shown in Figure 26-4. The positive terminal is identified with a plus (+) sign.

The *photodiode* also uses a PN junction and has a construction similar to that of the solar cell. It is used the same way as the photo cell, as a light-variable resistor. **Photodiodes** are semiconductor devices that are made primarily from silicon. They are constructed in two ways. One method is as a

simple PN junction (Figure 26-5). The other method inserts an intrinsic (undoped) layer between the P and N region (Figure 26-6) forming a PIN photodiode.

A PN junction photodiode operates on the same principles as a photovoltaic cell except that it is used to control current flow, not generate it. A reverse-bias voltage is applied across the photodiode, forming a wide depletion region. When light energy strikes the photodiode, it enters the depletion region, creating free electrons. The electrons are drawn toward the positive side of the bias source. A small current flows through the photodiode in the reverse direction. As the light energy increases, more free electrons are generated, creating more current flow.

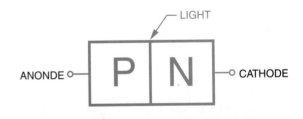

Figure 26-5 PN junction photodiode.

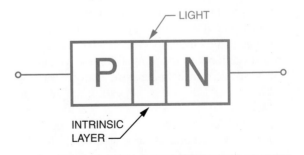

Figure 26-6 PIN junction photodiode.

A **PIN photodiode** has an intrinsic layer between the P and N regions. This effectively extends the depletion region. The wider depletion region allows the PIN photodiode to respond to lower light frequencies. The lower light frequencies have less energy and so must penetrate deeper into the depletion region before generating free electrons. The wider depletion region offers a greater chance of free electrons being generated. The PIN photodiode is more efficient over a wide range.

The PIN photodiode has a lower internal capacitance because of the intrinsic layer. This allows for a faster response to changes in light intensity. Also a more linear change in reverse current with light intensity is produced.

The advantage of the photodiode is fast response to light changes, the fastest of any photosensitive device. The disadvantage is a low output compared to other photosensitive devices.

Figure 26–7 shows a typical photodiode package. A glass window allows light energy to strike the photodiode. The schematic symbol is shown in Figure 26–8. A typical circuit is shown in Figure 26–9.

A **phototransistor** is constructed like other transistors with two PN junctions. It resembles a standard NPN transistor. It is used like a photodiode

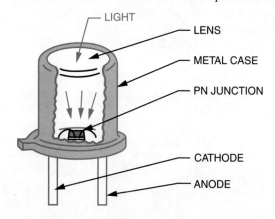

Figure 26–7 Photodiode package.

Figure 26–8 Schematic symbol for a photodiode.

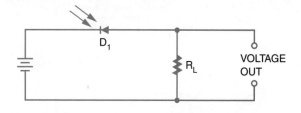

Figure 26–9 Voltage divider using a photodiode.

and packaged like a photodiode, except that it has three leads (emitter, base, and collector). Figure 26–10 shows the equivalent circuit. The transistor conduction depends on the conduction of the photodiode. The base lead is seldom used. When it is used, it is to adjust the turn-on point of the transistor.

Phototransistors can produce higher output current than photodiodes. Their response to light changes is not as fast as that of the photodiode. There is a sacrifice of speed for a higher output current.

Applications include phototachometers, photographic exposure controls, flame detectors, object counters, and mechanical positioners.

Figure 26–11 shows the schematic symbol for a phototransistor. Figure 26–12 shows a typical circuit application.

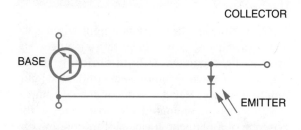

Figure 26–10 Equivalent circuit for a phototransistor.

Figure 26–11 Schematic symbol for a phototransistor.

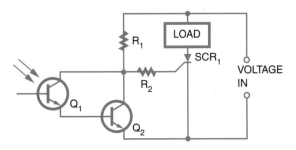

Figure 26–12 Darkness-on DC switch.

26–2 Questions

1. Explain how a photoconductive cell operates.
2. Explain how a solar cell functions.
3. What is the difference between the two types of photodiodes?
4. How is the phototransistor an improvement over the photodiode?
5. Draw and label the schematic symbols for a photoconductive cell, solar cell, photodiode, and phototransistor.

26–3 Light-Emitting Devices

Light-emitting devices produce light when subject to a current flow, converting electrical energy to light energy. A **light-emitting diode** (LED) is the most common semiconductor light-emitting device. Being a semiconductor component, it has an unlimited life span due to the absence of a filament, a cause of trouble in conventional lamps.

Any PN junction diode can emit light when subject to an electrical current flow. The light is produced when the free electrons combine with holes and the extra energy is released in the form of light. The frequency of the light emitted is determined by the type of semiconductor material used in construction of the diode. Regular diodes do not emit light because of the opaque material used in their packaging.

An LED is simply a PN junction diode that emits light when a current flows through its junction. The light is visible because the LED is packaged in semitransparent material. The frequency of the light emitted depends on the material used in the construction of the LED. Gallium arsenide (GaAs) produces light in the infrared region, which is invisible to the human eye. Gallium arsenide phosphide (GaAsP) emits a visible red light. By changing the phosphide content, different frequencies of light can be produced.

Figure 26–13 shows the construction of an LED. The P layer is made thin so that light energy near the PN junction only needs to travel a short distance through it.

After the LED is formed, it is mounted in a package designed for optimum emission of light. Figure 26–14 shows some of the more common LED packages. Most LEDs contain a lens that gathers and intensifies the light. The case may act as a filter with a color to enhance the natural light emitted.

In a circuit, an LED is connected with forward bias to emit light (Figure 26–15). The forward bias must exceed 1.2 volts before a forward current can flow. Because LEDs can be easily damaged by an excessive amount of voltage or current, a series resistor is connected to limit current flow.

The schematic symbol for an LED is shown in Figure 26–16. Figure 26–17 shows a properly biased

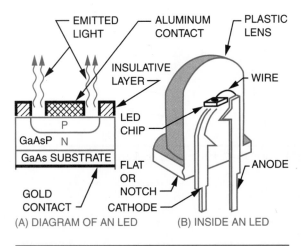

Figure 26–13 LED construction.

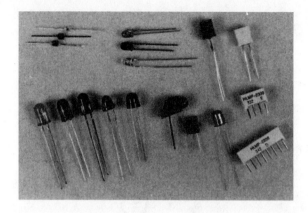

Figure 26–14 Common LED packages.

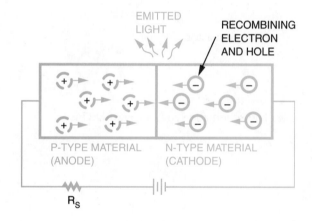

Figure 26–15 Forward-biased LED.

Figure 26–16 Schematic symbol for an LED.

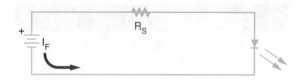

Figure 26–17 Properly biased LED circuit.

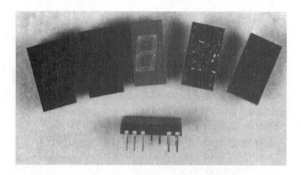

Figure 26–18 Seven-segment display of LEDs for digital readout.

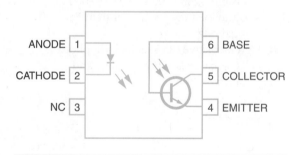

Figure 26–19 Commercial optical coupler.

circuit. The series resistor (R_S) is used to limit the forward current (I_F) based on the applied voltage.

Figure 26–18 shows LEDs arranged to form a seven-segment display used for digital readouts. Figure 26–19 shows an LED used in conjunction with a photodiode to form an **optical coupler.** Both devices are housed in the same package. An *optical coupler* consists of an LED and a photo-

transistor. They are coupled by the light beam produced by the LED. The signal to the LED can vary, which, in turn, varies the amount of light available. The phototransistor converts the varying light back into electrical energy. An optical coupler allows one circuit to pass a signal to another circuit while providing a high degree of electrical insulation.

26—3 Questions

1. Explain how an LED differs from a conventional diode.
2. How does an LED emit different colors of light?
3. How may the LED package enhance the light emitted?
4. Draw and label the schematic symbol for an LED.
5. What is the function of an optical coupler?

SUMMARY

■ Semiconductor devices that interact with light can be classified as light-detection devices, light-conversion devices, or light-emitting devices.

■ Light is electromagnetic radiation that is visible to the human eye.

■ The frequency range of light is:
—Infrared light—less than 400,000 gigahertz
—Visible light—400,000 to 750,000 gigahertz
—Ultraviolet light—greater than 750,000 gigahertz

■ Light-sensitive devices include photo cells, solar cells, photodiodes, and phototransistors.

■ Light-emitting devices include the LED (light-emitting diode).

■ An optical coupler combines a light-sensitive device with a light-emitting device.

■ The schematic symbols for light-sensitive devices are:

Photo cell

Solar cell

Photodiode

Phototransistor

■ The schematic symbol for an LED is:

Chapter 26 Self-Test

1. Which light-sensitive device has the fastest response time for changes in light intensity?

2. Which would lend itself to a wider range of applications, a photodiode or a phototransistor? Why?

3. How does the amount of current flowing in an LED affect the intensity of the light emitted?

Linear Electronic Circuits

27

POWER SUPPLIES

Objectives

After completing this chapter, the student will be able to:

- Explain the purpose of a power supply.
- Draw a block diagram of the circuits and parts of a power supply.
- Describe the three different rectifier configurations.
- Explain the function of a filter.
- Describe the two basic types of voltage regulator and how they operate.
- Explain the function of a voltage multiplier.
- Identify over-voltage and over-current protection devices.

Power supplies are used to supply voltage to a variety of circuits. Their basic principles are the same.

The primary function of the power supply is to convert alternating current (AC) to direct current (DC). The power supply may increase or decrease the incoming AC voltage by means of a transformer.

Once the voltage is at the desired level, it is converted to a DC voltage through a process called rectification. The rectified voltage still contains an AC signal, which is referred to as a ripple frequency. The ripple is removed with a filter.

To ensure that the output voltage remains at a constant level, a voltage regulator is used. The voltage regulator holds the output voltage at a constant level.

27–1 Transformers

Transformers are used in power supplies to isolate the power supply from the AC voltage source. They are also used to step up voltages if higher voltages are required or to step down voltages if lower voltages are required.

If transformers are used in power supplies, the AC power source is connected only to the primary of the transformer. This isolates the electrical circuits from the power source.

When selecting a transformer, its primary power rating is the first concern. The most common primary ratings are 110 to 120 volts and 220 to 240 volts. The next concern is its frequency. Some frequencies are 50 to 60 hertz, 400 hertz, and 10,000 hertz. The third concern is its secondary voltage and current ratings. The final concern is its power-handling capability, or volt-ampere rating. This is essentially the amount of power that can be delivered to the secondary of the transformer. It is given as volt-amperes because of the loads that can be placed on the secondary.

27–1 Questions

1. Why are transformers used in power supplies?
2. How is a transformer connected in a power supply?
3. What are the important considerations when selecting a transformer for a power supply?
4. How are transformers rated?

27–2 Rectifier Circuits

The **rectifier circuit** is the heart of the power supply. Its function is to convert the incoming AC voltage to a DC voltage. There are three basic types of rectifier circuit used with power supplies: **half-wave rectifiers, full-wave rectifiers,** and **bridge rectifiers.**

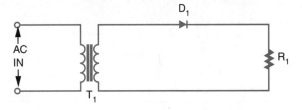

Figure 27–1 Basic half-wave rectifier.

Figure 27–1 shows a basic half-wave rectifier. The diode is located in series with the load. The current in the circuit flows in only one direction because of the diode.

Figure 27–2 shows a half-wave rectifier during the positive alternation of the sine wave. The diode is forward biased, allowing a current to flow through the load. This allows the positive alternation of the cycle to develop across the load.

Figure 27–3 shows the circuit during the negative alternation of the sine wave. The diode is now reverse biased and does not conduct. Because no current flows through the load, no voltage is dropped across the load.

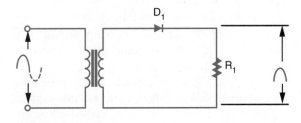

Figure 27–2 Half-wave rectifier during positive alternation.

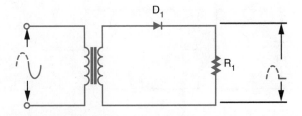

Figure 27–3 Half-wave rectifier during negative alternation.

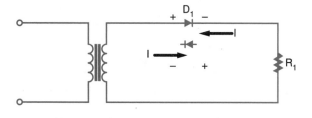

Figure 27–4 The diode determines the direction of current flow.

The half-wave rectifier operates during only one-half of the input cycle. The output is a series of positive or negative pulses, depending on how the diode is connected in the circuit. The frequency of the pulses is the same as the input frequency. The frequency of the pulses is called the *ripple frequency.*

The polarity of the output depends on which way the diode is connected in the circuit (Figure 27–4). The current flows through the diode from cathode to anode. When current flows through a diode, a deficiency of electrons exists at the anode end, making it the positive end of the diode. The polarity of the power supply can be reversed by reversing the diode.

There is a serious disadvantage with the half-wave rectifier because current flows during only half of each cycle. To overcome this disadvantage, a full-wave rectifier can be used.

Figure 27–5 shows a basic full-wave rectifier circuit. It requires two diodes and a center-tapped transformer. The center tap is grounded. The voltages at each end of the transformer are 180 degrees out of phase with each other.

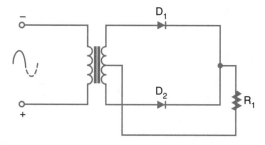

Figure 27–6 Full-wave rectifier during positive alternation.

Figure 27–6 shows a full-wave rectifier during the positive alternation of the input signal. The anode of diode D_1 is positive, and the anode of diode D_2 is negative. Diode D_1 is forward biased and conducts current. Diode D_2 is reverse biased and does not conduct. The current flows from the center tap of the transformer through the load and diode D_1, to the top of the secondary of the transformer. This permits the positive half of the cycle to be felt across the load.

Figure 27–7 shows the full-wave rectifier during the negative half of the cycle. The anode of diode D_2 becomes positive, and the anode of diode D_1 becomes negative. Diode D_2 is now forward biased and conducts. Diode D_1 is reverse biased and does not conduct. The current flows from the center tap of the transformer through the load and diode D_2 to the bottom of the secondary of the transformer.

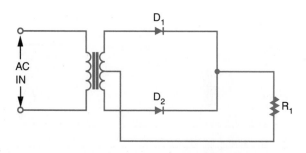

Figure 27–5 Basic full-wave rectifier circuit.

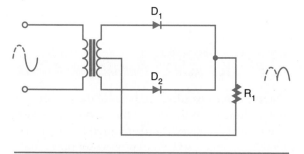

Figure 27–7 Full-wave rectifier during negative alternation.

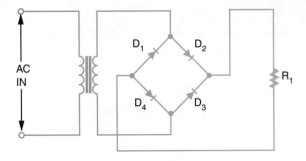

Figure 27–8 Bridge rectifier circuit.

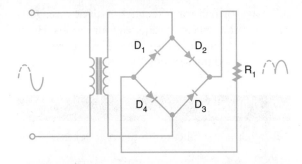

Figure 27–10 Bridge rectifier during negative alternation.

With a full-wave rectifier, the current flows during both half cycles. This means that the ripple frequency is twice the input frequency.

There is a disadvantage with the full-wave rectifier because the output voltage is half that of a half-wave rectifier for the same transformer. This disadvantage can be overcome by the use of a bridge rectifier circuit.

Figure 27–8 shows a bridge rectifier circuit. The four diodes are arranged so that the current flows in only one direction through the load.

Figure 27–9 shows the current flow during the positive alternation of the input signal. Current flows from the bottom of the secondary side of the transformer, up through diode D_4, through the load, through diode D_2 to the top of the secondary of the transformer. The entire voltage is dropped across the load.

Figure 27–10 shows the current flow during the negative alternation of the input signal. The top

of the secondary is negative, and the bottom is positive. The current flows from the top of the secondary, down through diode D_1, through the load and diode D_3 to the bottom of the secondary. Note that the current flows in the same direction through the load as during the positive alternation. Again the entire voltage is dropped across the load.

A bridge rectifier is a type of full-wave rectifier, because it operates on both half cycles of the input sine wave. The advantage of the bridge rectifier is that the circuit does not require a center-tapped secondary. This circuit does not require a transformer to operate. A transformer is used only to step up or step down the voltage or provide isolation.

To summarize the differences in rectifiers: The advantages of the half-wave rectifier are its simplicity and low cost. It requires one diode and a transformer. It is not very efficient, because only half of the input signal is used. It is also restricted to low-current applications.

The full-wave rectifier is more efficient than the half-wave rectifier. It operates on both alternations of the sine wave. The higher ripple frequency of the full-wave rectifier is easier to filter. A disadvantage is that it requires a center-tapped transformer. Its output voltage is lower than that of a half-wave rectifier for the same transformer because of the center tap.

The bridge rectifier can operate without a transformer. However, a transformer is needed to step up or step down the voltage. The output of the bridge rectifier is higher than that of either the

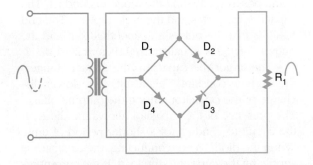

Figure 27–9 Bridge rectifier during positive alternation.

full-wave or half-wave rectifier. A disadvantage is that the bridge rectifier requires four diodes. However, the diodes are inexpensive compared to a center-tapped transformer.

27–2 Questions

1. What is the function of the rectifier in a power supply?
2. What are three configurations for connecting rectifiers to power supplies?
3. What are the differences in operation of the three configurations?
4. What are the advantages of one rectifier configuration over another?
5. Which rectifier configuration represents the best selection? Why?

27–3 Filter Circuits

The output of the rectifier circuit is a pulsating DC voltage. This is not suitable for most electronic circuits. Therefore, a filter follows the rectifier in most power supplies. The **filter** converts the pulsating DC voltage to a smooth DC voltage.

The simplest filter is a capacitor connected across the output of the rectifier (Figure 27–11). Figure 27–12 compares the outputs of a rectifier without and with the addition of a filter capacitor.

A capacitor affects the circuit in the following manner. When the anode of the diode is positive, current flows in the circuit. At the same time, the

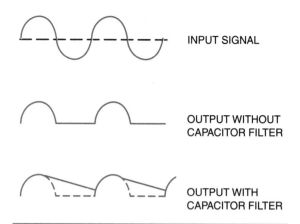

INPUT SIGNAL

OUTPUT WITHOUT CAPACITOR FILTER

OUTPUT WITH CAPACITOR FILTER

Figure 27–12 Output of a half-wave rectifier without and with a filter capacitor.

filter capacitor charges to the polarity indicated in Figure 27–11. After 90 degrees of the input signal, the capacitor is fully charged to the peak potential of the circuit.

When the input signal starts to drop in the negative direction, the capacitor discharges through the load. The resistance of the load controls the rate the capacitor discharges by the RC time constant. The discharge time constant is long compared to the cycle time. Therefore the cycle is completed before the capacitor can discharge. Thus, after the first quarter cycle, the current through the load is supplied by the discharging capacitor. As the capacitor discharges, the voltage stored in the capacitor decreases. However, before the capacitor can completely discharge, the next cycle of the sine wave occurs. This causes the anode of the diode to become positive, allowing the diode to conduct. The capacitor recharges and the cycle repeats. The end result is that the pulses are smoothed out and the output voltage is actually raised (Figure 27–13).

The larger the capacitor, the greater (longer) the RC time constant. This results in a slower discharge of the capacitor, raising the output voltage. The presence of the capacitor allows the diode in the circuit to conduct for a shorter period of time. When the diode is not conducting, the capacitor is supplying the current to the load. If the current required by the load is large, a very large capacitor must be used.

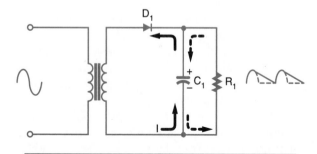

Figure 27–11 Half-wave rectifier with capacitor filter.

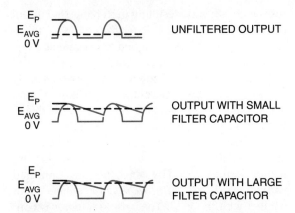

Figure 27–13 Effects of different filter capacitors on output of half-wave rectifier.

A capacitor filter across a full-wave or bridge rectifier behaves much like the capacitor filter in the half-wave rectifier just described. Figure 27–14 shows the output of a full-wave or bridge rectifier. The ripple frequency is twice that of the half-wave rectifier. When the capacitor filter is added to the output of the rectifier, the capacitor does not discharge very far before the next pulse occurs. The output voltage is high. If a large capacitor is used, the output equals the peak voltage of the input signal. Thus, the capacitor does a better job of filtering in a full-wave circuit than in a half-wave circuit.

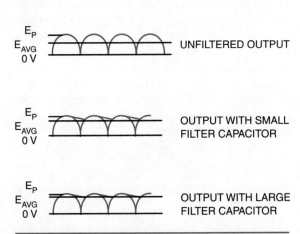

Figure 27–14 Effects of different filter capacitors on output of full-wave or bridge rectifier.

The purpose of the filter capacitor is to smooth out the pulsating DC voltage from the rectifier. The filter's performance is determined by the ripple remaining on the DC voltage. The ripple can be made lower by using a large capacitor or by increasing the load resistance. Typically, the load resistance is determined by the circuit design. Therefore, the size of the filter capacitor is determined by the amount of the ripple.

It should be realized that the filter capacitor places additional stress on the diodes used in the rectifier circuit. A half-wave and a full-wave rectifier followed by a filter capacitor are shown in Figure 27–15. The capacitor charges to the peak value of the secondary voltage and holds this value throughout the input cycle. When the diode becomes reverse biased, it shuts off and a maximum negative voltage is felt on the anode of the diode. The filter capacitor holds a maximum positive voltage on the cathode of the diode. The difference of potential across the diode is twice the peak value of the secondary. The diode must be selected to withstand this voltage.

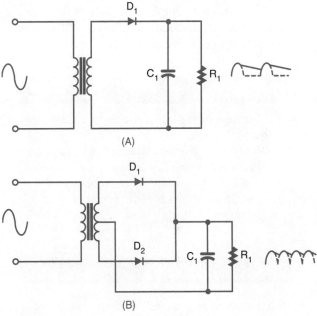

Figure 27–15 Half-wave rectifier (A) and full-wave rectifier (B) followed by a filter capacitor.

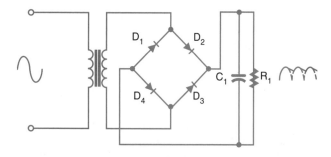

Figure 27–16 Bridge rectifier with capacitor filter.

The maximum voltage a diode can withstand when reverse biased is called the *peak-inverse-voltage (PIV)*. A diode should be selected that has a PIV higher than twice the peak value. Ideally, the diode should be operated at 80% of its rated value to allow for changes in the input voltage. This holds true for both the half-wave and the full-wave rectifier. This does not hold true for the bridge rectifier.

The diodes in a bridge rectifier are never exposed to more than the peak value of the secondary. In Figure 27–16, none of the diodes is exposed to more than the peak value of the input signal. The use of diodes with lower PIV ratings represents another advantage of the bridge rectifier.

27–3 Questions

1. What is the purpose of a filter in a power supply?
2. What is the simplest filter configuration?
3. What is the ripple frequency?
4. How is a filter capacitor selected?
5. What adverse effects result from the addition of a filter?

27–4 Voltage Regulators

Two factors can cause the output voltage of a power supply to vary. First, the input voltage to the power supply can vary, resulting in increases and decreases in the output voltage. Second, the load resistance can vary, resulting in a change in current demand.

Many circuits are designed to operate at a certain voltage. If the voltage varies, the operation of the circuit is affected. Therefore, the power supply must produce the same output regardless of load and input voltage changes. To accomplish this, a **voltage regulator** is added after the filter.

There are two basic types of voltage regulators: **shunt regulators** and **series regulators.** They are named for the method by which they are connected with the load. The *shunt regulator* is connected in parallel with the load. The *series regulator* is connected in series with the load. Series regulators are more popular than shunt regulators because they are more efficient and dissipate less power. The shunt regulator also acts as a control device, protecting the regulator from a short in the load.

Figure 27–17 shows a basic zener diode regulator circuit. This is a shunt regulator. The zener diode is connected in series with a resistor. The input voltage, unregulated DC voltage, is applied across both the zener diode and the resistor so as to make the zener diode reverse biased. The resistor allows a small current to flow to keep the zener diode in its zener breakdown region. The input voltage must be higher than the zener breakdown voltage of the diode. The voltage across the zener diode is equal to the zener diode's voltage rating. The voltage dropped across the resistor is equal to the difference between the zener diode's voltage and the input voltage.

The circuit shown in Figure 27–17 provides a constant output voltage for a changing input voltage. Any change in the voltage appears across the resistor. The sum of the voltage drops must equal the input voltage. By changing the output zener diode and the series resistor, the output voltage can be increased or decreased.

Figure 27–17 Basic zener diode regulator circuit.

The current through the load is determined by the load resistance and the output voltage. The load current plus the zener current flow through the series resistor. The series resistor must be chosen carefully so that the zener current keeps the zener diode in the breakdown zone and allows the current to flow.

When the load current increases, the zener current decreases and the load current and zener current together maintain a constant voltage. This allows the circuit to regulate for changes in output current as well as input voltage.

Figure 27–18 shows a shunt regulator circuit using a transistor. Note that transistor Q_1 is in parallel with the load. This protects the regulator in case a short develops with the load. More complex shunt regulators are available that use more than one transistor.

The series regulator is more popular than the shunt regulator. The simplest series regulator is a variable resistor in series with the load (Figure 27–19). The resistance is adjusted continuously to maintain a constant voltage across the load. When the DC voltage increases, the resistance increases, dropping more voltage. This maintains the voltage drop across the load by dropping the extra voltage across the series resistor.

The variable resistor can also compensate for changes in load current. If the load current increases, more voltage drops across the variable resistor. This results in less voltage dropped across the load. If the resistance can be made to decrease at the same instant that the current increases, the voltage dropped across the variable resistor can

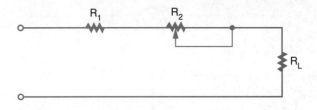

Figure 27–19 Series regulator using a variable resistor.

remain constant. This results in a constant output voltage despite changes in the load current.

In reality, it is too difficult to vary the resistance manually to compensate for voltage and current changes. It is more efficient to replace the variable resistor with a transistor (Figure 27–20). The transistor is connected so that the load's current flows through it. By changing the current on the base of the transistor, the transistor can be biased to conduct more or less current. Additional components are required to make the circuit self-adjusting (Figure 27–21). These components allow the transistor to compensate automatically for changes in the input voltage or load current.

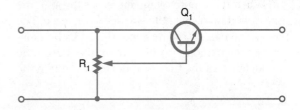

Figure 27–20 Transistor series regulator using a manually adjusted variable resistor.

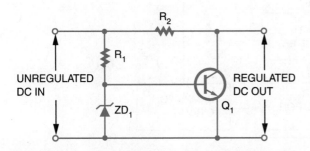

Figure 27–18 Shunt regulator using a transistor.

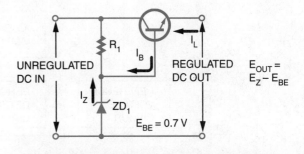

Figure 27–21 Self-adjusting series regulator.

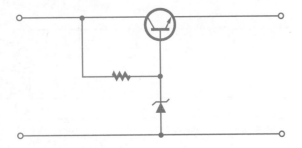

Figure 27–22 A series regulator.

Figure 27–22 shows a simple series regulator. The input is unregulated DC voltage, and the output is in a lower, regulated DC voltage. The transistor is connected as an emitter follower, meaning that there is a lack of phase inversion between the base and the emitter. The emitter voltage follows the base voltage. The load is connected between the emitter of the transistor and ground. The voltage at the base of the transistor is set by the zener diode. Therefore, the output voltage is equal to the zener voltage minus the 0.7 volt dropped across the base-emitter junction of the transistor.

When the input voltage increases through the transistor, the voltage out also tries to increase. The base voltage is set by the zener diode. If the emitter becomes more positive than the base, the conductance of the transistor decreases. When the transistor conducts less, it reacts the same way a large resistor would react if placed between the input and output. Most of the increase in input voltage is dropped across the transistor and there is only a small increase in the output voltage.

There is a disadvantage with the emitter-follower regulator because the zener diode must have a large power rating. Zener diodes with large power-handling capabilities are expensive.

A more popular type of series regulator is the feedback regulator. A feedback regulator consists of a feedback circuit to monitor the output voltage. If the output voltage changes, a control signal is developed. This signal controls the transistor's conductance. Figure 27–23 shows a block diagram of the feedback regulator. An unregulated DC voltage is applied to the input of the regulator. A lower, regulated DC output voltage appears at the output terminals of the regulator.

A sampling circuit appears across the output terminals. A *sampling circuit* is a voltage divider that sends a sample of the output voltage to the error-detection circuit. The voltage changes if the output voltage changes.

The error-detection circuit compares the sampled voltage with a reference voltage. In order to produce the reference voltage, a zener diode is used. The difference between the sample and reference voltages is the *error voltage*. The error voltage is amplified by an error amplifier. The error

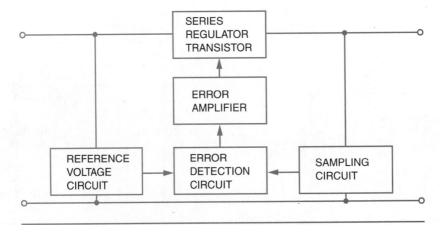

Figure 27–23 Block diagram of a feedback series regulator.

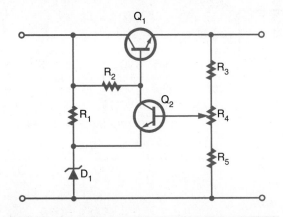

Figure 27–24 Basic feedback series regulator.

amplifier controls the conduction of the series transistor. The transistor conducts more or less to compensate for changes in the output voltage.

Figure 27–24 shows a feedback voltage regulator circuit. Resistors R_3, R_4, and R_5 form the sampling circuit. Transistor Q_2 acts as both the error-detection and error-amplification circuit. Zener diode D_1 and resistor R_1 produce the reference voltage. Transistor Q_1 is the series regulator transistor. Resistor R_2 is the collector load resistor for transistor Q_2 and the biasing resistor for transistor Q_1.

If the output voltage tries to increase, the sample voltage also increases. This increases the bias voltage on the base of transistor Q_2. The emitter voltage of transistor Q_2 is held constant by zener diode D_1. This results in transistor Q_2 conducting more and increasing the current through resistor R_2. In turn, the voltage on the collector of transistor Q_2 and the base of transistor Q_1 decreases. This decreases the forward bias of transistor Q_1 causing it to conduct less. When transistor Q_1 conducts less, less current flows. This causes a smaller voltage drop across the load and cancels the increase in voltage.

The output voltage can be adjusted accurately by varying potentiometer R_4. To increase the output voltage of the regulator, the wiper of potentiometer R_4 is moved in the more negative direction. This decreases the sample voltage on the base of transistor Q_2, decreasing the forward bias. This results in transistor Q_2 conducting less, caus-

ing the collector voltage of transistor Q_2 and base of transistor Q_1 to increase. This increases the forward bias on transistor Q_1, causing it to conduct more. More current flows through the load, which increases the voltage out.

There is a serious disadvantage with the series regulator because the transistor is in series with the load. A short in the load would result in a large current flowing through the transistor, which could destroy the transistor. A circuit is needed to keep the current passing through the transistor at a safe level.

Figure 27–25 shows a circuit that limits current through the series regulator transistor. It is shown added to the feedback series voltage regulator. Transistor Q_3 and resistor R_6 form the current-limiting circuit. In order for transistor Q_3 to conduct, the base-to-emitter junction must be forward biased by a minimum of 0.7 volt. When 0.7 volt is applied between the base and emitter, the transistor conducts. If R_6 is 1 ohm, the current necessary to develop 0.7 volt to the base of transistor Q_3 is:

$$I = \frac{E}{R}$$

$$I = \frac{0.7}{1}$$

$$I = 0.7 \text{ amp or } 700 \text{ mA}$$

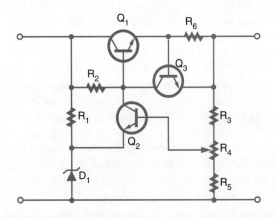

Figure 27–25 Feedback series regulator with current-limiting circuit.

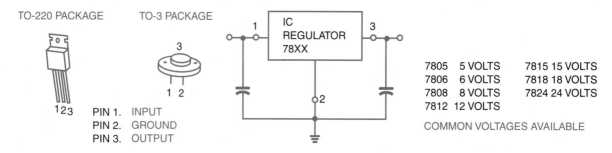

Figure 27–26 Three-terminal IC regulator.

When less than 700 milliamperes of current flows, transistor Q_3's base-to-emitter voltage is less than 0.7 volt, keeping it shut off. When transistor Q_3 is shut off, the circuit acts as if it does not exist. When the current exceeds 700 milliamperes, the voltage dropped across resistor R_6 increases above 0.7 volt. This results in transistor Q_3 conducting through resistor R_2. This decreases the voltage on the base of transistor Q_1 causing it to conduct less. The current cannot increase above 700 milliamperes. The amount of current that can be limited can be varied by changing the value of resistor R_6. Increasing the value of resistor R_6 lowers the limit on current value.

The feedback series regulator has another disadvantage, the number of components required. This problem can be overcome by the use of an integrated circuit (IC) regulator.

Modern IC regulators are low in cost and easy to use. Most IC regulators have only three terminals (input, output, and ground) and can be connected directly to the filtered output of a rectifier (Figure 27–26). The IC regulator provides a wide variety of output voltages of both positive and negative polarities. If a voltage is needed that is not a standard voltage, adjustable IC regulators are available.

In selecting an IC regulator, the voltage and load current requirements must be known along with the electrical characteristics of the unregulated power supply. IC regulators are classified by their output voltage. Fixed voltage regulators have three terminals and provide only one output voltage. They are available in both positive and negative voltages. Dual-polarity voltage regulators can supply both a positive and a negative voltage. Both fixed- and dual-polarity voltage regulators are available as adjustable voltage regulators. In using any of the IC voltage regulators, refer to the manufacturer's specification sheet.

27–4 Questions

1. What is the purpose of a voltage regulator in a power supply?
2. What are the two basic types of voltage regulator?
3. Which type of voltage regulator is used more?
4. Draw a schematic of a simple zener diode voltage regulator and explain how it operates.
5. Draw a block diagram of a series feedback regulator and explain how it operates.

27–5 Voltage Multipliers

In all previous cases, the DC voltage is limited to the peak value of the input sine wave. When higher DC voltages are required, a step-up transformer is used. However, higher DC voltages can be produced without a step-up transformer. Circuits that are capable of producing higher DC voltages without the benefit of a transformer are called **voltage multipliers.** Two voltage multipliers are the **voltage doubler** and the **voltage tripler.**

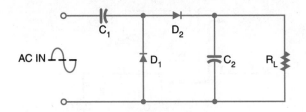

Figure 27–27 Half-wave voltage doubler.

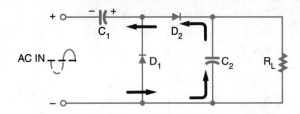

Figure 27–29 Half-wave voltage doubler during positive alternation of input signal.

Figure 27–27 shows a half-wave voltage doubler. It produces a DC output voltage that is twice the peak value of the input signal. Figure 27–28 shows the circuit during the negative alternation of the input signal. Diode D_1 conducts, and the current flows along the path shown. Capacitor C_1 charges to the peak value of the input signal. Because there is no discharge path, capacitor C_1 remains charged. Figure 27–29 shows the positive alternation of the input signal. At this point, capacitor C_1 is charged to the negative peak value. This keeps diode D_1 reverse biased and forward biases diode D_2. This allows diode D_2 to conduct, charging capacitor C_2. Because capacitor C_1 is charged to the maximum negative value, capacitor C_2 charges to twice the peak value of the input signal.

As the sine wave changes from the positive half cycle to the negative half cycle, diode D_2 is cut off. This is because capacitor C_2 holds diode D_2 reverse biased. Capacitor C_2 discharges through the load, holding the voltage across the load constant. Therefore, it also acts as a filter capacitor.

Capacitor C_2 recharges only during the positive cycle of the input signal, resulting in a ripple

frequency of 60 hertz (and the name half-wave voltage doubler). The half-wave voltage doubler is hard to filter because of the 60-hertz ripple frequency. Another disadvantage is that capacitor C_2 must have a voltage rating of at least twice the peak value of the AC input signal.

A full-wave voltage doubler overcomes some of the disadvantages of the half-wave voltage doubler. Figure 27–30 is a schematic of a circuit that operates as a full-wave voltage doubler. Figure 27–31 shows that, on the positive alternation of the input signal, capacitor C_1 charges through diode D_1 to the peak value of the AC input signal. Figure 27–32

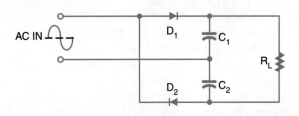

Figure 27–30 Full-wave voltage doubler.

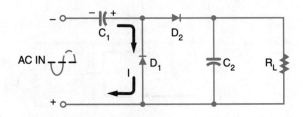

Figure 27–28 Half-wave voltage doubler during negative alternation of input signal.

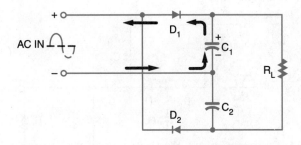

Figure 27–31 Full-wave voltage doubler during positive alternation of input signal.

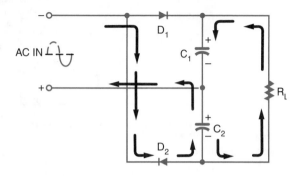

Figure 27–32 Full-wave voltage doubler during negative alternation of input signal.

shows that, on the negative alternation, capacitor C_2 charges through diode D_2 to the peak value of the input signal.

When the AC input signal is changing between the peaks of the alternations, capacitors C_1 and C_2 discharge in series through the load. Because each capacitor is charged to the peak value of the input signal, the total voltage across the load is two times the peak value of the input signal.

Capacitors C_1 and C_2 are charged during the peaks of the input signal. The ripple frequency is 120 hertz because both capacitors C_1 and C_2 are charged during each cycle. Capacitors C_1 and C_2 split the output voltage to the load, so each capacitor is subject to the peak value of the input signal.

Figure 27–33 shows the circuit of a voltage tripler. In Figure 27–34, the positive alternation of the input signal biases diode D_1 so that it conducts. This charges capacitor C_1 to the peak value of the input signal. Capacitor C_1 places a positive potential across diode D_2.

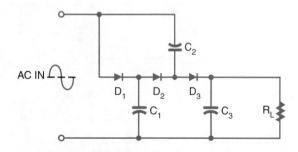

Figure 27–33 Voltage tripler.

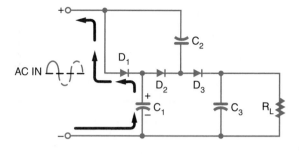

Figure 27–34 Voltage tripler during first positive alternation of input signal.

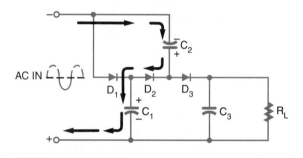

Figure 27–35 Voltage tripler during negative alternation of input signal.

Figure 27–35 shows the negative alternation of the input signal. Because diode D_2 is now forward biased, current flows through it to capacitor C_1 via capacitor C_2. This charges capacitor C_2 to twice the peak value because of the voltage stored in capacitor C_1.

Figure 27–36 shows the occurrence of the next positive alternation. It places a difference of

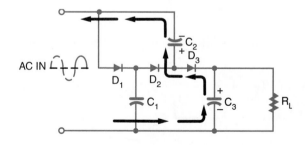

Figure 27–36 Voltage tripler during second positive alternation of input signal.

potential across capacitor C_2 that is three times the peak value. The top plate of capacitor C_2 has a positive peak value of two times the peak value. The anode of diode D_3 has a positive value of three times the peak value with respect to ground. This charges capacitor C_3 to three times the peak value. This is the voltage that is applied to the load.

27–5 Questions

1. What is the function of a voltage multiplier circuit?
2. Draw a schematic of a half-wave voltage doubler and explain how it operates.
3. Draw a schematic of a full-wave voltage doubler.
4. Draw a schematic of a voltage tripler.
5. What requirement must be placed on capacitors used in voltage doubler and tripler circuits?

27–6 Circuit-Protection Devices

To protect the load from failure of the power supply, an **over-voltage protection circuit** is used.

Figure 27–37 shows an over-voltage protection circuit called a *crowbar.* The SCR, placed in parallel with the load, is normally cut off (not conducting). If the output voltage rises above a predetermined level, the SCR turns on and places a short circuit across the load. With the short circuit across the load, very little current flows through the load. This fully protects the load. The short circuit across

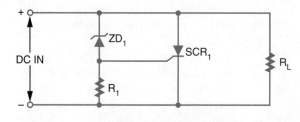

Figure 27–37 Crowbar overprotection circuit.

the load does not protect the power supply. The output of the power supply is shorted, thereby blowing the fuse of the power supply.

The zener diode establishes the voltage level at which the SCR turns on. It protects the load from voltage above the zener voltage. As long as the supply voltage is less than the zener diode voltage rating, the diode does not conduct. This keeps the SCR turned off.

If the supply voltage rises above the zener voltage due to a malfunction, the zener diode conducts. This creates a gate current to the SCR, turning it on and shorting out the load. It should be noted that the SCR must be large enough to handle the large short-circuit current.

Another protection device is a **fuse** (Figure 27–38). A *fuse* is a device that fails when an overload occurs. A fuse is essentially a small piece of wire between two metal terminals. A hollow glass cylinder holds the metal terminals apart and protects the wire. Typically, a fuse is placed in series with the primary of the power supply transformer. If a large current flows in the power supply, it causes the fuse wire to overheat and melt. This opens the circuit so that no more current can flow. The glass housing of the fuse allows a visual check to see if the fuse is blown.

Figure 27–38 Fuses used for protection of electronic circuits. *(From Shimizu,* Electronic Fabrication, *Delmar Publishers.)*

Fuses are classified as normal or slow blow. A *normal* fuse opens as soon as its current is exceeded. In some circuits this is an advantage, because it removes the overload very quickly. A *slow-blow* fuse can withstand a brief period of overloading before it blows. This brief period of overloading occurs because the fuse wire heats more slowly. If the overload is present for longer than a few seconds, it blows the fuse. A slow-blow fuse may contain a spring for pulling the fuse wire apart once it melts. Some circuits can withstand a surge in current. In these, a slow-blow fuse is preferable to a normal fuse.

The fuse is always installed after the switch on the hot (alive and energized) lead of the AC power source. This disconnects the transformer from the AC power source when the fuse blows. By installing it after the switch, power can be removed from the fuse holder for added safety in replacing a blown fuse. [Warning: A blown fuse should not be replaced until the fault is diagnosed and corrected.]

A disadvantage of the fuse is that it must be replaced each time it blows. A **circuit breaker** performs the same job but does not have to be replaced each time an overload occurs. Instead, the circuit breaker can be manually reset after the overload occurs (Figure 27–39). Circuit breakers are connected into the circuit the same way as fuses.

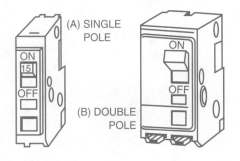

Figure 27–39 Circuit breakers used for protection of electronic circuits.

27—6 Questions

1. How does a crowbar over-voltage protection circuit operate?
2. How does a fuse operate when used in a circuit?
3. What are the different types of fuses?
4. Where is the fuse of any circuit-protection device located in a circuit?
5. What is the advantage of a circuit breaker over a fuse?

SUMMARY

■ The primary purpose of a power supply is to convert AC to DC.
■ Transformers are used in power supplies for isolation and to step up or step down the voltage.
■ A rectifier circuit converts incoming AC voltage to pulsating DC voltage.
■ The basic rectifier circuits are half-wave, full-wave, and bridge.
■ Half-wave rectifiers are simpler and less expensive than either full-wave or bridge rectifiers.
■ The full-wave rectifier is more efficient than the half-wave rectifier.
■ The bridge rectifier can operate without a transformer.
■ To convert pulsating DC voltage to a smooth DC voltage, a filter must follow the rectifier in the circuit.
■ A capacitor in parallel with the load is an effective filter.
■ A voltage regulator provides constant output regardless of load and input voltage changes.
■ The voltage regulator is located after the filter in the circuit.
■ The two basic types of regulator are the shunt regulator and the series regulator.
■ The series regulator is more efficient and therefore more popular than the shunt regulator.

■ Voltage multipliers are circuits capable of providing higher DC voltages than the input without the aid of a transformer.
■ Voltage doublers and voltage triplers are voltage multipliers.
■ A crowbar is a circuit designed for over-voltage protection.

■ A fuse protects a circuit from a current overload.
■ Fuses are classified as either normal or slow-blow.
■ Circuit breakers perform the same job as fuses but do not have to be replaced each time there is an overload.

Chapter 27 Self-Test

1. What are the four areas of concern when selecting a transformer for a power supply?
2. What is the purpose of the transformer in a power supply?
3. What purpose does a rectifier serve with a power supply?
4. What are the advantages and disadvantages between the full-wave rectifier circuit and the bridge rectifier circuit?
5. Describe the process of how a filter capacitor converts a pulsating DC voltage to a smooth DC voltage.
6. On what basis is the size of the filter capacitor selected?
7. How does a series regulator maintain the output voltage at a constant level?
8. What circuit characteristics must be known when selecting a regulator circuit?
9. What practical use do voltage multipliers serve?
10. What advantages does a full-wave voltage doubler have over a half-wave voltage doubler?
11. What type of device(s) is used for over-voltage protection?
12. What type of device(s) is used for over-current protection?

28

AMPLIFIER BASICS

Objectives

After completing this chapter, the student will be able to:

- Describe the purpose of an amplifier.
- Identify the three basic configurations of transistor amplifier circuits.
- Identify the classes of amplifiers.
- Describe the operation of direct coupled amplifiers, audio amplifiers, video amplifiers, RF amplifiers, IF amplifiers, and operational amplifiers.
- Draw and label schematic diagrams for the different types of amplifier circuits.

Amplifiers are electronic circuits that are used to increase the amplitude of an electronic signal. A circuit designed to convert a low voltage to a higher voltage is called a voltage amplifier. A circuit designed to convert a low current to a higher current is called a current amplifier.

28–1 Amplifier Configurations

In order for a transistor to provide **amplification,** it must be able to accept an input signal and produce an output signal that is greater than the input.

The input signal controls current flow in the transistor. This, in turn, controls the voltage through the load. The transistor circuit is designed to take voltage from an external power source (V_{CC}) and apply it to a load resistor (R_L) in the form of an output voltage. The output voltage is controlled by a small input voltage.

The transistor is used primarily as an amplifying device. However, there is more than one way of achieving this amplification. The transistor can be connected in three different circuit configurations.

The three circuit configurations are the **common-base circuit,** the **common-emitter circuit,** and the **common-collector circuit.** In each configuration, one of the transistor's leads serves as a common reference point and the other two leads serve as input and output connections. Each configuration can be constructed using either NPN or PNP transistors. In each, the transistor's emitter-base junction is forward biased, while the collector-base junction is reverse biased. Each configuration offers advantages and disadvantages.

In the common-base circuit (Figure 28–1) the input signal enters the emitter-base circuit, and the output leaves from the collector-base circuit. The base is the element common to both the input and output circuits.

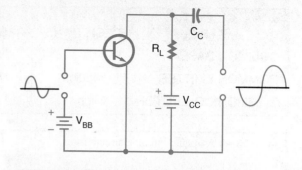

Figure 28–2 Common-emitter amplifier circuit.

In the common-emitter circuit (Figure 28–2) the input signal enters the base-emitter circuit, and the output signal leaves from the collector-emitter circuit. The emitter is common to both the input and output circuit. This method of connecting a transistor is the most widely used.

The third type of connection (Figure 28–3) is the common-collector circuit. In this configuration, the input signal enters the base-collector circuit, and the output signal leaves from the emitter-collector circuit. Here, the collector is common to both the input and output circuits. This circuit is used as an impedance-matching circuit.

Figure 28–4 charts the input-output resistance and voltage, current, and power gains for the three circuit configurations. Figure 28–5 shows the phase relationship of input and output waveforms for the three configurations. Note that the common-emitter configuration provides a phase reversal of the input-output signal.

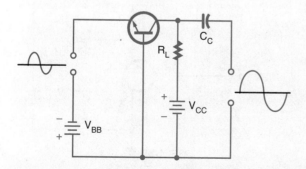

Figure 28–1 Common-base amplifier circuit.

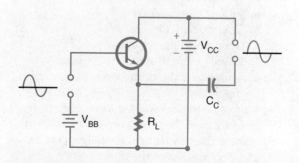

Figure 28–3 Common-collector amplifier circuit.

CIRCUIT TYPE	INPUT RESISTANCE	OUTPUT RESISTANCE	VOLTAGE GAIN	CURRENT GAIN	POWER GAIN
COMMON BASE	Low	High	High	Less than 1	Medium
COMMON EMITTER	Medium	Medium	Medium	Medium	High
COMMON COLLECTOR	High	Low	Less than 1	Medium	Medium

Figure 28–4 Amplifier circuit characteristics.

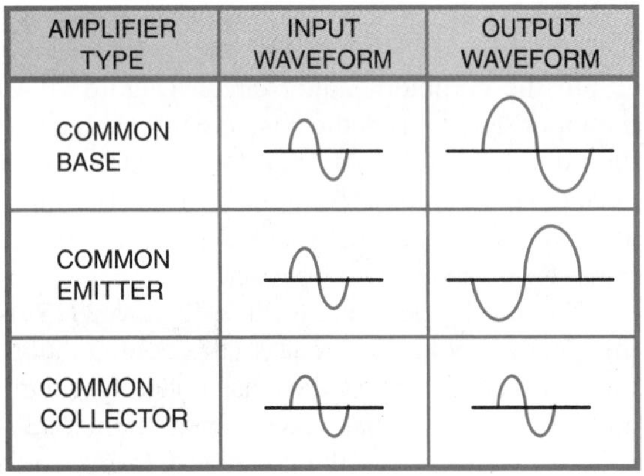

AMPLIFIER TYPE	INPUT WAVEFORM	OUTPUT WAVEFORM
COMMON BASE		
COMMON EMITTER		
COMMON COLLECTOR		

Figure 28–5 Amplifier circuit input-output phase relationships.

28–1 Questions

1. Draw schematic diagrams of the three basic configurations of transistor amplifier circuits.
2. List the characteristics of the:
   a. Common-base circuit
   b. Common-emitter circuit
   c. Common-collector circuit
3. Make a chart showing the input-output phase relationship of the three configurations.
4. Make a chart showing the input-output resistances of the three configurations.
5. Make a chart showing the voltage, current, and power gains of the three configurations.

28–2 Amplifier Biasing

The basic configurations of transistor amplifier circuits are the common base, the common emitter, and the common collector. All require two voltages for proper biasing. The base-emitter junction must be forward biased and the base-collector junction must be reverse biased. However, both bias voltages can be provided from a single source.

Because the common-emitter circuit configuration is the most often used, it is described in detail here. The same principles apply to the common-base and common-collector circuits.

Figure 28–6 shows a common-emitter transistor amplifier using a single voltage source. The circuit is schematically diagrammed in Figure 28–7. The voltage source is identified as $+V_{CC}$. The ground symbol is the negative side of the voltage source V_{CC}. The single voltage source provides proper

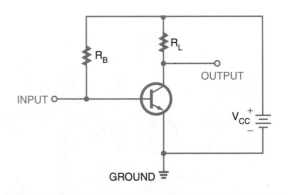

Figure 28–6 Common-emitter amplifier with single voltage source.

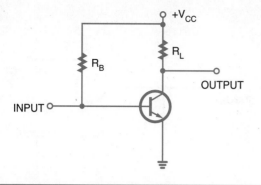

Figure 28–7 Schematic representation of common-emitter amplifier with single voltage source.

biasing for the base-emitter and base-collector junctions. The two resistors (R_B and R_L) are used to distribute the voltage for proper operation. Resistor R_L, the collector load resistor, is in series with the collector. When a collector current flows, a voltage develops across resistor R_L. The voltage dropped across resistor R_L and the voltage dropped across the transistor's collector-to-emitter junction must add up to the total voltage applied.

Resistor R_B, connected between the base and the voltage source, controls the amount of current flowing out of the base. The base current flowing through resistor R_B creates a voltage across it. Most of the voltage from the source is dropped across it. A small amount of voltage drops across the transistor's base-to-emitter junction, providing the proper forward bias.

The single voltage source can provide the necessary forward-bias and reverse-bias voltages. For an NPN transistor, the transistor's base and collector must be positive with respect to the emitter. Therefore, the voltage source can be connected to the base and the collector through resistors R_B and R_L. This circuit is often called a base-biased circuit, because the base current is controlled by resistor R_B and the voltage source.

The input signal is applied between the transistor's base and emitter or between the input terminal and ground. The input signal either aids or opposes the forward bias across the emitter junction. This causes the collector current to vary, which then causes the voltage across R_L to vary.

The output signal is developed between the output terminal and ground.

The circuit shown in Figure 28–6 is unstable, because it cannot compensate for changes in the bias current with no signal applied. Temperature changes cause the transistor's internal resistance to vary, which causes the bias currents to change. This shifts the transistor operating point, reducing the gain of the transistor. This process is referred to as *thermal instability*.

It is possible to compensate for temperature changes in a transistor amplifier circuit. If a portion of the unwanted output signal is fed back to the circuit input, the signal opposes the change. This is referred to as **degenerative** or **negative feedback** (Figure 28–8). In a circuit using degenerative feedback, the base resistor R_B is connected directly to the collector of the transistor. The current flowing through resistor R_B is determined by the voltage at the collector. If the temperature increases, the collector current increases, and the voltage across R_L increases. The collector-to-emitter voltage decreases, reducing the voltage applied to R_B. This reduces the base current, which causes the collector current to decrease. This is referred to as a *collector feedback circuit*.

Figure 28–9 shows another type of feedback. The circuit is similar to the one in Figure 28–7 except that a resistor (R_E) is connected in series with the emitter lead. Resistors R_B and R_E and the transistor's emitter-base junction are connected in series with the source voltage V_{CC}.

An increase in temperature causes the collector current to increase. The emitter current then

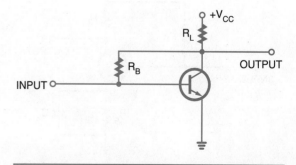

Figure 28–8 Common-emitter amplifier with collector feedback.

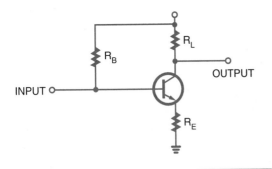

Figure 28–9 Common-emitter amplifier with emitter feedback.

also increases, causing the voltage drop across resistor R_E to increase and the voltage across resistor R_B to decrease. The base current then decreases, which reduces both the collector current and the emitter current. Because the feedback is generated at the transistor's emitter, the circuit is called an *emitter feedback circuit.*

There is a problem with this type of circuit because an AC input signal develops across resistor R_E as well as across the load resistor R_L and the transistor. This reduces the overall gain of the circuit. With the addition of a capacitor across the emitter resistor R_E (Figure 28–10), the AC signal bypasses resistor R_E. The capacitor is often referred to as a *bypass capacitor.*

The bypass capacitor prevents any sudden voltage changes from appearing across resistor R_E by offering a lower impedance to the AC signal. The bypass capacitor holds the voltage across

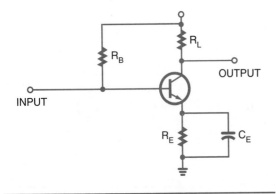

Figure 28–10 Emitter feedback with bypass capacitor.

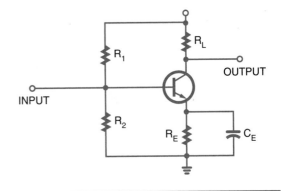

Figure 28–11 Common-emitter amplifier with voltage-divider feedback.

resistor R_E steady, while at the same time not interfering with the feedback action provided by resistor R_E.

A voltage-divider feedback circuit (Figure 28–11) offers even more stability. This circuit is the one most widely used. Resistor R_B is replaced with two resistors, R_1 and R_2. The two resistors are connected in series across the source voltage V_{CC}. The resistors divide the source voltage into two voltages, forming a voltage divider.

Resistor R_2 drops less voltage than resistor R_1. The voltage at the base, with respect to ground, is equal to the voltage developed across resistor R_2. The purpose of the voltage divider is to establish a constant voltage from the base of the transistor to ground. The current flow through resistor R_2 is toward the base. Therefore, the end of resistor R_2 attached to the base is positive with respect to ground.

Because the emitter current flows up through resistor R_E, the voltage dropped across resistor R_E is more positive at the end that is attached to the emitter. The voltage developed across the emitter-base junction is the difference between the two positive voltages developed across resistor R_2 and resistor R_E. For the proper forward bias to occur, the base must be slightly more positive than the emitter.

When the temperature increases, the collector and emitter currents also increase. The increase in the emitter current causes an increase in the voltage drop across the emitter resistor R_E. This results in the emitter becoming more positive with respect to ground. The forward bias across the emitter-base junction is then reduced, causing the base current to decrease. The reduction in the base current reduces the collector and emitter currents. The opposite action takes place if the temperature decreases: The base current increases, causing the collector and emitter currents to increase.

The amplifier circuits discussed so far have been biased so that all of the applied AC input signal appears at the output. Except for a higher voltage, the output signal is the same as the input signal. An amplifier that is biased so that the current flows throughout the entire cycle is operating as a **class A amplifier** (Figure 28–12).

An amplifier that is biased so that the output current flows for less than a full cycle but more than a half cycle is operating as a **class AB amplifier.** More than half but less than the full AC input signal is amplified in the class AB mode (Figure 28–13).

An amplifier that is biased so that the output current flows for only half of the input cycle is operating as a **class B amplifier.** Only one-half of the AC input signal is amplified in the class B mode (Figure 28–14).

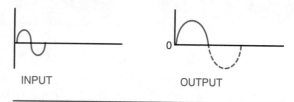

Figure 28–14 Class B amplifier output.

An amplifier that is biased so that the output current flows for less than half of the AC input cycle is operating as a **class C amplifier.** Less than one alternation is amplified in the class C mode (Figure 28–15).

Class A amplifiers are the most linear of the types mentioned. They produce the least amount of distortion. They also have lower output ratings and are the least efficient. They find wide application where the full signal must be maintained, as in the amplification of audio signals in radios and televisions. However, because of the high power-handling capabilities required for class A operation, transistors are usually operated in the class AB or class B mode.

Class AB, B, and C amplifiers produce a substantial amount of distortion. This is because they amplify only a portion of the input signal. To amplify the full AC input signal, two transistors are needed, connected in a push-pull configuration (Figure 28–16). Class B amplifiers are used as output stages of stereo systems and public address amplifiers, and in many industrial controls. Class C amplifiers are used for high-power amplifiers and transmitters where only a single frequency is amplified, such as the RF (radio frequency) carrier used for radio and television transmission.

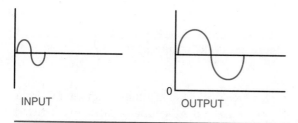

Figure 28–12 Class A amplifier output.

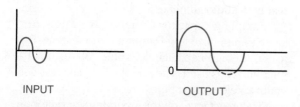

Figure 28–13 Class AB amplifier output.

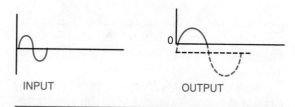

Figure 28–15 Class C amplifier output.

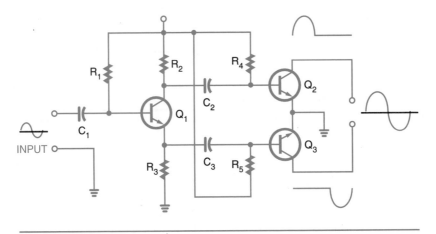

Figure 28–16 Push-pull amplifier configuration.

28–2 Questions

1. Draw a schematic diagram of a common-emitter transistor amplifier using a single voltage source.
2. How are temperature changes compensated for in a transistor amplifier?
3. Draw a schematic diagram of a voltage-divider feedback circuit.
4. List the classes of amplifiers and identify their outputs.
5. List the applications of each class of amplifier.

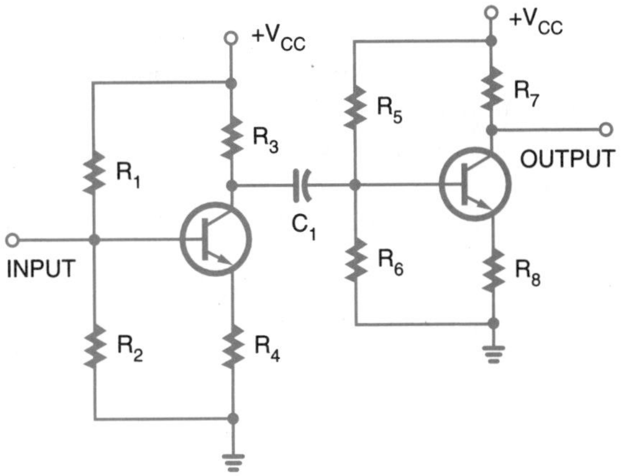

Figure 28–17 RC coupling.

28–3 Amplifier Coupling

To obtain greater amplification, transistor amplifiers may be connected together. However, to prevent one amplifier's bias voltage from affecting the operation of the second amplifier, a **coupling** technique must be used. The coupling method used must not disrupt the operation of either circuit. Coupling methods used include resistance-capacitance coupling, impedance coupling, transformer coupling, and direct coupling.

Resistive-capacitive coupling or *RC* coupling consists of two resistors and a capacitor connected as shown in Figure 28–17. Resistor R_3 is the collec-

tor load resistor of the first stage. Capacitor C_1 is the DC-blocking and AC-coupling capacitor. Resistors R_5 and R_6 are the input load resistor and DC return resistor for the base-emitter junction of the second stage. Resistance-capacitive coupling is used primarily in audio amplifiers.

Coupling capacitor C_1 must have a low reactance to minimize low-frequency attenuation. Typically, a high-capacitance value of 10 to 100 microfarads is used. The coupling capacitor is generally an electrolytic type.

The reactance of the coupling capacitor increases as the frequency decreases. The low-

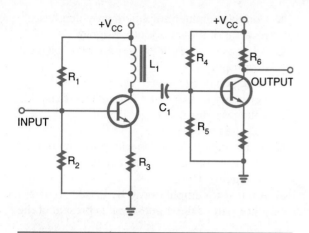

Figure 28–18 Impedance coupling.

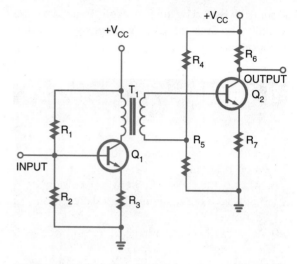

Figure 28–19 Transformer coupling.

frequency limit is determined by the size of the coupling capacitor. The high-frequency limit is determined by the type of transistor used.

The *impedance-coupling* method is similar to the RC-coupling method, but an inductor is used in place of the collector load resistor for the first stage of amplification (Figure 28–18).

Impedance coupling works just like RC coupling. The advantage is that the inductor has a very low DC resistance across its windings. The AC output signal is developed across the inductor just as across the load resistor. However, the inductor consumes less power than the resistor, increasing the overall efficiency of the circuit.

The disadvantage of impedance coupling is that the inductance reactance increases with the frequency. The voltage gain varies with the frequency. This type of coupling is ideal for single-frequency amplification when a very narrow band of frequencies must be amplified.

In a *transformer-coupled* circuit, the two amplifier stages are coupled through the transformer (Figure 28–19). The transformer can effectively match a high-impedance source to a low-impedance load. The disadvantage is that transformers are large and expensive. Also, like the inductor-coupled amplifier, the transformer-coupled amplifier is useful only for a narrow band of frequencies.

When very low frequencies or a DC signal must be amplified, the *direct-coupling* technique

must be used (Figure 28–20). Direct-coupled amplifiers provide a uniform current or voltage gain over a wide range of frequencies. This type of amplifier can amplify frequencies from zero (DC) hertz to many thousands of hertz. However, direct-coupled amplifiers find their best applications with low frequencies.

A drawback of direct-coupled amplifiers is that they are not stable. Any change in the output current of the first stage is amplified by the second stage. This occurs because the second stage is

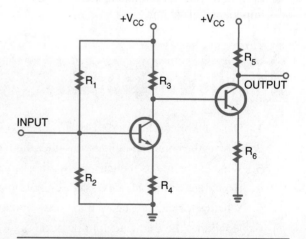

Figure 28–20 Direct coupling.

essentially biased by the first stage. To improve the stability requires the use of expensive precision components.

28–3 Questions

1. What are the four main methods of coupling transistor amplifiers?
2. Where is resistance-capacitive coupling used primarily?
3. What is the difference between resistance-capacitive coupling and impedance coupling?
4. What is the disadvantage of transformer coupling?
5. What coupling method is used for DC and low-frequency signals?

SUMMARY

- Amplifiers are electronic circuits used to increase the amplitude of an electronic signal.
- The transistor is used primarily as an amplifying device.
- The three transistor amplifier configurations are common base, common collector, and common emitter.
- Common-collector amplifiers are used for impedance matching.
- Common-emitter amplifiers provide phase reversal of the input-output signal.
- All transistor amplifiers require two voltages for proper biasing.
- A single voltage source can provide the necessary forward-bias and reverse-bias voltages using a voltage-divider arrangement.
- A voltage-divider feedback arrangement is the most commonly used biasing arrangement.
- A transistor amplifier can be biased so that all or part of the input signal is present at the output.
- Class A amplifiers are biased so that the output current flows throughout the cycle.
- Class AB amplifiers are biased so that the output current flows for less than the full but more than half of the input cycle.
- Class B amplifiers are biased so that the output current flows for only half of the input cycle.
- Class C amplifiers are biased so that the output current flows for less than half of the input cycle.
- Coupling methods used to connect one transistor to another include resistance-capacitance coupling, impedance coupling, transformer coupling, and direct coupling.
- Direct-coupled amplifiers are used for high gain at low frequencies or amplification of a DC signal.

Chapter 28 Self-Test

1. Describe briefly how a transistor is used to provide amplification.
2. Why is the common-emitter amplifier the most widely used transistor amplifier configuration?
3. What factor affects the gain of the transistor, and what must be done to compensate for it?
4. How does an amplifier's class of operation affect the biasing of the amplifier?
5. What factor must be taken into consideration when connecting one amplifier to another?
6. How does the frequency of operation of an amplifier affect the coupling method used in connecting amplifiers together?

Chapter

29

AMPLIFIER APPLICATIONS

Objectives

After completing this chapter, the student will be able to:

- Describe the operation of:
  - —direct coupled amplifiers
  - —audio amplifiers
  - —video amplifiers
  - —RF amplifiers
  - —IF amplifiers
  - —operational amplifiers
- Identify schematic diagrams for the different types of amplifier circuits.

An amplifier may be defined as an electronic circuit designed to increase the amplitude of an electronic signal. Amplifier circuits are one of the most basic circuits in electronics. Amplifier circuits make signal level greater, sound louder, and provide the circuit with gain. Gain is the ability of an electronic circuit to increase the amplitude of an electronic signal. This chapter looks at some unique types of amplifier circuits.

29–1 Direct-Coupled Amplifiers

Direct-coupled or **DC amplifiers** are used for high gain at low frequencies or for amplification of a direct-current signal. The DC amplifier is also used to eliminate frequency loss through a coupling network. Applications of the DC amplifier include computers, measuring and test instruments, and industrial control equipment.

A simple DC amplifier is shown in Figure 29–1. The common-emitter amplifier is the one most frequently used. The circuit is shown with voltage-divider bias and emitter feedback. This type of circuit does not use a coupling capacitor. The input is applied directly to the base of the transistor. The output is taken from the collector.

The DC amplifier can provide both voltage and current gain. However, it is used primarily as a voltage amplifier. The voltage gain is uniform for both AC and DC signals.

In most applications, one stage of amplification is not enough. Two or more stages are required to obtain a higher gain. Two or more stages connected together are referred to as *multiple-stage amplifier*. Figure 29–2 shows a two-stage amplifier. The input signal is amplified by the first stage. The amplified signal is then applied to the base of the transistor in the second stage. The overall gain of the circuit is the product of the voltage gains of the two stages. For example, if both the

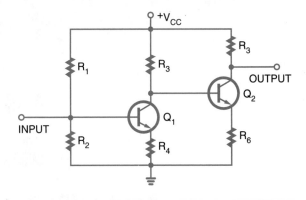

Figure 29–2 Two-stage DC amplifier.

first and second stages have a voltage gain of 10, the overall gain of the circuit is 100.

Figure 29–3 shows another type of two-stage DC amplifier. Both an NPN and a PNP transistor are used. The type of circuit is called a *complementary amplifier*. The circuit functions the same way as the circuit shown in Figure 29–2. The difference is that the second-stage transistor is a PNP transistor. The PNP transistor is reversed so that the emitter and collector are biased properly.

Figure 29–4 shows two transistors connected together to function as a single unit. This circuit arrangement is called a *darlington arrangement*. Transistor Q_1 is used to control the conduction of transistor Q_2. The input signal applied to the base of transistor Q_1 controls the base of transistor Q_2. The darlington arrangement may be a single

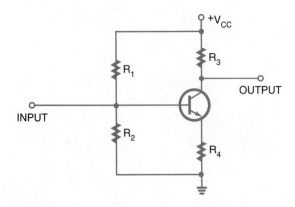

Figure 29–1 Simple DC amplifier.

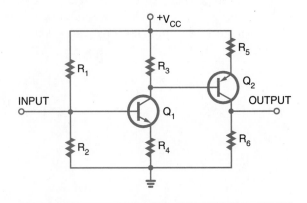

Figure 29–3 Complementary DC amplifier.

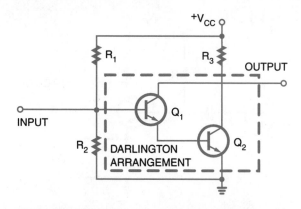

Figure 29–4 Darlington arrangement.

package with three leads: emitter (E), base (B), and collector (C). It is used as a simple DC amplifier but offers a very high voltage gain.

The main disadvantage of multiple-stage amplifiers is their high thermal instability. In circuits requiring three or four stages of DC amplification, the final output stage may not amplify the original DC or AC signal but one that has been greatly distorted. The same problem exists with the darlington arrangement.

In applications where both high gain and temperature stability are required, another type of amplifier is necessary. This type is called a **differential amplifier** (Figure 29–5). It is unique in that it has two separate inputs and can provide either

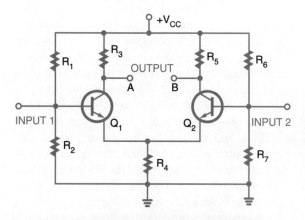

Figure 29–5 Differential amplifier.

one or two outputs. If a signal is applied to the input of transistor Q_1, an amplified signal is produced between output A and ground as in an ordinary amplifier. However, a small signal is also developed across resistor R_4 and is applied to the emitter of transistor Q_2. Transistor Q_2 functions as a common-base amplifier and amplifies the signal on its base. An amplified output signal is produced between output B and ground. The output produced at B is 180 degrees out of phase with output A. This makes the differential amplifier much more versatile than a conventional amplifier.

The differential amplifier is generally not used to obtain an output between either output and ground. The output signal is usually obtained between output A and output B. Because the two outputs are 180 degrees out of phase, a substantial output voltage is developed between the two points. The input signal can be applied to either input.

The differential amplifier has a high degree of temperature stability, because transistors Q_1 and Q_2 are mounted close together and therefore are equally affected by temperature changes. Also, the collector currents of transistors Q_1 and Q_2 tend to increase and decrease by the same amount, so that the output voltage remains constant.

The differential amplifier is used extensively in integrated circuits and in electronic equipment generally. It is used to amplify and/or compare the amplitudes of both DC and AC signals. It is possible to connect one or more differential amplifiers together to obtain a higher overall gain. In some cases, the differential amplifier is used as a first stage, with conventional amplifiers used in succeeding circuits. Because of their versatility and temperature stability, differential amplifiers are the most important type of direct-coupled amplifier.

29–1 Questions

1. When are direct-coupled amplifiers used in a circuit?
2. What kind of amplifier are direct-coupled amplifiers used for primarily?

3. Draw schematic diagrams of the following circuits:
   a. Complementary amplifier
   b. Darlington arrangement
   c. Differential amplifier
4. How does a differential amplifier differ from a conventional amplifier?
5. Where are differential amplifiers primarily used?

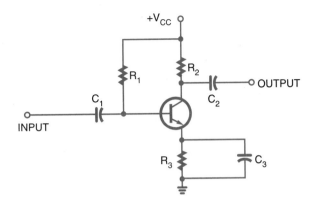

Figure 29–6 Voltage amplifier.

29–2 Audio Amplifiers

Audio amplifiers amplify AC signals in the frequency range of approximately 20 to 20,000 hertz. They may amplify the whole audio range or only a small portion of it.

Audio amplifiers are divided into two categories: **voltage amplifiers** and **power amplifiers.** *Voltage amplifiers* are primarily used to produce a high voltage gain. *Power amplifiers* are used to deliver a large amount of power to a load. For example, a voltage amplifier is typically used to increase the voltage level of a signal sufficiently to drive a power amplifier. The power amplifier then supplies a high output to drive a load such as a loudspeaker or other high-power device. Typically, voltage amplifiers are biased to operate as class A amplifiers, and power amplifiers are biased to operate as class B amplifiers.

Figure 29–6 shows a simple voltage amplifier. The circuit shown is a common-emitter circuit. It is biased class A to provide a minimum amount of distortion. The amplifier can provide a substantial voltage gain over a wide frequency range. Because of the coupling capacitors, the circuit cannot amplify a DC signal.

Two or more voltage amplifiers can be connected together to provide higher amplification. The stages may be RC coupled or transformer coupled. Transformer coupling is more efficient. The transformer is used to match the input and output impedance of the two stages. This keeps the second stage from loading down the first stage. *Loading down* is the condition when a device creates too large a load and severely affects the output by drawing too much current. The transformer used to link the two stages together is called an *interstage transformer.*

Once a sufficient voltage level is available, a power amplifier is used to drive the load. Power amplifiers are designed to drive specific loads and are rated in watts. Typically a load may vary from 4 to 16 ohms.

Figure 29–7 shows a two-transistor power amplifier circuit, called a *push-pull amplifier.* The top half of the circuit is a mirror image of the bottom half. Each half is a single transistor amplifier. The output voltage is developed across the primary of the transformer during alternate half cycles of the input signal. Both transistors are biased either class AB or class B. The input to a push-pull amplifier requires complementary signals. That is, one signal must be inverted compared to the other. However, both signals must have the same amplitude and the same frequency. The circuit that produces the complementary signal is called a *phase splitter.* A single-transistor phase splitter is shown in Figure 29–8. The complementary outputs are taken from the collector and the emitter of the transistor. The phase splitter is operated as a class A amplifier to provide minimum distortion. The coupling capacitors are necessary to offset the differences between the DC collector and emitter voltages.

A push-pull amplifier that does not require a phase splitter is called a *complementary push-pull*

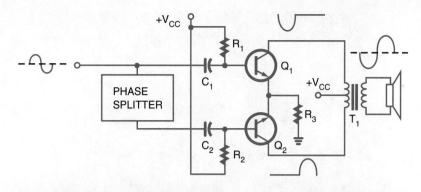

Figure 29–7 Push-pull power amplifier.

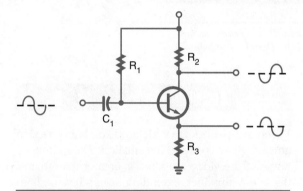

Figure 29–8 Phase splitter.

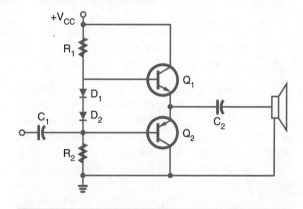

Figure 29–9 Complementary push-pull power amplifier.

amplifier. It uses an NPN and a PNP transistor to accomplish the push-pull action (Figure 29–9). The two transistors are connected in series with their emitters together. When each transistor is properly biased, there is 0.7 volt between the base and emitter or 1.4 volts between the two bases. The two diodes help to keep the 1.4-volt difference constant. The output is taken from between the two emitters through a coupling capacitor.

For amplifiers greater than 10 watts, it is difficult and expensive to match NPN and PNP transistors to ensure that they have the same characteristics. Figure 29–10 shows a circuit that uses two NPN transistors for the output-power transistors. The power transistors are driven by lower power NPN and PNP transistors while the upper set of transistors is connected in a darlington

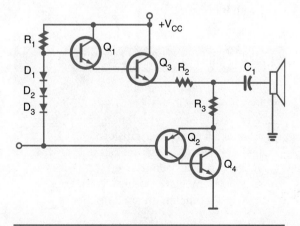

Figure 29–10 Quasi-complementary power amplifier.

Figure 29–11 Types of heat sinks available. *(From Herman,* Electronics for Industrial Electricians, *by Delmar Publishers.)*

configuration. The lower set of transistors uses a PNP and an NPN transistor. Operating as a single unit, they respond like a PNP transistor. This type of amplifier is called a *quasi-complementary amplifier*. It operates like a complementary amplifier but does not require high-power complementary output transistors.

Because of the large amounts of power generated by power amplifiers, some components get hot. To assist in the removal of this heat buildup, a heat sink is used. A *heat sink* is a device that provides a large area from which the heat can radiate. Figure 29–11 shows different types of heat sinks used with transistors.

29–2 Questions

1. For what frequency range are audio amplifiers used?
2. What are the two types of audio amplifiers?
3. What is an interstage transformer?
4. Draw schematic diagrams of the following:
   a. Push-pull amplifier
   b. Complementary push-pull amplifier
   c. Quasi-complementary amplifier

29–3 Video Amplifiers

Video amplifiers are wideband amplifiers used to amplify video (picture) information. The frequency range of the video amplifier is greater than that of the audio amplifier, extending from a few hertz to 5 or 6 megahertz. For example, a television requires a uniform bandwidth of 60 hertz to 4 megahertz. Radar requires a bandwidth of 30 hertz to 2 megahertz. In circuits that use sawtooth or pulse waveforms, it is necessary to cover a range of frequencies from one-tenth of the lowest frequency to ten times the highest frequency. The extended range is necessary because nonsinusoidal waveforms contain many harmonics and they must all be amplified equally.

Because video amplifiers require good uniformity in frequency response, only direct or RC coupling is used. Direct coupling provides the best frequency response, whereas RC coupling has economic advantages. The RC-coupled amplifier also has a flat response in the middle frequency range that is suitable for video amplifiers. *Flat response* is the term used to indicate that the gain of an amplifier varies only slightly within a stated frequency range. The response curve plotted for such an amplifier is almost a straight line; hence the term "flat response."

A factor that limits the high-frequency response in a transistor amplifier is the shunt capacitance of the circuit. A small capacitance exists between the junctions of the transistor. The capacitance is determined by the size of the junction and the spacing between the transistor's leads. The capacitance is further affected by the junction bias. A forward-biased base-emitter junction has a greater capacitance than a reverse-biased collector-base junction.

To reduce the effects of shunt capacitance and increase the frequency response in transistor video amplifiers, peaking coils are used. Figure 29–12 shows the *shunt-peaking* method. A small inductor is placed in series with the load resistor. At the low- and mid-frequency range, the peaking coil will have little effect on the amplifier response. At the

higher frequencies, the inductor resonates with the circuit's capacitance, which results in an increase in the output impedance and boosts the gain.

Another method is to insert a small inductor in series with the interstage coupling capacitor. This method is called *series peaking* (Figure 29–13). The peaking coil effectively isolates the input and output capacitance of the two stages.

Often series and shunt peaking are combined to gain the advantages of both (Figure 29–14). This combination can extend the bandwidth to over 5 megahertz.

The most common use of video amplifiers is in television receivers (Figure 29–15). Transistor Q_1 is connected as an emitter-follower. Input to transistor Q_1 is from the video detector. The video detector recovers the video signal from the intermediate frequency. In the collector circuit of transistor Q_2 is a shunt-peaking coil (L_1). In the signal-output path is a series-peaking coil (L_2). The video signal is then coupled to the picture tube through coupling capacitor C_5.

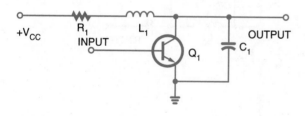

Figure 29–12 Shunt peaking.

29–3 Questions

1. What is a video amplifier?
2. What is the frequency range of a video amplifier?

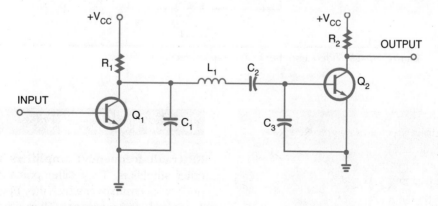

Figure 29–13 Series peaking.

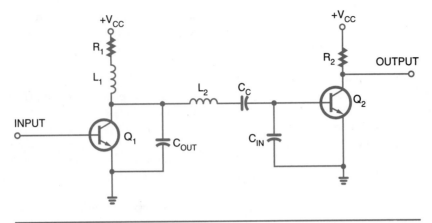

Figure 29–14 Series-shunt peaking.

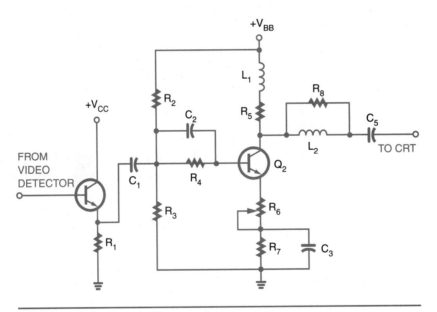

Figure 29–15 Video amplifier in a television receiver.

3. What coupling techniques are used for video amplifiers?
4. Define the following:
   a. Shunt peaking
   b. Series peaking
5. Where are video amplifiers used?

29–4 RF and IF Amplifiers

RF (radio-frequency) amplifiers are similar to other amplifiers. They differ primarily in the frequency spectrum over which they operate, which is 10,000 to 30,000 megahertz. There are two classes of RF amplifier: untuned and tuned. In an *untuned* *amplifier,* a response is desired over a large RF range, and the main function is amplification. In a *tuned amplifier,* high amplification is desired over a small range of frequencies or a single frequency. Normally, when RF amplifiers are mentioned, they are assumed to be tuned unless otherwise specified.

In receiving equipment, the RF amplifier serves to amplify the signal and select the proper frequency. In transmitters, the RF amplifier serves to amplify a single frequency for application to the antenna. Basically, the receiver RF amplifier is a voltage amplifier, and the transmitter RF amplifier is a power amplifier.

In a receiver circuit, the RF amplifier must provide sufficient gain, produce low internal noise, provide good selectivity, and respond well to the selected frequencies.

Figure 29–16 shows an RF amplifier used for an AM radio. Capacitors C_1 and C_4 tune the antenna and the output transformer T_1 to the same frequency. The input signal is magnetically coupled to the base of transistor Q_1. Transistor Q_1 operates as a class A amplifier. Capacitor C_4 and transformer T_1 provide a high voltage gain at the resonant frequency for the collector load circuit. Transformer T_1 is tapped to provide a good impedance match for the transistor.

Figure 29–17 shows an RF amplifier used in a television VHF tuner. The circuit is tuned by coils L_{1A}, L_{1B}, and L_{1C}. When the channel selector is turned, a new set of coils is switched into the circuit. This provides the necessary bandwidth response for each channel. The input signal is developed across the tuned circuit consisting of L_{1A}, C_1, and C_2. Transistor Q_1 operates as a class A amplifier. The collector-output circuit is a double-tuned transformer. Coil L_{1B} is tuned by capacitor C_4, and coil L_{1C} is tuned by capacitor C_7. Resistor R_2 and capacitor C_6 form a decoupling filter to prevent any RF from entering the power supply to interact with other circuits.

In an AM radio, the incoming RF signal is converted to a constant **IF (intermediate frequency) signal.** A fixed-tuned IF amplifier is then used to increase the signal to a usable level. The IF amplifier is a single-frequency amplifier. Typically, two or more IF amplifiers are used to increase the signal to the proper level. The sensitivity of a receiver is determined by its signal to noise (S/N) ratio. The higher the gain, the better the sensitivity. Figure 29–18 shows a typical IF amplifier in an AM radio. The IF frequency is 455,000 hertz. Figure 29–19 shows an IF amplifier in a television receiver. Figure 29–20 compares the frequencies of radio and television receivers.

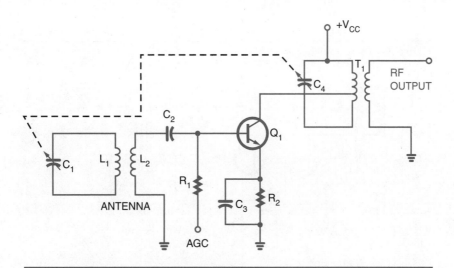

Figure 29–16 RF amplifier in an AM radio.

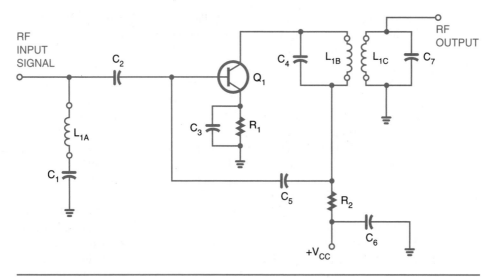

Figure 29–17 RF amplifier in a television VHF tuner.

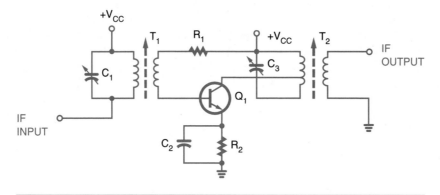

Figure 29–18 IF amplifier in an AM radio.

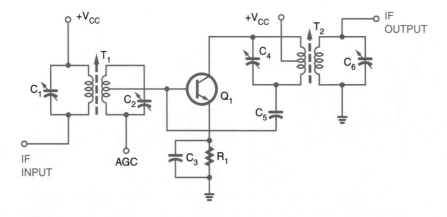

Figure 29–19 IF amplifier in a television receiver.

TYPE	RECEIVED RF	COMMON IF	BANDWIDTH
AM Radio	535-1605 kHz	455 kHz	10 kHz
FM Radio	88-108 MHz	10.7 MHz	150 kHz
Television	54-88 MHz		
Channels 2-6	174-216 MHz	41-47 MHz	6 MHz
Channels 7-13	470-890 MHz		
Channels 14-83			

Figure 29–20 Comparison of radio and television frequencies.

29—4 Questions

1. How do RF amplifiers differ from other amplifiers?
2. What are the two types of RF amplifiers?
3. Where are RF amplifiers used?
4. What is an IF amplifier?
5. What is significant about an IF amplifier?

29—5 Operational Amplifiers

Operational amplifiers are usually called **op-amps.** An *op-amp* is a very high-gain DC amplifier. Typically, op-amps have an output gain in the range of 20,000 to 1,000,000 times the input. Figure 29–21 shows the schematic symbol used for op-amps. The negative (–) input is called the *inverting input* and the positive (+) is called the *noninverting input*.

Figure 29–22 is a block diagram of an op-amp. The op-amp consists of three stages. Each stage is an amplifier with some unique characteristic.

The input stage is a differential amplifier. It allows the op-amp to respond only to the differences between input signals. Also, the differential amplifier amplifies only the differential input voltage and is unaffected by signals common to both inputs. This is referred to as *common-mode rejection.* Common-mode rejection is useful when measuring a small signal in the presence of 60-hertz noise. The 60-hertz noise common to both inputs is rejected, and the op-amp amplifies only the small difference between the two inputs. The differential amplifier has a low-frequency response that extends down to a DC level. This means that the differential amplifier can respond not only to low-frequency AC signals but to DC signals as well.

The second stage is a high-gain voltage amplifier. This stage is composed of several darlington-pair transistors. This stage provides a voltage gain of 200,000 or more and supplies most of the op-amp's gain.

The last stage is an output amplifier. Typically, this is a complementary emitter-follower amplifier. It is used to give the op-amp a low out-put impedance. The op-amp can deliver several milliamperes of current to a load.

Generally, op-amps are designed to be powered by a dual-voltage power supply in the range of ±5 to ±15 volts. The positive power source delivers +5 to +15 volts with respect to ground. The negative power source delivers –5 to –15 volts with respect to ground. This allows the output voltage to swing from positive to negative with respect to

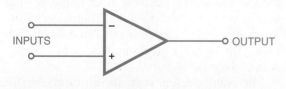

Figure 29–21 Schematic symbol for an op-amp.

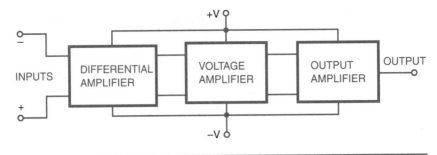

Figure 29–22 Block diagram of an op-amp.

ground. However, in certain cases, op-amps may be operated from a single voltage source.

A schematic diagram of a typical op-amp is shown in Figure 29–23. The op-amp shown is called a 741. This op-amp does not require frequency compensation, is short-circuit protected, and has no latch-up problems. It provides good performance for a low price and is the most commonly used op-amp. A device that contains two 741 op-amps in a single package is called a 747 op-amp. Because no coupling capacitors are used, the circuit can amplify DC signals as well as AC signals.

The op-amp's normal mode of operation is a closed loop. The **closed-loop mode** uses feedback, as compared to the **open-loop mode,** which does not use feedback. A lot of degenerative feedback is used with the closed-loop mode. This reduces the overall gain of the op-amp but provides better stability.

In closed-loop operation, the output signal is applied back to one of the input terminals as a feedback signal. This feedback signal opposes the input signal. There are two basic closed-loop circuits: inverting and noninverting. The inverting configuration is more popular.

Figure 29–24 shows the op-amp connected as an *inverting amplifier.* The input signal is applied to the op-amp's inverting (−) input through resistor R_1. The feedback is provided through resistor R_2. The signal at the inverting input is determined by both the input and output voltages.

A minus sign indicates a negative output signal when the input signal is positive. A plus sign indicates a positive output signal when the input signal is negative. The output is 180 degrees out of phase

with the input. Depending on the ratio of resistor R_2 to R_1, the gain of the inverting amplifier can be less than, equal to, or greater than 1. When the gain is equal to 1, it is called a *unity-gain amplifier* and is used to invert the polarity of the input signal.

Figure 29–25 shows an op-amp connected as a *noninverting amplifier.* The output signal is in phase with the input. The input signal is applied to the op-amp's noninverting input. The output voltage is divided between resistors R_1 and R_2 to produce a voltage that is fed back to the inverting (−) input. The voltage gain of a noninverting input amplifier is always greater than 1.

The gain of an op-amp varies with frequency. Typically, the gain given by a specification sheet is the DC gain. As the frequency increases, the gain decreases. Without some means to increase the bandwidth, the op-amp is only good for amplifying a DC signal. To increase the bandwidth, feedback is used to reduce the gain. By reducing the gain, the bandwidth increases by the same amount. In this way, the 741 op-amp bandwidth can be increased to 1 megahertz.

Besides its use for comparing, inverting, or noninverting a signal, the op-amp has several other applications. It can be used to add several signals together as shown in Figure 29–26. This is referred to as a *summing amplifier.* The negative feedback holds the inverting input of the op-amp very close to ground potential. Therefore, all the input signals are electrically isolated from each other. The output of the amplifier is the inverted sum of the input signal.

In a summing amplifier, the resistor chosen for the noninverting input-to-ground is equal to the

Operational Amplifiers/Buffers

LM741/LM741A/LM741C/LM741E Operational Amplifier

General Description

The LM741 series are general purpose operational amplifiers which feature improved performance over industry standards like the LM709. They are direct, plug-in replacements for the 709C, LM201, MC1439 and 748 in most applications.

The amplifiers offer many features which make their application nearly foolproof: overload pro-

tection on the input and output, no latch-up when the common mode range is exceeded, as well as freedom from oscillations.

The LM741C/LM741E are identical to the LM741/LM741A except that the LM741C/ LM741E have their performance guaranteed over a 0°C to +70°C temperature range, instead of −55°C to +125°C.

Schematic and Connection Diagrams (Top Views)

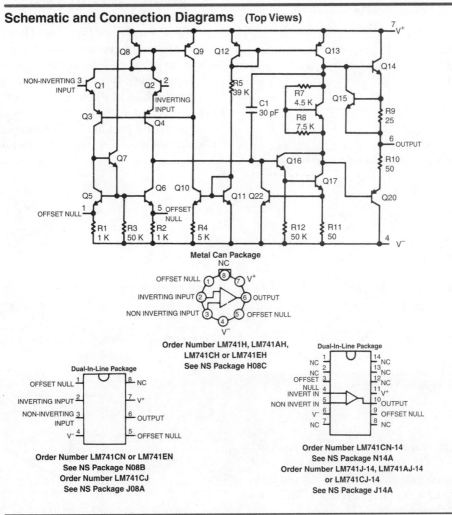

Figure 29–23 Schematic diagram of an op-amp. *(Courtesy of National Semiconductor Corporation.)*

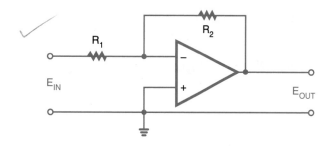

Figure 29–24 Op-amp connected as an inverting amplifier.

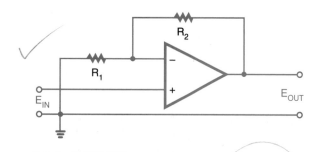

Figure 29–25 Op-amp connected as a noninverting amplifier.

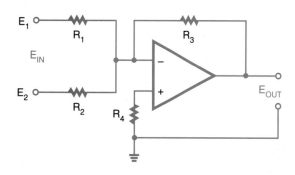

Figure 29–26 Op-amp connected as a summing amplifier.

total parallel resistance of the input and feedback resistances. If the feedback resistance is increased, the circuit can provide gain. If different input resistances are used, the input signals can be added together with different gains.

Summing amplifiers are used when mixing audio signals together. Potentiometers are used for the input resistances to adjust the strength of each of the input signals.

Op-amps can also be used as active filters. Filters that use resistors, inductors, and capacitors are called **passive filters. Active filters** are inductorless filters using integrated circuits. The advantage of active filters is the absence of inductors, which is handy at low frequencies because of the size of inductors.

There are some disadvantages when using op-amps as active filters because they require a power supply, can generate noise, and can oscillate due to thermal drift or aging components.

Figure 29–27 shows an op-amp used as a high-pass filter. A high-pass filter rejects low frequencies while passing frequencies above a certain cut-off frequency. Figure 29–28 shows an op-amp used as a low-pass filter. A low-pass filter passes low frequencies while blocking frequencies above the cut-off frequency. Figure 29–29 shows an op-amp used as a band-pass filter. A band-pass filter passes frequencies around some central frequency while attenuating both higher and lower frequencies.

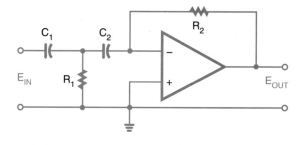

Figure 29–27 Op-amp connected as a high-pass filter.

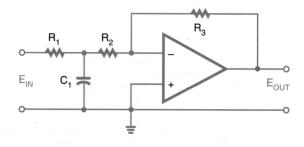

Figure 29–28 Op-amp connected as a low-pass filter.

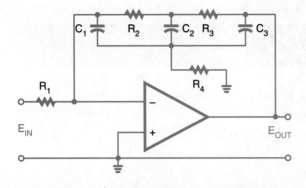

Figure 29–29 Op-amp connected as a band-pass filter.

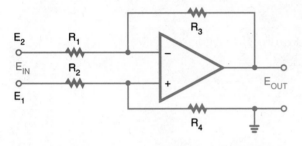

Figure 29–30 Op-amp connected as a difference amplifier.

A **difference amplifier** subtracts one signal from another signal. Figure 29–30 shows a basic difference amplifier. This circuit is called a *subtractor* because it subtracts the value of E_2 from the value of E_1.

29–5 Questions

1. What is an op-amp?
2. Draw a block diagram of an op-amp.
3. Briefly explain how an op-amp works.
4. What is the normal mode of operation for an op-amp?
5. What type of gain can be obtained with an op-amp?
6. Draw schematic diagrams of the following:
   a. Inverting amplifier
   b. Summing amplifier
   c. High-pass filter
   d. Band-pass filter
   e. Difference amplifier

SUMMARY

- Direct-coupled amplifiers are used primarily as voltage amplifiers.
- A differential amplifier has two separate inputs and may provide either one or two outputs.
- Audio amplifiers amplify AC signals in the audio range of 20 to 20,000 hertz.
- The two types of audio amplifiers are voltage amplifiers and power amplifiers.
- Video amplifiers are wideband amplifiers used to amplify video information.
- Video frequencies extend from a few hertz to 5 or 6 megahertz.
- RF amplifiers operate from 10,000 to 30,000 megahertz.
- The two types of RF amplifiers are tuned and untuned.
- Op-amps may provide output gains of 20,000 to 1,000,000 times the input.
- The two basic closed-loop modes are the inverting configuration and the non-inverting configuration.

Chapter 29 Self-Test

1. Under what conditions would a DC amplifier be used?

2. How is the problem of temperature instability resolved with high-gain DC amplifiers?

3. What are the main differences between audio voltage amplifiers and audio power amplifiers?

4. What is the practical advantage of using a quasi-complementary power amplifier over a complementary push-pull amplifier?

5. How does a video amplifier differ from an audio amplifier?

6. What factor is involved in limiting the output of high-frequency video amplifiers?

7. What is the purpose of an RF amplifier?

8. How are IF amplifiers used in a circuit?

9. Identify the three stages of an op-amp, and describe its function.

10. For what type of application is the op-amp used?

Chapter

30

OSCILLATORS

Objectives

After completing this chapter, the student will be able to:

- Describe an oscillator and its purpose.
- Identify the main requirements of an oscillator.
- Explain how a tank circuit operates and its relationship to an oscillator.
- Draw a block diagram of an oscillator.
- Identify LC, crystal, and RC sinusoidal oscillator circuits.
- Identify nonsinusoidal relaxation oscillator circuits.
- Draw examples of sinusoidal and nonsinusoidal oscillators.

An oscillator is a nonrotating device for producing alternating current. Oscillators are used extensively in the field of electronics: in radios and televisions, communication systems, computers, industrial controls, and timekeeping devices. Without the oscillator, very few electronic circuits would be possible.

30–1 Fundamentals of Oscillators

An **oscillator** is a circuit that generates a repetitive AC signal. The frequency of the AC signal may vary from a few hertz to many millions of hertz. The oscillator is an alternative to the mechanical generator used to produce electrical power. The advantages of the oscillator are the absence of moving parts and the range over which the AC signal can be produced. The output of an oscillator may be a sinusoidal, rectangular, or sawtooth waveform, depending on the type of oscillator used. The main requirement of an oscillator is that the output be uniform; that is, the output must not vary in either frequency or amplitude.

When an inductor and a capacitor are connected in parallel, they form what is called a **tank circuit.** When a tank circuit is excited by an external DC source, it oscillates; that is, it produces a back-and-forth current flow. If it were not for the resistance of the circuit, the tank circuit would oscillate forever. However, the resistance of the tank circuit absorbs energy from the current, and the oscillations of the circuit are dampened.

For the tank circuit to maintain its oscillation, the energy that is dissipated must be replaced. The energy that is replaced is referred to as **positive feedback.** Positive feedback is the feeding back into the tank circuit of a portion of the output signal to sustain oscillation. The feedback must be in phase with the signal in the tank circuit.

Figure 30–1 shows a block diagram of an oscillator. The basic oscillator can be broken down into three sections. The frequency-determining oscillator circuit is usually an LC tank circuit. An amplifier increases the output signal of the tank circuit. A feedback circuit delivers the proper amount of energy to the tank circuit to sustain oscillation. The oscillator circuit is essentially a closed loop that uses DC power to maintain AC oscillations.

30–1 Questions

1. What is an oscillator?
2. How does a tank circuit operate?
3. What makes a tank circuit continue to oscillate?
4. Draw and label a block diagram of an oscillator.
5. What are the functions of the basic parts of an oscillator?

30–2 Sinusoidal Oscillators

Sinusoidal oscillators are oscillators that produce a sine-wave output. They are classified according to their frequency-determining components. The three basic types of sinusoidal oscillators are LC oscillators, crystal oscillators, and RC oscillators.

LC oscillators use a tank circuit of capacitors and inductors, connected either in series or parallel, to determine the frequency. Crystal oscillators are like LC oscillators except that crystal oscillators maintain a higher degree of stability. LC and crystal oscillators are used in the radio frequency (RF) range. They are not suitable for low-frequency applications. For low-frequency applications, RC oscillators are used. RC oscillators use a resistance-capacitance network to determine the oscillator frequency.

Three basic types of LC oscillator are the Hartley oscillator, the Colpitts oscillator, and the Clapp oscillator. Figures 30–2 and 30–3 show the two basic types of Hartley oscillator. The tapped inductor in the tank circuit identifies these circuits as Hartley oscillators. The disadvantage of the series-fed Hartley (Figure 30–2) is that DC current flows through a portion of the tank circuit. The shunt-fed Hartley (Figure 30–3) overcomes the problem of DC current in the tank circuit by using a coupling capacitor in the feedback line.

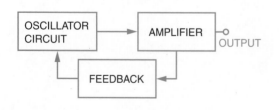

Figure 30–1 Block diagram of an oscillator.

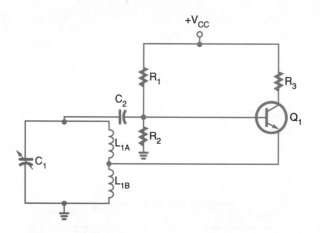

Figure 30–2 Series-fed Hartley oscillator.

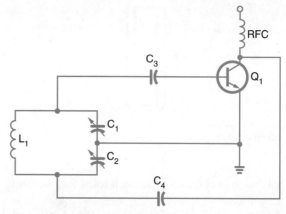

Figure 30–4 Colpitts oscillator.

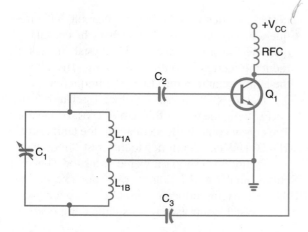

Figure 30–3 Shunt-fed Hartley oscillator.

Crystals are materials that can convert mechanical energy to electrical energy when pressure is applied or can convert electrical energy to mechanical energy when a voltage is applied. When an AC voltage is applied to a crystal, the crystal stretches and compresses, creating mechanical vibrations that correspond to the frequency of the AC signal.

Crystals, because of their structure, have a natural frequency of vibration. If the AC signal applied matches the natural frequency, the crystal vibrates more. If the AC signal is different from the crystal's natural frequency, little vibration is produced. The crystal's mechanical frequency of

The Colpitts oscillator (Figure 30–4) is like the shunt-fed Hartley except that two capacitors are substituted for the tapped inductor. The Colpitts is more stable than the Hartley and is more often used.

The Clapp oscillator (Figure 30–5) is a variation of the Colpitts oscillator. The main difference is the addition of a capacitor in series with the inductor in the tank circuit. The capacitor allows tuning of the oscillator frequency.

Temperature changes, aging of components, and varying load requirements cause oscillators to become unstable. When stability is a requirement, crystal oscillators are used.

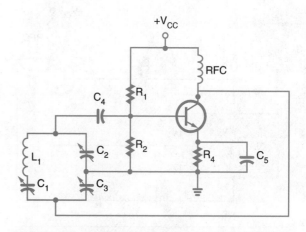

Figure 30–5 Clapp oscillator.

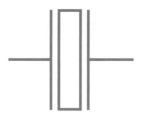

Figure 30–6 Schematic symbol for crystal.

vibration is constant, making it ideal for oscillator circuits.

The most common materials used for crystals are Rochelle salt, tourmaline, and quartz. Rochelle salt has the most electrical activity, but it fractures easily. Tourmaline has the least electrical activity, but it is the strongest. Quartz is a compromise: It has good electrical activity and is strong. Quartz is the most commonly used crystal in oscillator circuits.

The crystal material is mounted between two metal plates, with pressure applied by a spring so that the metal plates make electrical contact with the crystal. The crystal is then placed in a metal package. Figure 30–6 shows the schematic symbol used to represent the crystal. The letters Y or $XTAL$ identify crystals in schematics.

Figure 30–7 shows a shunt-fed Hartley oscillator with the addition of a crystal. The crystal is con-

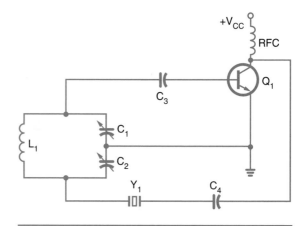

Figure 30–8 Colpitts crystal oscillator.

nected in series with the feedback circuit. If the frequency of the tank circuit drifts from the crystal frequency, the impedance of the crystal increases, reducing feedback to the tank circuit. This allows the tank circuit to return to crystal frequency.

Figure 30–8 shows a Colpitts oscillator connected the same way as the Hartley crystal oscillator. The crystal controls the feedback to the tank circuit. The LC tank circuit is tuned to the crystal frequency.

Figure 30–9 shows a Pierce oscillator. This circuit is similar to the Colpitts oscillator except that the tank-circuit inductor is replaced with a crystal. The crystal controls the tank-circuit impedance,

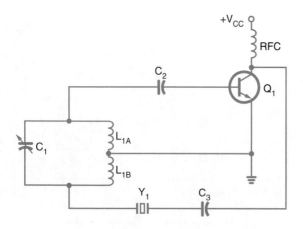

Figure 30–7 Crystal shunt-fed Hartley oscillator.

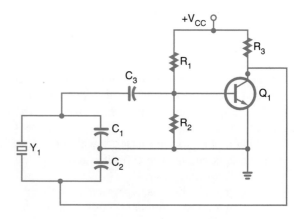

Figure 30–9 Pierce oscillator.

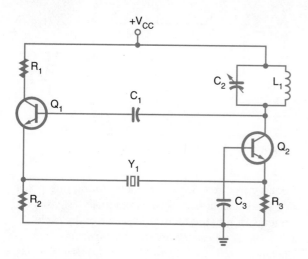

Figure 30–10 Butler oscillator.

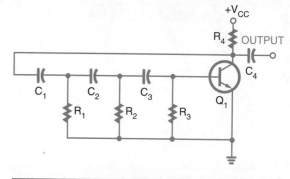

Figure 30–11 Phase-shift oscillator.

which determines the feedback and stabilizes the oscillator.

Figure 30–10 shows a Butler oscillator. This is a two-transistor circuit. It uses a tank circuit, and the crystal determines the frequency. The tank circuit must be tuned to the crystal frequency or the oscillator does not work. The advantage of the Butler oscillator is that a small voltage exists across the crystal, reducing stress on the crystal. By replacing the tank-circuit components, the oscillator can be tuned to operate on one of the crystal's overtone frequencies.

RC oscillators use resistance-capacitance networks to determine the oscillator frequency. There are two basic types of RC oscillators that produce sinusoidal waveforms: the phase-shift oscillator and the Wien-bridge oscillator.

A *phase-shift oscillator* is a conventional amplifier with a phase-shifting RC feedback network (Figure 30–11). The feedback must shift the signal 180 degrees. Because the capacitance reactance changes with a change in frequency, it is the frequency-sensitive component. Stability is improved by reducing the amount of phase shift across each RC network. However, there is a power loss across the combined RC network. The transistor must have enough gain to offset these losses.

A *Wien-bridge oscillator* is a two-stage amplifier with a lead-lag network and voltage divider (Figure 30–12). The lead-lag network consists of a series RC network ($R_1 C_1$) and a parallel network. It is called a *lead-lag network* because the output phase angle leads for some frequencies and lags for other frequencies. At the resonant frequency, the phase shift is zero, and the output voltage is maximum. Resistors R_3 and R_4 form the voltage-divider network, which is used to develop the degenerative feedback. *Regenerative feedback* is applied to the base and *degenerative feedback* is applied to the emitter of oscillator transistor Q_1. The output of transistor Q_1 is capacitively coupled

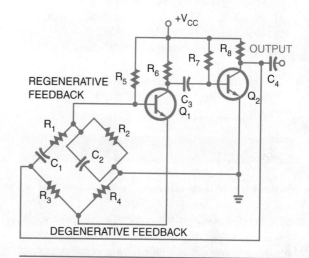

Figure 30–12 Wien-bridge oscillator.

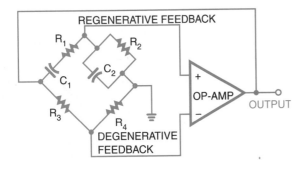

Figure 30–13 IC Wien-bridge oscillator.

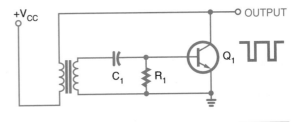

Figure 30–14 Blocking oscillator.

to the base of transistor Q_2 where it is amplified and shifted and required 180 degrees. The output is coupled by capacitor C_3 to the bridge network.

Figure 30–13 shows an integrated circuit Wienbridge oscillator. The inverting and noninverting inputs of the op-amp are ideal for use as a Wienbridge oscillator. The gain of the op-amp is high, which offsets the circuit losses.

30–2 Questions

1. What are the three types of sinusoidal oscillator?
2. Draw and label schematic diagrams of the three types of LC oscillator.
3. What is the difference between a Colpitts oscillator and a Hartley oscillator?
4. How can the stability of an LC oscillator be improved?
5. What are the two types of RC oscillator used for producing sinusoidal waves?

30–3 Nonsinusoidal Oscillators

Nonsinusoidal oscillators are oscillators that do not produce a sine-wave output. There is no specific nonsinusoidal waveshape. The nonsinusoidal oscillator output may be a square, sawtooth, rectangular, or triangular waveform, or a combination of two waveforms. A common characteristic of all nonsinusoidal oscillators is that they are a form of relaxation oscillator. A **relaxation oscillator** stores

energy in a reactive component during one phase of the oscillation cycle and gradually releases the energy during the relaxation phase of the cycle.

Blocking oscillators and *multivibrators* are relaxation oscillators. Figure 30–14 shows a blocking oscillator circuit. The reason for the name is that the transistor is easily driven into the blocking (cut-off) mode. The blocking condition is determined by the discharge from capacitor C_1. Capacitor C_1 is charged through the emitter-base junction of transistor Q_1. However, once capacitor C_1 is charged, the only discharge path is through resistor R_1. The RC time constant of resistor R_1 and capacitor C_1 determines how long the transistor is blocked or cut off and also determines the oscillator frequency. A long time constant produces a low frequency; a short time constant produces a high frequency.

If the output is taken from an RC network in the emitter circuit of the transistor, the output is a sawtooth waveshape (Figure 30–15). The RC network determines the frequency of oscillation and produces the sawtooth output. Transistor Q_1 is forward biased by resistor R_1. As transistor Q_1 conducts, capacitor C_1 charges quickly. The positive

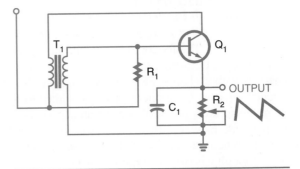

Figure 30–15 Sawtooth waveform generated by a blocking oscillator.

potential on the top plate of capacitor C_1 reverse biases the emitter junction, turning off transistor Q_1. Capacitor C_1 discharges through resistor R_2, producing the trailing portion of the sawtooth output. When capacitor C_1 discharges, transistor Q_1 is

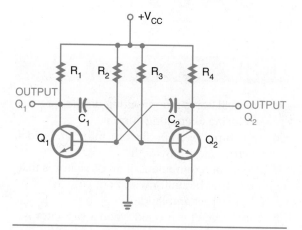

Figure 30–16 Free-running multivibrator.

again forward biased and conducts, repeating the action.

Capacitor C_1 and resistor R_2 determine the frequency of oscillation. By making resistor R_2 variable, the frequency can be adjusted. If resistor R_2 offers high resistance, a long RC time constant results, producing a low frequency of oscillation. If resistor R_2 offers low resistance, a short RC time constant results, producing a high frequency of oscillation.

A **multivibrator** is a relaxation oscillator that can function in either of two temporarily stable conditions and is capable of rapidly switching from one temporary state to the other.

Figure 30–16 shows a basic free-running multivibrator circuit. It is basically an oscillator consisting of two stages coupled so that the input signal to each stage is taken from the output of the other. One stage conducts while the other stage is cut off, until a point is reached where the stages reverse their conditions. The circuit is free-running

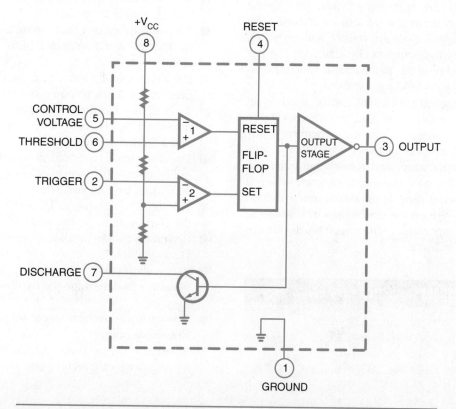

Figure 30–17 Block diagram of a 555 timer integrated circuit.

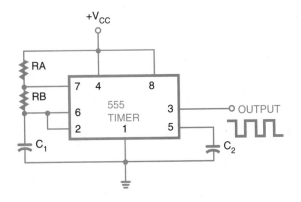

Figure 30–18 A stable multivibrator using a 555 timer.

because of regenerative feedback. The frequency of oscillation is determined by the coupling circuit.

An **astable multivibrator** is one type of free-running multivibrator. The output of an astable multivibrator is rectangular. By varying the RC time constants of the coupling circuits, rectangular pulses of any desired width can be obtained. By changing the values of the resistor and capacitor, the operating frequency can be changed. The frequency stability of the multivibrator is better than that of the typical blocking oscillator.

An integrated circuit that can be used as an astable multivibrator is the 555 timer (Figure 30–17). This integrated circuit can perform many functions. It consists of two comparators, a flip-flop, an output stage, and a discharge transistor. Figure 30–18 shows a schematic diagram in which the 555 timer is used as an astable multivibrator. The output frequency is determined by resistors R_A and R_B and capacitor C_1. This circuit finds wide application in industry.

30–3 Questions

1. Draw the more commonly used nonsinusoidal waveforms.
2. What is a relaxation oscillator?
3. Give two examples of relaxation oscillators.
4. Draw a schematic diagram of a blocking oscillator.
5. Draw a schematic diagram of a 555 timer used as an astable multivibrator.

SUMMARY

■ An oscillator is a nonrotating device for producing alternating current.

■ The output of an oscillator can be sinusoidal, rectangular, or sawtooth.

■ The main requirement of an oscillator is that the output be uniform and not vary in frequency or amplitude.

■ A tank circuit is formed when a capacitor is connected in parallel with an inductor.

■ A tank circuit oscillates when an external voltage source is applied.

■ The oscillations of a tank circuit are dampened by the resistance of the circuit.

■ For a tank circuit to maintain oscillation, positive feedback is required.

■ An oscillator has three basic parts: a frequency-determining device, an amplifier, and a feedback circuit.

■ The three basic types of sinusoidal oscillators are LC oscillators, crystal oscillators, and RC oscillators.

■ The three basic types of LC oscillators are the Hartley, the Colpitts, and the Clapp.

■ Crystal oscillators provide more stability than LC oscillators.

■ RC oscillators use resistance-capacitance networks to determine the oscillator frequency.

■ Nonsinusoidal oscillators do not produce a sine-wave output.

■ Nonsinusoidal oscillator outputs include square, sawtooth, rectangular, and triangular

waveforms and combinations of two waveforms.
■ A relaxation oscillator is the basis of all nonsinusoidal oscillators.

■ A relaxation oscillator stores energy in a reactive component during one part of the oscillation cycle.
■ Examples of relaxation oscillators are blocking oscillators and multivibrators.

Chapter 30 Self-Test

1. Identify the parts of an oscillator, and explain what each part does in making the oscillator function.
2. Explain how a tank circuit can sustain an oscillation.
3. What are the major types of sinusoidal oscillators?
4. How are crystals used in oscillator circuits?
5. How does a nonsinusoidal oscillator differ from a sinusoidal oscillator?
6. What types of components make up nonsinusoidal oscillators?

31

WAVESHAPING CIRCUITS

Objectives

After completing this chapter, the student will be able to:

■ Identify ways in which waveform shapes can be changed.

■ Explain the frequency-domain concept in waveform construction.

■ Define *pulse width, duty cycle, rise* and *fall time, undershoot, overshoot,* and *ringing* as they relate to waveforms.

■ Explain how differentiators and integrators work.

■ Describe clipper and clamper circuits.

■ Describe the differences between monostable and bistable multivibrators.

■ Draw schematic diagrams of waveshaping circuits.

In electronics it is sometimes necessary to change the shape of a waveform. Sine waves may need to be changed to square waves, rectangular waveforms may need to be changed to pulse waveforms, and pulse waveforms may need to be changed to square or rectangular waveforms. Waveforms can be analyzed by two methods. Analyzing a waveform by its amplitude per unit of time is called a *time-domain analysis*. Analyzing a waveform by the sine waves that make it up is called a *frequency-domain analysis*. This concept assumes that all periodic waveforms are composed of sine waves.

31–1 Nonsinusoidal Waveforms

Figure 31–1 shows three basic waveforms that are represented by **time-domain.** The three waveforms shown are sine wave, square-wave, and sawtooth wave. Although the three waveforms are different, they all have the same period of frequency. By using various electronic circuits, these waveforms can be changed from one shape to another.

A *periodic waveform* is one with the same waveform for all cycles. According to the **frequency-domain concept,** all periodic waveforms are made up of sine waves. In other words, any periodic wave can be formed by superimposing a number of sine waves having different amplitudes, phases, and frequencies. Sine waves are important because they are the only waveform that cannot be distorted by RC, RL, or LC circuits.

The sine wave that has the same frequency as the periodic waveform is called the *fundamental frequency.* The fundamental frequency is also called the *first harmonic.* **Harmonics** are multiples of the fundamental frequency. The second harmonic is twice the fundamental, the third harmonic is three times the fundamental, and so on. Figure 31–2 shows the fundamental frequency of

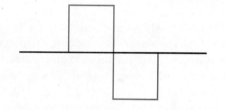

(Fundamental)	1st Harmonic	1000 Hz
	2nd Harmonic	2000 Hz
	3rd Harmonic	3000 Hz
	4th Harmonic	4000 Hz
	5th Harmonic	5000 Hz

Figure 31–2 Chart of fundamental frequency of 1000 hertz and some of its harmonics.

1000 hertz and some of its harmonics. Harmonics can be combined in an infinite number of ways to produce any periodic waveform. The type and number of harmonics included in the periodic waveform depend on the shape of the waveform.

For example, Figure 31–3 shows a square wave. Figure 31–4 shows how a square wave is formed by the combination of the fundamental frequency with an infinite number of odd harmonics

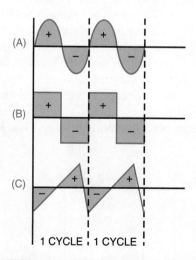

Figure 31–1 Three basic waveforms: (A) sine wave, (B) square wave, (C) sawtooth wave.

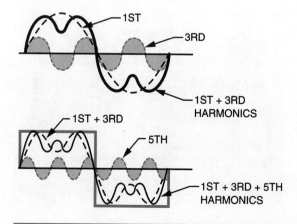

Figure 31–3 Square wave.

Figure 31–4 Formation of a square wave by the frequency-domain method.

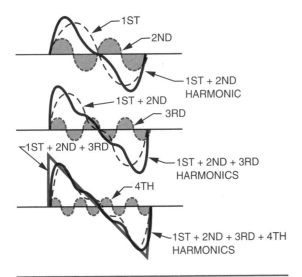

Figure 31–5 Formation of a sawtooth wave by the frequency-domain method.

that cross the zero reference line in phase with the fundamental.

Figure 31–5 shows the formation of a sawtooth waveform. It consists of the fundamental frequency plus even and odd harmonics crossing the zero reference line 180 degrees out of phase with the fundamental.

An oscilloscope displays waveforms in the time domain. A *spectrum analyzer* (Figure 31–6)

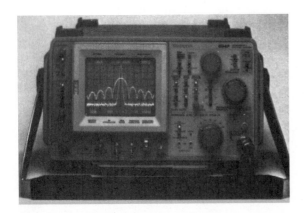

Figure 31–6 Spectrum analyzer. *(Courtesy of Tektronix, Inc.)*

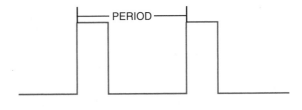

Figure 31–7 Period of a waveform.

displays waveforms in the frequency domain. Frequency-domain analysis can be used to determine how a circuit affects a waveform.

Periodic waveforms are waveforms that occur at regular intervals. The *period* of a waveform is measured from any point on one cycle to the same point on the next cycle (Figure 31–7).

The **pulse width** is the length of the pulse (Figure 31–8). The **duty cycle** is the ratio of the pulse width to the period. It can be represented as a percentage indicating the amount of time that the pulse exists during each cycle.

$$\text{duty cycle} = \frac{\text{pulse width}}{\text{period}}$$

All pulses have rise and fall times. The **rise time** is the time it takes from the pulse to rise from 10% to 90% of its maximum amplitude. The **fall time** is the time it takes for a pulse to fall from 90% to 10% of its maximum amplitude (Figure 31–9).

Overshoot, undershoot, and **ringing** are conditions common to high-frequency pulses (Figure 31–10). *Overshoot* occurs when the leading edge of a waveform exceeds its normal maximum value. *Undershoot* occurs when the trailing edge exceeds its normal minimum value. (The **leading**

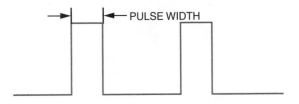

Figure 31–8 Pulse width of a waveform.

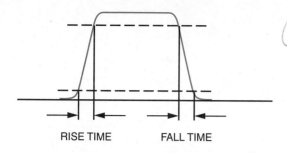

Figure 31–9 The rise and fall times of a waveform are measured at 10% and 90% of the waveform's maximum amplitude.

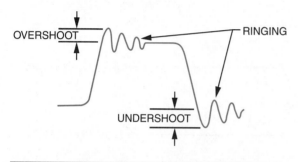

Figure 31–10 Overshoot, undershoot, and ringing.

edge is the front edge of the waveform; the **trailing edge** is the back edge of the waveform.) Both conditions are followed by damped oscillations known as *ringing*. These conditions are undesirable but exist because of imperfect circuits.

31—1 Questions

1. Define the *frequency-domain concept*.
2. How are the following waveforms constructed according to the frequency-domain concept?
   a. Square wave
   b. Sawtooth wave
3. What is a periodic waveform?
4. What is a duty cycle?
5. Draw examples of overshoot, undershoot, and ringing as they apply to a waveform.

31—2 Waveshaping Circuits

An RC network can change the shape of complex waveforms so that the output barely resembles the input. The amount of distortion is determined by the RC time constant. The type of distortion is determined by the component the output is taken across. If the output is taken across the resistor, the circuit is called a **differentiator.** A *differentiator* is used to produce a pip or peaked waveform from square or rectangular waveforms for timing or synchronizing circuits. It is also used to produce trigger or marker pulses. If the output is taken across the capacitor, the circuit is called an **integrator.** An integrator is used for waveshaping in radio, television, radar, and computers.

Figure 31–11 shows a differentiator circuit. Recall that complex waveforms are made of the fundamental frequency plus a large number of harmonics. When a complex waveform is applied to a differentiator, each frequency is affected differently. The ratio of the capacitive reactance (X_C) to R is different for each harmonic. This results in each harmonic being shifted in phase and reduced in amplitude by a different amount. The net result is distortion of the original waveform. Figure 31–12

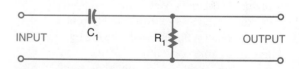

Figure 31–11 Differentiator circuit.

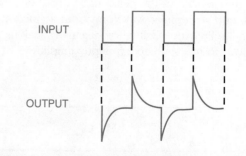

Figure 31–12 Result of applying a square wave to a differentiator circuit.

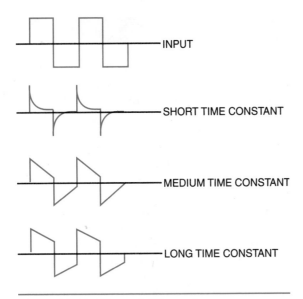

Figure 31–13 Effects of different time constants on a differentiator circuit.

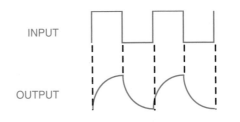

Figure 31–15 Result of applying a square wave to an integrator circuit.

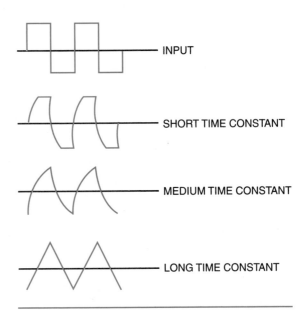

Figure 31–16 Effects of different time constants on an integrator circuit.

shows what happens to a square wave applied to a differentiator. Figure 31–13 shows the effects of different RC time constants.

An integrator circuit is similar to a differentiator except that the output is taken across the capacitor (Figure 31–14). Figure 31–15 shows the result of applying a square wave to an integrator. The integrator distorts the waveform in a different way than the differentiator. Figure 31–16 shows the effects of different RC time constants.

Another type of circuit that can change the shape of a waveform is a **clipping,** or **limiter circuit** (Figure 31–17). A clipping circuit can be used to square off the peaks of an applied signal, obtain a rectangular waveform from a sine wave signal, eliminate positive or negative portions of a waveform, or keep an input amplitude at a

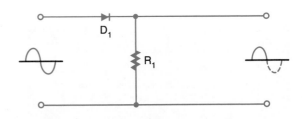

Figure 31–17 Basic series diode clipping circuit.

constant level. The diode is forward biased and conducts during the positive portion of the input signal. During the negative portion of the input signal, the diode is reverse biased and does not conduct. Figure 31–18 shows the effect of reversing the

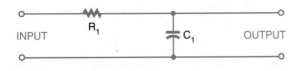

Figure 31–14 Integrator circuit.

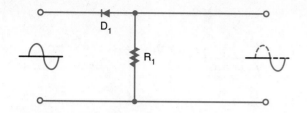

Figure 31–18 Result of reversing the diode in a clipping circuit.

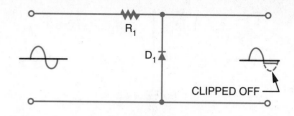

Figure 31–21 Shunt diode clipping circuit.

diode: The positive portion of the input signal is clipped off. The circuit is essentially a half-wave rectifier.

By using a bias voltage, the amount of signal that is clipped off can be regulated. Figure 31-19 shows a biased series clipping circuit. The diode cannot conduct until the input signal exceeds the bias source. Figure 31-20 shows the result of reversing the diode and the bias source.

A shunt clipping circuit performs the same function as the series clipper (Figure 31–21). The difference is that the output is taken across the diode. This circuit clips off the negative portion of the input signal. Figure 31–22 shows the effect of reversing the diode. A shunt clipper can also be biased to change the clipping level as shown in Figures 31–23 and 31–24.

If both the positive and the negative peaks must be limited, two biased diodes are used (Figure 31–25). This prevents the output signal from exceeding predetermined values for both peaks. With the elimination of both peaks, the remaining signal is generally square-shaped. Therefore,

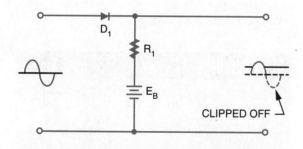

Figure 31–19 Biased series diode clipping circuit.

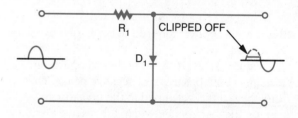

Figure 31–22 Result of reversing the diode in a shunt diode clipping circuit.

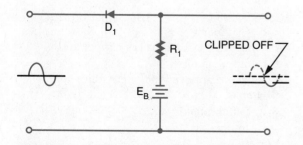

Figure 31–20 Result of reversing diode and bias source in a biased series clipping circuit.

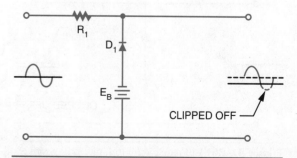

Figure 31–23 Biased shunt diode clipping circuit.

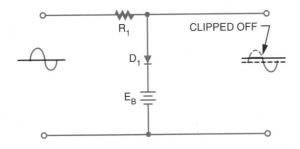

Figure 31–24 Effect of reversing the diode and bias source in a shunt diode clipping circuit.

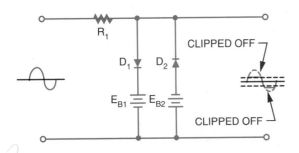

Figure 31–25 Clipping circuit used to limit both the positive and negative peaks.

this circuit is often referred to as a square-wave generator.

Figure 31–26 shows another clipping circuit that limits both positive and negative peaks. Therefore, the output is clamped to the breakdown voltage of the zeners. Between the two extremes, neither zener diode will conduct, and the input signal is passed to the output.

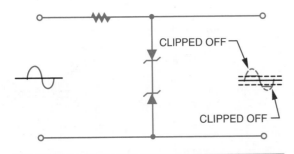

Figure 31–26 Alternate circuit to limit both positive and negative peaks.

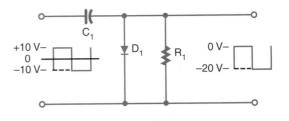

Figure 31–27 Diode clamper circuit.

Sometimes it is desirable to change the DC reference level of a waveform. The *DC reference level* is the starting point from which measurements are made. A **clamping circuit** can be used to clamp the top or bottom of the waveform to a given DC voltage. Unlike a clipper or limiter circuit, a clamping circuit does not change the shape of the waveform. A diode clamper (Figure 31–27) is also called a DC restorer. This circuit is commonly used in radar, television, telecommunications, and computers. In the circuit shown, a square wave is applied to an input signal. The purpose of the circuit is to clamp the top of the square wave to 0 volts, without changing the shape of the waveform.

31–2 Questions

1. Draw schematic diagrams of the following RC networks:
   a. Differentiator
   b. Integrator
2. What are the functions of integrator and differentiator circuits?
3. Draw schematic diagrams of the following circuits:
   a. Clipper
   b. Clamper
4. What are the functions of clipper and clamper circuits?
5. What are applications of the following circuits:
   a. Differentiator
   b. Integrator
   c. Clipper
   d. Clamper

31—3 Special-Purpose Circuits

The prefix *mono-* means *one*. A **monostable multivibrator** has only one stable state. It is also called a **one-shot multivibrator** because it produces one output pulse for each input pulse. The output pulse is generally longer than the input pulse. Therefore, this circuit is also called pulse stretcher. Typically, the circuit is used as a gate in computers, electronic control circuits, and communication equipment.

Figure 31–28 shows a schematic diagram of a monostable multivibrator. The circuit is normally in its stable state. When it receives an input trigger pulse, it switches to its unstable state. The length of time the circuit is in the unstable state is determined by the RC time constant of resistor R_2 and capacitor C_1. Capacitor C_2 and resistor R_5 form a differentiator circuit, which is used to convert the input pulse to a positive and negative spike. Diode D_1 allows only the negative spike to pass through to turn on the circuit.

A **bistable multivibrator** is a multivibrator having two stable states (*bi-* meaning *two*). This circuit requires two inputs to complete one cycle. A pulse at one input sets the circuit to one of its stable states. A pulse at the other input resets it to its

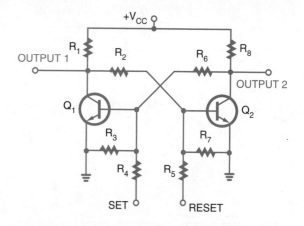

Figure 31–29 Basic flip-flop circuit.

other stable state. This circuit is often called a **flip-flop** because of its mode of operation.

A basic flip-flop circuit produces a square or rectangular waveform for use in gating or timing signals or for on-off switching operations in binary counter circuits (Figure 31–29). A binary counter circuit is essentially two transistor amplifiers with the output of each transistor coupled to the input of the other transistor. When an input signal is applied to the set input, transistor Q_1 turns on, which turns transistor Q_2 off. When transistor Q_2 turns off, it places a positive potential on the base of transistor Q_1, holding it on. If a pulse is not applied to the reset input, it causes transistor Q_2 to conduct, turning off transistor Q_1. Turning transistor Q_1 off holds transistor Q_2 on.

Discrete versions of the flip-flop find little application today. However, integrated circuit versions of the flip-flop find wide application. It is perhaps the most important circuit in digital electronics, used for frequency division, data storage, counting, and data manipulation.

Another bistable circuit is the **Schmitt trigger** (Figure 31–30). One application of the Schmitt trigger is to convert a sine-wave, sawtooth, or other irregularly shaped waveform to a square or rectangular wave. The circuit differs from a conventional bistable multivibrator in that one of the coupling networks is replaced by a common-emitter resistor (R_3). This provides additional

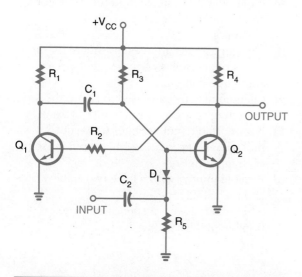

Figure 31–28 Monostable multivibrator.

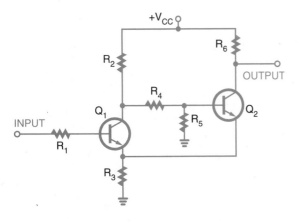

Figure 31–30 Basic Schmitt trigger circuit.

regeneration for quicker action and straighter leading and trailing edges on the output waveform.

31–3 Questions

1. What is a monostable multivibrator?
2. Draw a schematic diagram of a one-shot multivibrator.
3. What is a bistable multivibrator?
4. Draw a schematic diagram of a flip-flop.
5. How does a Schmitt trigger differ from a conventional bistable multivibrator?

SUMMARY

■ Waveforms can be changed from one shape to another using various electronic circuits.
■ The frequency-domain concept holds that all periodic waveforms are made of sine waves.
■ Periodic waveforms have the same waveshape in all cycles.
■ Sine waves are the only waveform that cannot be distorted by RC, RL, or LC circuits.
■ According to the frequency-domain concept, waveforms consist of the fundamental frequency plus combinations of even or odd harmonics or both.
■ A square wave consists of the fundamental plus an infinite number of odd harmonics.
■ A sawtooth wave consists of the fundamental plus even and odd harmonics crossing the zero reference line 180 degrees out of phase with the fundamental.
■ Periodic waveforms are measured from any point on a cycle to the same point on the next cycle.
■ The pulse width is the length of the pulse.
■ The duty cycle is the ratio of the pulse width to the period.
■ The rise time of a pulse is the time it takes to fall from 10% to 90% of its maximum amplitude.
■ The fall time of a pulse is the time it takes to fall from 90% to 10% of its maximum amplitude.
■ Overshoot, undershoot, and ringing are undesirable in a circuit and exist because of imperfect circuits.
■ An RC circuit can be used to change the shape of a complex waveform.
■ If the output is taken across the resistor in an RC circuit, the circuit is called a differentiator.
■ If the output is taken across the capacitor in an RC circuit, the circuit is called an integrator.
■ Clipping circuits are used to square off the peaks of an applied signal or to keep an amplitude constant.
■ Clamping circuits are used to clamp the top or bottom of a waveform to a DC voltage.
■ A monostable multivibrator (one-shot multivibrator) produces one output pulse for each input pulse.
■ Bistable multivibrators have two stable states and are called flip-flops.
■ A Schmitt trigger is a special-purpose bistable multivibrator.

Chapter 31 Self-Test

1. Describe the frequency-domain concept of waveforms.
2. Why do such problems as overshoot, undershoot, and ringing occur in waveshaping circuits?
3. Describe where integrator and differentiator waveshaping circuits are used.
4. How can the DC reference level of a signal be changed?
5. Explain the difference between monostable and bistable circuit functions.
6. Of what significance is the flip-flop?

Digital Electronic Circuits

32

BINARY NUMBER SYSTEM

Objectives

After completing this chapter, the student will be able to:

- Describe the binary number system.
- Identify the place value for each bit in a binary number.
- Convert binary numbers to decimal numbers.
- Convert decimal numbers to binary numbers.
- Convert decimal numbers to 8421 BCD code.
- Convert 8421 BCD code numbers to decimal numbers.

A number system is nothing more than a code. For each distinct quantity, there is an assigned symbol. When the code is learned, counting can be accomplished. This leads to arithmetic and higher forms of mathematics.

The simplest number system is the **binary number system.** The *binary system* contains only two digits, 0 and 1. These digits have the same value as in the decimal number system.

The binary number system is used in digital and microprocessor circuits because of its simplicity. Binary data are represented by binary digits, called **bits.** The term *bit* is derived from *binary digit.*

32–1 Binary Numbers

The decimal system is called a base-10 system because it contains ten digits, 0 through 9. The binary system is a base-two system because it contains two digits, 0 and 1. The position of the 0 or 1 in a binary number indicates its value within the number. This is referred to as its place value or weight. The place value of the digits in a binary number increases by powers of two.

$$\begin{array}{ccccccc} & \multicolumn{6}{c}{\text{Place Value}} \\ & 32 & 16 & 8 & 4 & 2 & 1 \\ \text{Power of 2:} & 2^5 & 2^4 & 2^3 & 2^2 & 2^1 & 2^0 \end{array}$$

Counting in binary starts with the numbers 0 and 1. When each digit has been used in the 1's place, another digit is added in the 2's place and the count continues with 10 and 11. This uses up all combinations of two digits, so a third digit is added in the 4's place and the count continues with 100, 101, 110, and 111. Now a fourth digit is needed in the 8's place to continue, and so on. Figure 32–1 shows the binary counting sequence.

To determine the largest value that can be represented by a given number of places in base 2, use the following formula:

$$\text{Highest number} = 2^n - 1$$

where: n represents the number of bits (or number of place values used)

EXAMPLE Two bits (two place values) can be used to count from 0 to 3 because:

$$2^n - 1 = 2^2 - 1 = 4 - 1 = 3$$

Four bits (four place values) are needed to count from 0 to 15 because:

$$2^n - 1 = 2^4 - 1 = 16 - 1 = 15$$

32–1 Questions

1. What is the advantage of the binary number system over the decimal number system for digital circuits?

DECIMAL AND BINARY EQUIVALENTS TABLE

Decimal Number	Binary Number				
	2^4 16	2^3 8	2^2 4	2^1 2	2^0 1
0	0	0	0	0	0
1	0	0	0	0	1
2	0	0	0	1	0
3	0	0	0	1	1
4	0	0	1	0	0
5	0	0	1	0	1
6	0	0	1	1	0
7	0	0	1	1	1
8	0	1	0	0	0
9	0	1	0	0	1
10	0	1	0	1	0
11	0	1	0	1	1
12	0	1	1	0	0
13	0	1	1	0	1
14	0	1	1	1	0
15	0	1	1	1	1
16	1	0	0	0	0
17	1	0	0	0	1
18	1	0	0	1	0
19	1	0	0	1	1
20	1	0	1	0	0
21	1	0	1	0	1
22	1	0	1	1	0
23	1	0	1	1	1
24	1	1	0	0	0
25	1	1	0	0	1

Figure 32–1 Decimal numbers and equivalent binary numbers. *(From Sullivan,* Modern Electronics Mathematics, *by Delmar Publishers.)*

2. How is the largest value of a binary number determined for a given number of place values?
3. What is the maximum value of a binary number with:
   a. 4 bits
   b. 8 bits
   c. 12 bits
   d. 16 bits

32–2 Binary and Decimal Conversion

As stated, a binary number is a weighted number with a place value. The value of a binary number can be determined by adding the product of each digit and its place value. The method for evaluating a binary number is shown by the following example:

EXAMPLE

$$
\begin{array}{cccccc}
\text{Place Value} \\
32 & 16 & 8 & 4 & 2 & 1 \\
\end{array}
$$

Binary number: 1   0   1   1   0   1

Value:

$$
\begin{aligned}
1 \times 32 &= 32 \\
0 \times 16 &= 0 \\
1 \times 8 &= 8 \\
1 \times 4 &= 4 \\
0 \times 2 &= 0 \\
+ 1 \times 1 &= 1 \\
\hline
101101_2 &= 45_{10}
\end{aligned}
$$

The number 45 is the decimal equivalent of the binary number 101101.

Fractional numbers can also be represented in binary form by placing digits to the right of the binary zero point, just as decimal numbers are placed to the right of the decimal zero point. All digits to the right of the zero point have weights that are negative powers of two, or fractional place values.

$$
\begin{array}{cc}
\text{Power} & \text{Place} \\
\text{of 2} & \text{value} \\
2^5 = 32 \\
2^4 = 16 \\
2^3 = 8 \\
2^2 = 4 \\
2^1 = 2 \\
2^0 = 1 \\
\end{array}
$$

decimal point

$$
2^{-1} = \frac{1}{2^1} = \frac{1}{2} = 0.5
$$

$$
2^{-2} = \frac{1}{2^2} = \frac{1}{4} = 0.25
$$

$$
2^{-3} = \frac{1}{2^3} = \frac{1}{8} = 0.125
$$

$$
2^{-4} = \frac{1}{2^4} = \frac{1}{16} = 0.0625
$$

EXAMPLE Determine the decimal value of the binary number 111011.011.

Binary number	Place value	Value
1	$\times 32$	$= 32$
1	$\times 16$	$= 16$
1	$\times 8$	$= 8$
0	$\times 4$	$= 0$
1	$\times 2$	$= 2$
1	$\times 1$	$= 1$
0	$\times 0.5$	$= 0$
1	$\times 0.25$	$= 0.25$
+ 1	$\times 0.125$	$= 0.125$

$$
111011.011_2 = 59.375
$$

In working with digital equipment, it is often necessary to convert from binary to decimal form and vice versa. The most popular way to convert decimal numbers to binary numbers is to progressively divide the decimal number by 2, writing down the remainder after each division. The remainders, taken in reverse order, form the binary number.

EXAMPLE To convert 11 to a binary number, progressively divide by 2. (Least Significant Bit)

$$
\begin{aligned}
11 \div 2 &= 5 \text{ with a remainder of 1}\quad \text{LSB} \\
5 \div 2 &= 2 \text{ with a remainder of 1} \\
2 \div 2 &= 1 \text{ with a remainder of 0} \\
1 \div 2 &= 0 \text{ with a remainder of 1}
\end{aligned}
$$

(1/2 = 0 means that 2 will no longer divide into 1, so 1 is the remainder) Decimal 11 is equal to 1011 in binary.

The process can be simplified by writing the numbers in an orderly fashion as shown for converting 25 to a binary number.

EXAMPLE

$$
\begin{array}{cccc}
2 & 25 & & \text{LSB} \\
2 & 12 & 1 \\
2 & 6 & 0 \\
2 & 3 & 0 \\
2 & 1 & 1 \\
& 0 & 1 \\
\end{array}
$$

Decimal number 25 is equal to binary number 11001.

Fractional numbers are done a little differently: The number is multiplied by 2 and the carry is recorded as the binary fraction.

EXAMPLE To convert decimal 0.85 to a binary fraction, progressively multiply by 2.

LSB

$0.85 \times 2 = 1.70 = 0.70$ with a carry of 1
$0.70 \times 2 = 1.40 = 0.40$ with a carry of 1
$0.40 \times 2 = 0.80 = 0.80$ with a carry of 0
$0.80 \times 2 = 1.60 = 0.60$ with a carry of 1
$0.60 \times 2 = 1.20 = 0.20$ with a carry of 1
$0.20 \times 2 = 0.40 = 0.40$ with a carry of 0

Continue to multiply by 2 until the needed accuracy is reached. Decimal 0.85 is equal to 0.110110 in binary form.

EXAMPLE Convert decimal 20.65 to a binary number. Split 20.65 into an integer of 20 and a fraction of 0.65 and apply the methods previously shown.

```
2   20          LSB
  2   10        0
    2    5      0
      2    2    1
        2    1  0
          0    1
```
Decimal 20 = Binary 10100

and

LSB

$0.65 \times 2 = 1.30 = 0.30$ with a carry of 1
$0.30 \times 2 = 0.60 = 0.60$ with a carry of 0
$0.60 \times 2 = 1.20 = 0.20$ with a carry of 1
$0.20 \times 2 = 0.40 = 0.40$ with a carry of 0
$0.40 \times 2 = 0.80 = 0.80$ with a carry of 0
$0.80 \times 2 = 1.60 = 0.60$ with a carry of 1
$0.60 \times 2 = 1.20 = 0.20$ with a carry of 1

Decimal 0.65 = Binary 0.1010011

Combining the two numbers results in 20.65_{10} = 10100.1010011_2. This 12-bit number is an approximation, because the conversion of the fraction was terminated after 7 bits.

32—2 Questions

1. What is the value of each position in an 8-bit binary number?
2. What is the value of each position to the right of the decimal point for 8 places?
3. Convert the following binary numbers to decimal numbers:
   a. 1001
   b. 11101111
   c. 11000010
   d. 10101010.1101
   e. 10110111.0001
4. What is the process for converting decimal numbers to binary digits?
5. Convert the following decimal numbers to binary form:
   a. 27
   b. 34.6
   c. 346
   d. 321.456
   e. 7465

32—3 BCD Code

An *8421 code* is a **binary-coded-decimal (BCD)** code consisting of four binary digits. It is used to represent the digits 0 through 9. The 8421 designation refers to the binary weight of the 4 bits.

Powers of 2:	2^3	2^2	2^1	2^0
Binary weight:	8	4	2	1

The main advantage of this code is that it permits easy conversion between decimal and binary form. This is the predominant BCD code used and is the one referred to unless otherwise stated.

Each decimal digit (0 through 9) is represented by a binary combination as follows:

Decimal	8421 code
0	0000
1	0001
2	0010
3	0011
4	0100
5	0101

6	0110
7	0111
8	1000
9	1001

Although sixteen numbers (2^4) can be represented by four binary positions, the six code combinations above decimal 9 (1010, 1011, 1100, 1101, 1110, and 1111) are invalid in the 8421 code.

To express any decimal number in the 8421 code, replace each decimal digit by the appropriate 4-bit code.

EXAMPLE Convert the following decimal numbers into a BCD code: 5, 13, 124, 576, 8769.

$$5 = 0101$$
$$13 = 0001\ 0011$$
$$124 = 0001\ 0010\ 0100$$
$$576 = 0101\ 0111\ 0110$$
$$8769 = 1000\ 0111\ 0110\ 1001$$

To determine a decimal number from an 8421 code number, break the code into groups of 4 bits. Then write the decimal digit represented by each 4-bit group.

EXAMPLE Find the decimal equivalent for each of the following BCD codes: 10010101, 1001000, 1100111, 1001100101001, 1001100001110110.

$$1001\ 0101 = 95$$
$$0100\ 1000 = 48$$
$$0110\ 0111 = 67$$
$$0001\ 0011\ 0010\ 1001 = 1329$$
$$1001\ 1000\ 0111\ 0110 = 9876$$

Note: If there is an insufficient number of bits in the group furthest to the left, zeros are implied.

32—3 Questions

1. What is the 8421 code and how is it used?
2. Convert the following decimal numbers into BCD code:
   a. 17
   b. 100
   c. 256
   d. 778
   e. 8573
3. Convert the following BCD codes into decimal numbers:
   a. 1000 0010
   b. 0111 0000 0101
   c. 1001 0001 0011 0100
   d. 0001 0000 0000 0000
   e. 0100 0110 1000 1001

SUMMARY

- The binary number system is the simplest number system.
- The binary number system contains two digits, 0 and 1.
- The binary number system is used to represent data for digital and computer systems.
- Binary data are represented by binary digits called bits.
- The term *bit* is derived from *binary digit*.
- The place value of each higher digit's position in a binary number is increased by a power of 2.
- The largest value that can be represented by a given number of places in base 2 is $2^n - 1$, where n represents the number of bits.
- The value of a binary digit can be determined by adding the product of each digit and its place value.
- Fractional numbers are represented by negative powers of 2.
- To convert from a decimal number to a binary number, divide the decimal number by 2, writing down the remainder after each division. The remainders, taken in reverse order, form the binary number.
- The 8421 code, a binary-coded-decimal (BCD) code, is used to represent digits 0 through 9.
- The advantage of the BCD code is ease of converting between decimal and binary forms of a number.

Chapter 32 Self-Test

1. What represents the decimal equivalent of 0 through 27 in binary?
2. How many binary bits are required to represent the decimal number 100?
3. Describe the process for converting a decimal number to a binary number.
4. Convert the following binary numbers to their decimal equivalents.
   a. 100101.001011
   b. 111101110.11101110
   c. 10000001.00000101
5. Describe the process for converting decimal numbers to BCD.
6. Convert the following BCD numbers to their decimal equivalents.
   a. 0100 0001 0000 0110
   b. 1001 0010 0100 0011
   c. 0101 0110 0111 1000

Chapter

33

BASIC LOGIC GATES

Objectives

After completing this chapter, the student will be able to:

- Identify and explain the function of the basic logic gates.
- Draw the symbols for the basic logic gates.
- Develop truth tables for the basic logic gates.

All digital equipment, whether simple or complex, is constructed of only a few basic circuits. These circuits, referred to as *logic elements*, perform some logic function on binary data.

There are two basic types of logic circuits: decision-making and memory. *Decision-making logic circuits* monitor binary inputs and produce an output based on the status of the inputs and the characteristics of the logic circuit. *Memory circuits* are used to store binary data.

33–1 AND Gate

The **AND gate** is a logic circuit that has two or more inputs and a single output. The AND gate produces an output of 1, only when all its inputs are 1's. If any of the inputs are 0's, the output is 0.

Figure 33–1 shows the standard symbol used for AND gates. An AND gate can have any number of inputs greater than one. Shown in the figure are symbols representing the more commonly used gates of two, three, four, and eight inputs.

The operation of the AND gate is summarized by the table in Figure 33–2. Such a table, called a *truth table,* shows the output for each possible input. The inputs are designated A and B. The output is designated Y. The total number of possible combinations in the truth table is determined by the following formula:

$$N = 2^n$$

where: N = the total number of possible combinations

n = the total number of input variables

EXAMPLE

For two input variables, $N = 2^2 = 4$

For three input variables, $N = 2^3 = 8$

For four input variables, $N = 2^4 = 16$

For eight input variables, $N = 2^8 = 256$

The AND gate performs the basic operation of multiplication. Multiplication is known as the **AND** function. The output of an AND gate is represented by the equation $Y = A \cdot B$ or $Y = AB$. The AND function is represented by the dot between the two variables A and B.

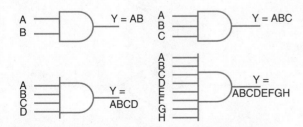

Figure 33–1 Logic symbol for an AND gate.

INPUTS		OUTPUT
A	B	Y
0	0	0
1	0	0
0	1	0
1	1	1

Figure 33–2 Truth table for a two-input AND gate.

33–1 Questions

1. Under what conditions does an AND gate produce a 1 output?
2. Draw the symbol used to represent a two-input AND gate.
3. Develop a truth table for a three-input AND gate.
4. What logical operation is performed by an AND gate?
5. What is the algebraic output of an AND gate?

33–2 OR Gate

An **OR gate** produces a 1 output if any of its inputs are 1's. The output is a 0 if all the inputs are 0's. The output of a two-input OR gate is shown in the truth table in Figure 33–3. The total number of possible combinations is expressed by $N = 2^2 = 4$. The truth table shows all four combinations.

INPUTS		OUTPUT
A	B	Y
0	0	0
1	0	1
0	1	1
1	1	1

Figure 33–3 Truth table for a two-input OR gate.

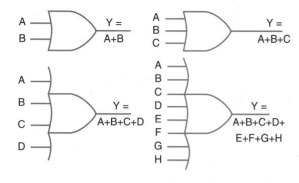

Figure 33–4 Logic symbol for an OR gate.

An OR gate performs the basic operation of addition. The algebraic expression for the output of an OR gate is Y = A + B. The plus sign designates the OR function.

Figure 33–4 shows the logic symbol for an OR gate. The inputs are labeled A and B, and the output is labeled Y. An OR gate can have any number of inputs greater than one. Shown in the figure are OR gates with two, three, four, and eight inputs.

33–2 Questions

1. What conditions produce a 1 output for an OR gate?
2. Draw the symbol used to represent a two-input OR gate.
3. Develop a truth table for a three-input OR gate.
4. What operation is performed by an OR gate?
5. What algebraic expression represents the output of an OR gate?

33–3 NOT Gate

The simplest logic circuit is the **NOT circuit.** It performs the function called inversion, or complementation, and is commonly referred to as an *inverter.*

The purpose of the inverter is to make the output state the opposite of the input state. The two states associated with logic circuits are 1 and 0. A

INPUTS	OUTPUT
A	Y
0	1
1	0

Figure 33–5 Truth table for an inverter.

1 state can also be referred to as a *high,* to indicate that the voltage is higher than in the 0 state. A 0 state can also be referred to as a *low,* to indicate that the voltage is lower than in the 1 state. If a 1, or high, is applied to the input of an inverter, a low, or 0, appears on its output. If a 0, or low, is applied to the input, a 1, or high, appears on its output.

The operation of an inverter is summarized in Figure 33–5. The input to an inverter is labeled A and the output is labeled $\overline{A}$ (read "A NOT" or "NOT A"). The bar over the letter A indicates the complement of A. Because the inverter has only one input, only two input combinations are possible.

The symbol used to represent an inverter or NOT function is shown in Figure 33–6. The triangle portion of the symbol represents the circuit, and the circle or "bubble" represents the circuit inversion or complementary characteristic. The choice of symbol depends on where the inverter is used. If the inverter uses a 1 as the qualifying input, the symbol in Figure 33–6A is used. If the inverter uses a 0 as the qualifying input, the symbol in Figure 33–6B is used.

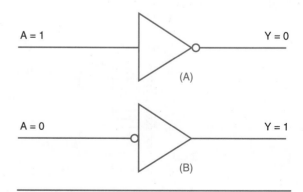

Figure 33–6 Logic symbol for an inverter.

33–3 Questions

1. What operation is performed by a NOT circuit?
2. Develop a truth table for the NOT circuit.
3. Draw the symbols used to represent the NOT circuit.
4. Why are two different symbols used to represent the NOT circuit?

33–4 NAND Gate

A **NAND gate** is a combination of an inverter and an AND gate. It is called a NAND gate from the NOT-AND function it performs. The NAND gate is the most commonly used logic function. This is because it can be used to construct an AND gate, OR gate, inverter, or any combination of these functions.

The logic symbol for a NAND gate is shown in Figure 33–7. Also shown is its equivalency to an AND gate and an inverter. The bubble on the output end of the symbol means to invert the AND function.

Figure 33–8 shows the truth table for a two-input NAND gate. Notice that the output of the NAND gate is the complement of the output of an AND gate. Any 0 in the input yields a 1 output.

The algebraic formula for NAND-gate output is $Y = \overline{AB}$, where Y is the output and A and B are the inputs. NAND gates are available with two, three, four, eight, and thirteen inputs.

INPUTS		OUTPUT
A	B	Y
0	0	1
1	0	1
0	1	1
1	1	0

Figure 33–8 Truth table for a two-input NAND gate.

NAND gates are the most widely available gates on the market. The availability and flexibility of the NAND gate allows it to be used for other types of gates. Figure 33–9 shows how a NAND gate can be used to generate other logic functions.

33–4 Questions

1. What is a NAND gate?
2. Why is the NAND gate so often used in circuits?
3. Draw the logic symbol used to represent a NAND gate.
4. What is the algebraic expression for a NAND gate?
5. Develop a truth table for a three-input NAND gate.

33–5 NOR Gate

A **NOR gate** is a combination of an inverter and an OR gate. Its name derives from its NOT-OR function. Like the NAND gate, the NOR gate can also be used to construct an AND gate, an OR gate, and an inverter.

The logic symbol for the NOR gate is shown in Figure 33–10. Also shown is its equivalency to an OR gate and an inverter. The bubble on the output of the symbol means to invert the OR function.

Figure 33–11 shows the truth table for a two-input NOR gate. Notice that the output is the complement of the OR-function output. A 1 occurs only

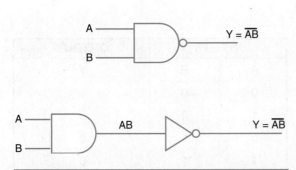

Figure 33–7 Logic symbol for a NAND gate.

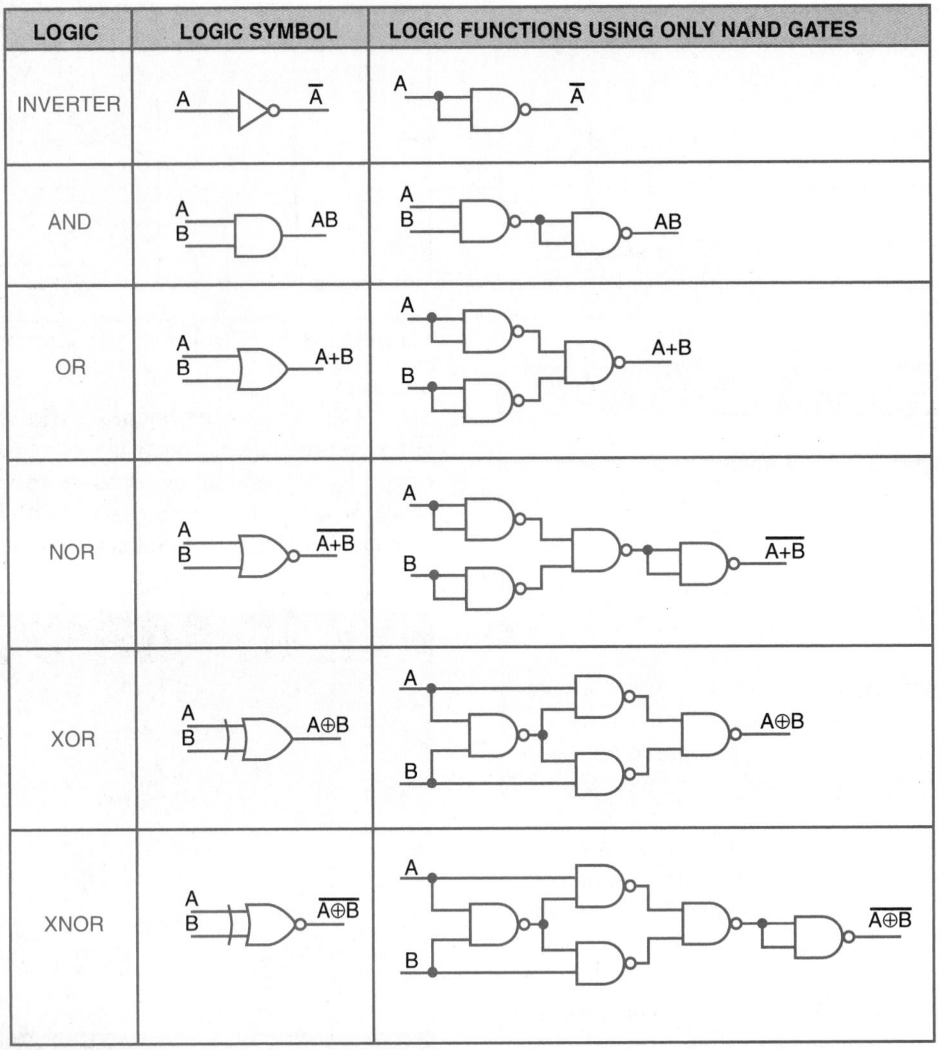

LOGIC	LOGIC SYMBOL	LOGIC FUNCTIONS USING ONLY NAND GATES
INVERTER		
AND		
OR		
NOR		
XOR		
XNOR		

Figure 33–9 Using the NAND gate to generate other logic functions.

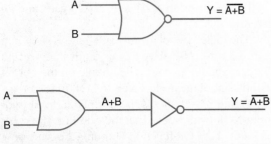

Figure 33–10 Logic symbol for a NOR gate.

INPUTS		OUTPUT
A	B	Y
0	0	1
1	0	0
0	1	0
1	1	0

Figure 33–11 Truth table for a two-input NOR gate.

when 0 is applied to both inputs. A 1 input produces a 0 output.

The algebraic expression for NOR-gate output is $Y = \overline{A + B}$, where Y is the output and A and B are the inputs. NOR gates are available with two, three, four, and eight inputs.

33–5 Questions

1. What is a NOR gate?
2. Why is the NOR gate useful in designing digital circuits?
3. Draw the symbol used to represent a NOR gate.
4. What is the algebraic expression for a NOR gate?
5. Develop a truth table for a three-input NOR gate.

33–6 Exclusive OR and NOR Gates

A less common but still important gate is called an **exclusive OR gate,** abbreviated as XOR. An XOR gate has only two inputs, unlike the OR gate, which may have several inputs. However, the XOR is similar to the OR gate in that it generates a 1 output if either input is a 1. The exclusive OR is different when both inputs are 1's or 0's. In that case, the output is a 0.

The symbol for an XOR gate is shown in Figure 33–12. Also shown is the equivalent logic circuit. Figure 33–13 shows the truth table for an exclusive OR gate. The algebraic output is written as $Y = A \oplus B$. It is read as "Y equals A exclusive or B."

The complement of the XOR gate is the XNOR (**exclusive NOR**) gate. Its symbol is shown in Figure 33–14. The bubble on the output implies inversion or complement. Also shown is the equivalent logic circuit. Figure 33–15 shows the truth table for an exclusive NOR gate. The alge-

Figure 33–12 Logic symbol for an exclusive OR gate.

INPUTS		OUTPUT
A	B	Y
0	0	0
1	0	1
0	1	1
1	1	0

Figure 33–13 Truth table for an exclusive OR gate.

Figure 33–14 Logic symbol for an exclusive NOR gate.

INPUTS		OUTPUT
A	B	Y
0	0	1
1	0	0
0	1	0
1	1	1

Figure 33–15 Truth table for an exclusive NOR gate.

braic output is written as $Y = \overline{A \oplus B}$, read as "Y equals A exclusive nor B."

33–6 Questions

1. What is the difference between an OR gate and an XOR gate?
2. Draw the symbol used to represent an XOR gate.
3. Develop a truth table for an XOR gate.
4. Draw the symbol used to represent an XNOR gate.
5. Write the algebraic expressions for XOR and XNOR gates.

SUMMARY

■ An AND gate produces a 1 output when all of its inputs are 1's.

■ An AND gate performs the basic operation of multiplication.

■ An OR gate produces a 1 output if any of its inputs are 1's.

■ An OR gate performs the basic operation of addition.

■ A NOT gate performs the function called inversion or complementation.

■ A NOT gate coverts the input state to an opposite output state.

■ A NAND gate is a combination of an AND gate and an inverter.

■ A NAND gate produces 1 output when any of the inputs are 0's.

■ A NOR gate is a combination of an OR gate and an inverter.

■ A NOR gate produces a 1 output only when both inputs are 0's.

■ An exclusive OR (XOR) gate produces a 1 output only if both inputs are different.

■ An exclusive NOR (XNOR) gate produces a 1 output only when both inputs are the same.

Chapter 33 Self-Test

1. Draw the schematic symbol for a six-input AND gate.
2. Develop the truth table for a four-input AND gate.
3. Draw the schematic symbol for a six-input OR gate.
4. Develop the truth table for a four-input OR gate.
5. What is the purpose for the NOT circuit?
6. How does an inverter for an input signal differ from the inverter for an output signal?
7. Draw the schematic symbol for an eight-input NAND gate.
8. Develop the truth table for a four-input NAND gate.
9. Draw the schematic symbol for an eight-input NOR gate.
10. Develop the truth table for a four-input NOR gate.
11. What is the significance of the XOR gate?
12. The XNOR gate has what maximum number of inputs?

Chapter

34

SIMPLIFYING LOGIC CIRCUITS

Objectives

After completing this chapter, the student will be able to:

■ Explain the function of Veitch diagrams.

■ Describe how to use a Veitch diagram to simplify Boolean expressions.

Digital circuits are being used more and more in electronics, not only in computers, but also in applications such as measurement, automatic control, robotics, and in situations requiring decisions. All of these applications require complex switching circuits that are formed from the five basic logic gates: the AND, OR, NAND, and NOR gates and the inverter.

The significant point about these logic gates is that they only have two operating conditions. They are either ON (1) or OFF (0). When logic gates are interconnected to form more complex circuits, it is necessary to obtain the simplest circuit possible.

Boolean algebra offers a means of expressing complex switching functions in equation form. A Boolean expression is an equation that expresses the output of a logic circuit in terms of its input. Veitch diagrams provide a fast and easy way to reduce a logic equation to its simplest form.

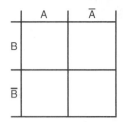 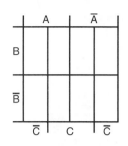 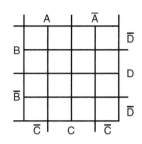

Figure 34–1 Two-, three-, four-variable Veitch diagrams.

34–1 Veitch Diagrams

Veitch diagrams provide a fast and easy method for reducing a complicated expression to its simplest form. They can be constructed for two, three, or four variables. Figure 34–1 shows several Veitch diagrams.

To use a Veitch diagram, follow these steps, as illustrated in the example.

1. Draw the diagram based on the number of variables.
2. Plot the logic functions by placing an X in each square representing a term.
3. Obtain the simplified logic function by looping adjacent groups of X's in groups of eight, four, or two. Continue to loop until all X's are included in a loop.
4. "OR" the loops with one term per loop. (Each expression is pulled off the Veitch diagram and ORed using the "+" symbol, e.g., ABC + BCD.)
5. Write the simplified expression.

EXAMPLE Reduce $AB + \overline{A}B + A\overline{B} = Y$ to its simplest form.

Step 1. Draw the Veitch diagram. There are two variables, A and B, so use the two-variable chart.

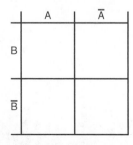

Step 2. Plot the logic function by placing an X in each square representing a term.

$$AB + \overline{A}B + A\overline{B}$$

1st	2nd	3rd
term	term	term

Plot 1st term AB

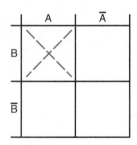

Plot 2nd term $\overline{A}B$

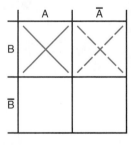

Plot 3rd term $A\overline{B}$

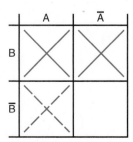

Step 3. Loop adjacent groups of X's in the largest group possible.

Start by analyzing chart for largest groups possible. The largest group possible here is two.

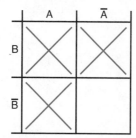

One possible group is the one indicated by the dotted line.

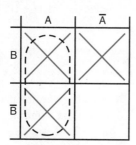

Another group is the one indicated by this dotted line.

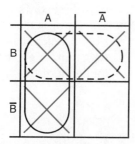

Step 4. "OR" the groups: either A or B = A + B

Step 5. The simplified expression for AB + $\overline{A}$B + A$\overline{B}$ = Y is A + B = Y obtained from the Veitch diagram.

EXAMPLE Find the simplified expression for

$$ABC + AB\overline{C} + A\overline{B}\,\overline{C} + \overline{A}\,\overline{B}\,\overline{C} = Y$$

Step 1. Draw a three-variable Veitch diagram.

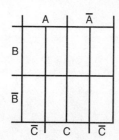

Step 2. Place an X for each term on the Veitch diagram.

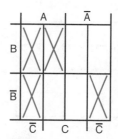

Step 3. Loop the groups.

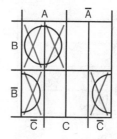

Step 4. Write the term for each loop, one term per expression:

$$AB, \quad \overline{B}\,\overline{C}$$

Step 5. The simplified expression is AB + $\overline{B}\,\overline{C}$ = Y.

Notice the unusual looping on the two bottom squares. The four corners of the Veitch diagram are considered connected as if the diagram were formed into a ball.

EXAMPLE Find the simplified expression for:

$$\overline{A}BCD + \overline{A}\,\overline{B}CD + \overline{A}\,\overline{B}\,\overline{C}D + A\overline{B}\,\overline{C}\,\overline{D} +$$
$$A\overline{B}\,\overline{C}D + \overline{A}B\overline{C}D = Y$$

Step 1. Draw a four-variable Veitch diagram.

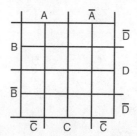

Step 2. Place an X for each term on the Veitch diagram.

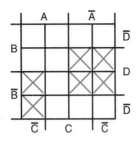

Step 3. Loop the groups.

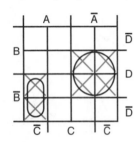

Step 4. Write the term for each loop, one term per expression: $\overline{A}D$, $A\overline{B}\,\overline{C}$

Step 5. "OR" the terms to form the simplified expression: $\overline{A}D + A\overline{B}\,\overline{C} = Y$

SUMMARY

■ Veitch diagrams provide a fast and easy way to reduce complicated expressions to their simplest form.

■ Veitch diagrams can be constructed from two, three, or four variables.

■ The simplest logic expression is obtained from a Veitch diagram by looping groups of two, four, or eight X's and "OR"ing the looped terms.

Chapter 34 Self-Test

1. Describe the procedure for using the Veitch diagram to simplify logic circuits.
2. Simplify the following Boolean expression by using a Veitch diagram.
   $\overline{A}BC\overline{D} + AB\overline{C}D + \overline{A}CD + A\overline{B}C + \overline{A}\,\overline{B} + AB\overline{C}\,\overline{D} = Y$

35

SEQUENTIAL LOGIC CIRCUITS

Objectives

After completing this chapter, the student will be able to:

- Describe the function of a flip-flop.
- Identify the basic types of flip-flop.
- Draw the symbols used to represent flip-flops.
- Describe how flip-flops are used in digital circuits.
- Describe how a counter and shift register operate.
- Identify the different types of counters and shift registers.
- Draw the symbols used to represent counters and shift registers.
- Identify applications of counters and shift registers.

Sequential logic circuits consist of circuits requiring timing and memory devices. The basic building block for sequential logic circuits is the flip-flop. Flip-flops can be wired together to form counters, shift registers, and memory devices.

The flip-flop belongs to a category of digital circuits called multivibrators. A multivibrator is a regenerative circuit with two active devices. It is designed so that one device conducts while the other device is cut off. Multivibrators can store binary numbers, count pulses, synchronize arithmetic operations, and perform other essential functions in digital systems.

There are three types of multivibrators: bistable, monostable, and astable. The bistable multivibrator is called a flip-flop.

35–1 Flip-Flops

A **flip-flop** is a bistable multivibrator whose output is either a high or low voltage, a 1 or a 0. The output stays high or low until an input called a trigger is applied.

The basic flip-flop is the *RS flip-flop*. It is formed by two cross-coupled NOR or NAND gates (Figure 35–1). The **RS flip-flop** has two outputs, Q and $\overline{Q}$, and two controlling inputs, R (Reset) and S (Set). The outputs are always opposite or complementary: If Q = 1, then $\overline{Q}$ = 0, and vice-versa.

To understand the operation of the circuit, assume that the Q output, R input, and S input are all low. The low on the Q output is connected to one of the inputs of gate 2. The S input is low. The output of gate 2 is high. This high is coupled to the input of gate 1, holding its output to a low. When the Q output is low, the flip-flop is said to be in the RESET state. It remains in this state indefinitely, until a high is applied to the S input of gate 2. When a high is applied to the S input of gate 2, the output of gate 2 becomes a low and is coupled to the input of gate 1. Because the R input of gate 1 is a low, the output changes to a high. The high is coupled back to the input of gate 2, ensuring that the $\overline{Q}$ output remains a low. When the Q output is high, the flip-flop is said to be in the SET state. It remains in the SET state until a high is applied to the R input, causing the flip-flop to RESET.

INPUTS		OUTPUTS	
S	R	Q	$\overline{Q}$
0	0	NC	NC
0	1	0	1
1	0	1	0
1	1	0	0

NC = No Change

Figure 35–2 Truth table for an RS flip-flop.

An "illegal" or "unallowed" condition occurs when a high is applied to both the R and S inputs simultaneously. In this case, the Q and $\overline{Q}$ outputs both try to go low, but Q and $\overline{Q}$ cannot be in the same state at the same time without violating the definition of flip-flop operation. When the highs on the R and S inputs are removed simultaneously, both of the outputs attempt to go high. Because there is always some difference in the gates, one gate dominates and becomes high. This forces the other gate to remain low. An unpredictable mode of operation exists and therefore the output state of the flip-flop cannot be determined.

Figure 35–2 shows the truth table for operation of an RS flip-flop. Figure 35–3 is a simplified symbol used to represent an RS flip-flop.

Another type of flip-flop is called a **clocked flip-flop.** It is different from the RS flip-flop in that an additional input is required for operation. The third input is called the clock or trigger. Figure 35–4 shows a logic diagram for a clocked flip-flop. A

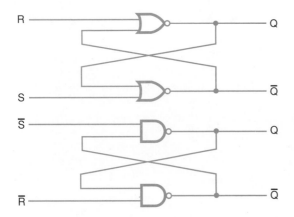

Figure 35–1 Basic flip-flop circuit.

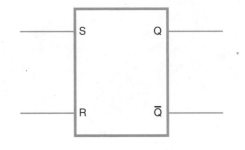

Figure 35–3 Logic symbol for an RS flip-flop.

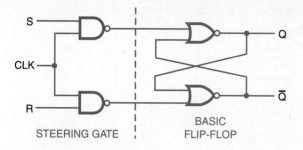

STEERING GATE | BASIC FLIP-FLOP

Figure 35–4 Logic circuit for a clocked RS flip-flop.

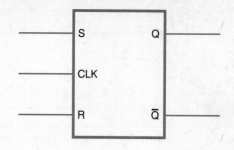

Figure 35–5 Logic symbol for a clocked RS flip-flop.

high at either input of the flip-flop portion activates the flip-flop, causing it to change states. The portion labeled "steering gate" steers or directs the clock pulses to either input gate.

The clocked flip-flop is controlled by the logic state of the S and R inputs when a clock pulse is present. A change in the state of the flip-flop occurs only when the leading edge of the clock pulse is applied. The leading edge of the clock pulse is a positive-going transition (low to high). This means that the pulse goes from a zero voltage level to a positive voltage level. This is referred to as positive edge-triggered (the edge of the pulse is what triggers the circuit).

As long as the clock input is low, the S and R inputs can be changed without affecting the state of the flip-flop. The only time that the effects of the S and R inputs are felt is when a clock pulse occurs. This is referred to as synchronous operation. The flip-flop operates in step with the clock. Synchronous operation is important in computers and calculators when each step must be performed in an exact order. Figure 35–5 shows the logic symbol used to represent a clocked RS flip-flop.

A **D flip-flop** is useful where only one data bit (1 or 0) is to be stored. Figure 35–6 shows the logic diagram for a D flip-flop. It has a single data input and a clock input. The D *flip-flop* is also referred to as a *delay flip-flop*. The D input is delayed one clock pulse from getting to the output (Q). Sometimes the D flip-flop has a PS (preset) input and CLR (clear) input. The preset input sets output Q to a 1 when a low or 0 is applied to it. The clear input clears the Q output to a 0 when it is enabled by a low or 0. D flip-flops are wired together to form shift registers and storage registers. These registers are widely used in digital systems.

The **JK flip-flop** is the most widely used flip-flop. It has all the features of the other types of flip-flops. The logic diagram and symbol for the JK flip-flop is shown in Figure 35–7. J and K are the inputs. The significant feature of the JK flip-flop is that when both the J and K inputs are high, repeated clock pulses cause the output to toggle or change state. The two asynchronous inputs, PS (preset) and CLR (clear), override the synchronous inputs, the J and K data inputs, and the clock input. JK flip-flops are widely used in many digital

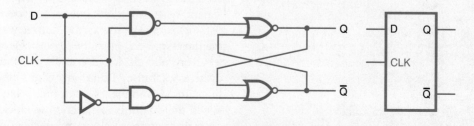

Figure 35–6 Logic circuit and symbol for the D flip-flop.

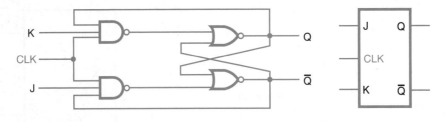

Figure 35–7 Logic circuit and symbol for the JK flip-flop.

circuits, especially counter circuits. Counters are found in almost every digital system.

A *latch* is a device that serves as a temporary buffer memory. It is used to hold data after the input signal is removed. The D flip-flop is a good example of a latch device. Other types of flip-flops can also be used.

A latch is used when inputting to a seven-segment display. Without a latch, the information being displayed is removed when the input signal is removed. With the latch, the information is displayed until it is updated.

Figure 35–8 shows a 4-bit latch. The unit has four D flip-flops enclosed in a single IC package. The E (enable) inputs are similar to the clock input of the D flip-flop. The data are latched when the enable line drops to a low, or 0. When the enable is high, or 1, the output follows the input. This means that the output will change to whatever state the input is in; e.g., if the input is high, the output will become high; if the input is low, the output will become low. This condition is referred to as a *transparent latch.*

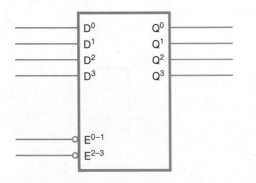

Figure 35–8 Four-bit latch.

35–2 Counters

A **counter** is a logic circuit that can count a sequence of numbers or states when activated by a clock input. The output of a counter indicates the binary number stored in the counter at any given time. The number of counts or states through which a counter progresses before returning to its original state (recycling) is called the **modulus** of the counter.

A flip-flop can act as a simple counter when connected as shown in Figure 35–9. Assuming that initially the flip-flop is reset, the first clock pulse causes it to set (Q = 1). The second clock pulse causes it to reset (Q = 0). Because the flip-flop has set and reset, two clock pulses have occurred.

Figure 35–10 shows the output waveform of the flip-flop. Notice that the Q output is high (1) after every odd clock pulse and low (0) after every even clock pulse. Therefore, when the output is high, an odd number of clock pulses has occurred. When the output is low, either no clock pulses or an even number of clock pulses has occurred. In this case, it is not known which has occurred.

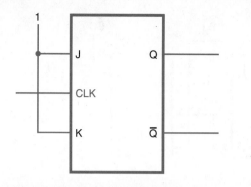

Figure 35–9 JK flip-flop set up for counting.

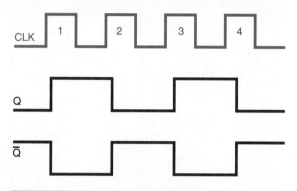

Figure 35–10 Input and output waveforms of JK flip-flop set up as a counter.

A single flip-flop produces a limited counting sequence, 0 or 1. To increase the counting capacity, additional flip-flops are needed. The maximum number of binary states in which a counter can exist depends on the number of flip-flops in the counter. This can be expressed as:

$$N = 2^n$$

where: N = maximum number of counter states
n = number of flip-flops in counter

Binary counters fall into two categories based on how the clock pulses are used to sequence the counter. The two categories are asynchronous and synchronous.

Asynchronous means not occurring at the same time. With respect to counter operations, asynchronous means that the flip-flops do not

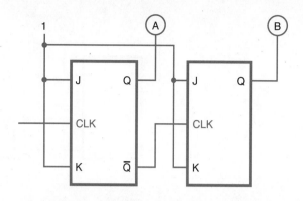

Figure 35–11 Two-stage counter.

change states at the same time. This is because the clock pulse is not connected to the clock input of each stage. Figure 35–11 shows a two-stage counter connected for asynchronous operation. Each flip-flop in a counter is referred to as a *stage*.

Notice that the $\overline{Q}$ output of the first stage is coupled to the clock input of the second stage. The second stage only changes state when triggered by the transition of the output of the first stage. Because of the delay through the flip-flop, the second flip-flop does not change state at the same time the clock pulse is applied. Therefore, the two flip-flops are never simultaneously triggered, which results in asynchronous operation.

Asynchronous counters are commonly referred to as *ripple counters*. The input clock pulse is first felt by the first flip-flop. The effect is not felt by the second flip-flop immediately because of the delay through the first flip-flop. In a multiple-stage counter, the delay is felt through each flip-flop, so that the effect of the input clock pulse "ripples" through the counter. Figure 35–12 shows a three-stage binary counter and timing chart for each of the stages. A truth table is also shown to show the counting sequence.

Synchronous means occurring at the same time. A *synchronous counter* is a counter in which each stage is clocked at the same time. This is accomplished by connecting the clock input to each stage of the counter (Figure 35–13). A synchronous counter is also called a *parallel counter* because the clock input is connected in parallel to each flip-flop.

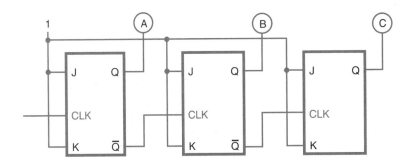

NUMBER OF CLOCK PULSES	BINARY COUNT SEQUENCE			DECIMAL COUNT
	C	B	A	
0	1	1	1	7
1	1	1	0	6
2	1	0	1	5
3	1	0	0	4
4	0	1	1	3
5	0	1	0	2
6	0	0	1	1
7	0	0	0	0
8	1	1	1	7

COUNT SEQUENCE

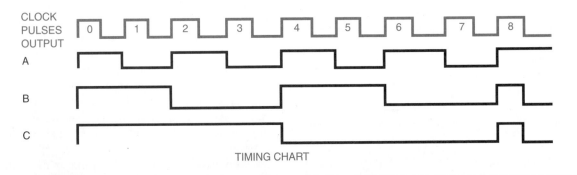

TIMING CHART

Figure 35–12 Three-stage binary counter.

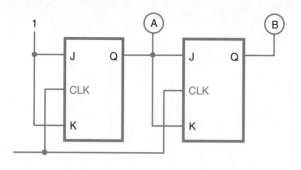

Figure 35–13 Two-stage synchronous counter.

A synchronous counter operates as follows. Initially the counter is reset with both flip-flops in the 0 state. When the first clock pulse is applied, the first flip-flop toggles and the output goes high. The second flip-flop does not toggle because of the delay from the input to the actual changing of the output state. Therefore, there is no change in the second flip-flop output state. When the second clock pulse is applied, the first flip-flop toggles and the output goes low. Because there is a high from the output of the first stage, the second stage toggles and its output goes high. After four clock pulses, the counter recycles to its original state. Figure 35–14 shows a timing chart for this sequence of events with a two-stage synchronous counter.

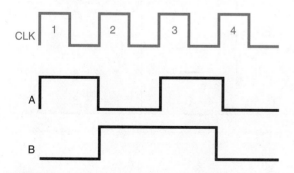

Figure 35–14 Input and output waveforms for the two-stage synchronous counter.

Figure 35–15 shows a three-stage binary counter and timing chart. Figure 35–16 shows a four-stage synchronous counter and logic symbol.

One application of a counter is frequency division. A single flip-flop produces an output pulse for every two input pulses. Therefore, it is essentially a divide-by-two device with an output one-half the frequency of the input. A two-stage binary counter is a divide-by-four device with an output equal to one-fourth the input clock frequency. A four-stage binary counter is a divide-by-sixteen device with the output equal to one-sixteenth the input clock frequency (Figure 35–17).

A binary counter with n stages divides the clock frequency by a factor of 2^n. A three-stage counter divides the frequency by eight (2^3), a four-stage counter by sixteen (2^4), a five-stage counter by thirty-two (2^5), and so on. Notice that the modulus of the counter is the same as the division factor.

Decade counters have a modulus of ten, or ten states in their counting sequence. A common decade counter is the BCD (8421) counter, which produces a binary-coded-decimal sequence (Figure 35–18). The AND and OR gates detect the occurrence of the ninth state and cause the counter to recycle on the next clock pulse. The symbol for a decade counter is shown in Figure 35–19.

An **up-down counter** can count in either direction through a certain sequence. It is also referred to as a bidirectional counter. The counter can be reversed at any point in the counting sequence. Its symbol is shown in Figure 35–20. An up-down counter can consist of any number of stages. Figure 35–21 shows the logic diagram for a BCD up-down counter. The inputs to the JK flip-flops are enabled by the up-down input qualifying the up or down set of the AND gates.

Counters can be stopped after any sequence of counting by using a logic gate or combination of logic gates. The output of the gate is fed back to the input of the first flip-flop in a ripple counter.

If a 0 is fed back to the JK input of the first flip-flop (Figure 35–22) it prevents the first flip-flop from toggling, thereby stopping the count.

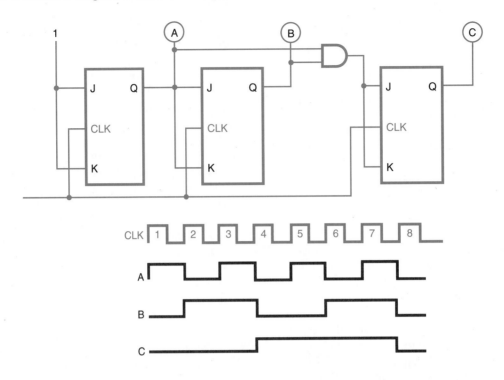

NUMBER OF CLOCK PULSES	BINARY COUNT SEQUENCE			DECIMAL COUNT
	C	B	A	
0	0	0	0	0
1	0	0	1	1
2	0	1	0	2
3	0	1	1	3
4	1	0	0	4
5	1	0	1	5
6	1	1	0	6
7	1	1	1	7
8	0	0	0	8

Figure 35–15 Three-stage binary counter and timing chart.

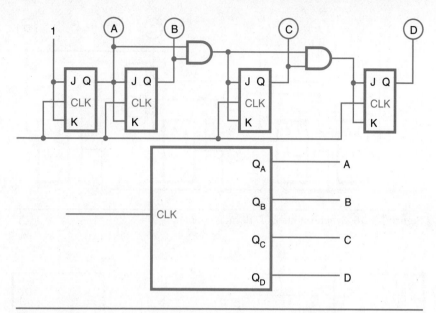

Figure 35–16 Logic symbol for a four-stage synchronous counter.

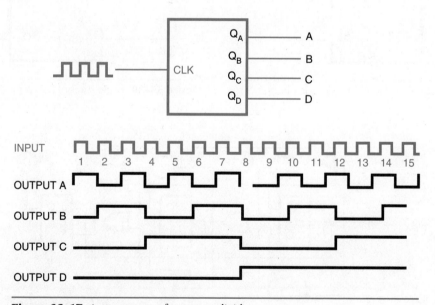

Figure 35–17 A counter as a frequency divider.

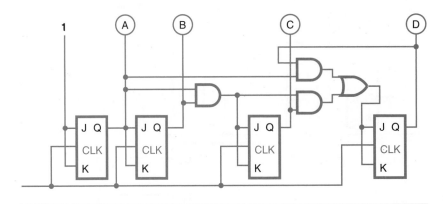

Figure 35–18 Synchronous BCD decade counter.

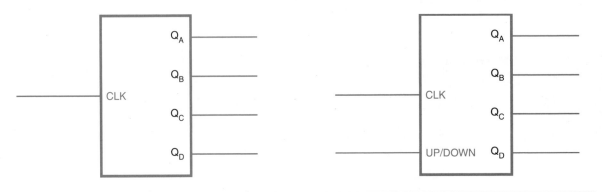

Figure 35–19 Logic symbol for a decade counter.

Figure 35–20 Logic symbol for an up-down counter.

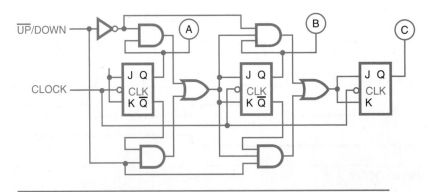

Figure 35–21 Logic diagram for a BCD up-down counter.

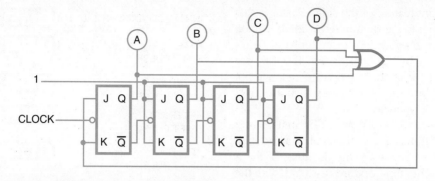

Figure 35–22 A low, applied to the JK input of the first flip-flop, prevents it from toggling, stopping the count.

35–2 Questions

1. What function does a counter serve?
2. How many counting sequences are available with an eight-stage counter?
3. How does an asynchronous counter operate?
4. How does a synchronous counter differ from an asynchronous counter?
5. How can a counter be made to stop at any count desired?

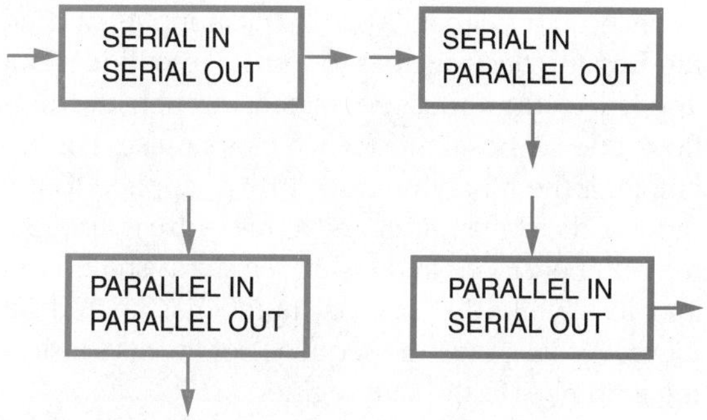

35–3 Shift Registers

A **shift register** is a sequential logic circuit widely used to store data temporarily. Data can be loaded into and removed from a shift register in either a parallel or serial format. Figure 35–23 shows the four different methods of loading and reading data in a shift register. Because of its ability to move data one bit at a time from one storage medium to another, the shift register is valuable in performing a variety of logic operations.

Shift registers are constructed of flip-flops wired together. Flip-flops have all the functions necessary for a register: They can be reset, preset, toggled, or steered to a 1 or 0 level. Figure 35–24 shows a basic shift register constructed from four flip-flops. It is called a 4-bit shift register because it consists of four binary storage elements.

Figure 35–23 Methods of loading and reading data in a shift register.

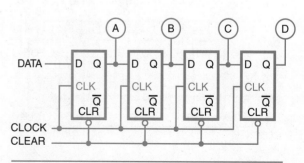

Figure 35–24 Shift register constructed from four flip-flops.

A significant feature of the shift register is that it can move data to the right or left a number of bit positions. This is equivalent to multiplying or dividing a number by a specific factor. The data are shifted one bit position at a time for each input clock pulse. The clock pulses have full control over the shift register operations.

Figure 35–25 shows a typical 4-bit shift register constructed of JK flip-flops. The serial data and their complements are applied to the JK inputs of the A flip-flop. The other flip-flops are cascaded, with the outputs of one connected to the inputs of the next. The toggles of all the flip-flops are connected together, and the clock pulse is applied to this line. Because all the flip-flops are toggled together, this is a synchronous circuit. Also, the clear inputs of each flip-flop are tied together to form a reset line. Data applied to the input are shifted through the flip-flops one bit position for each clock pulse. For example, if the binary number 1011 is applied to the input of the shift register and a shift pulse is applied, the number stored in the shift register is shifted out and lost while the external number is shifted in. Figure 35–26 shows the sequence of events for storing a number in the shift register.

One of the most common applications of a shift register is serial-to-parallel or parallel-to-serial data conversion. Figure 35–27 shows how a shift register can be loaded by a parallel input. For parallel operations, the input data are preset into the shift register. Once the data are in the shift register, they can be shifted out serially, as discussed earlier.

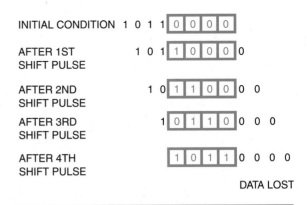

Figure 35–26 Storing a number in a shift register.

For serial-to-parallel data conversion, the data are initially shifted into the shift register with clock pulses. Once the data are in the shift register, the outputs of the individual flip-flops are monitored simultaneously, and the data are routed to their destination.

Shift registers can perform arithmetic operations such as multiplication or division. Shifting a binary number stored in the shift register to the right has the same effect as dividing the number by some power of 2. Shifting the binary number stored in the shift register to the left has the same effect as multi-

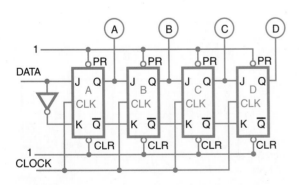

Figure 35–25 A typical shift register constructed of JK flip-flops.

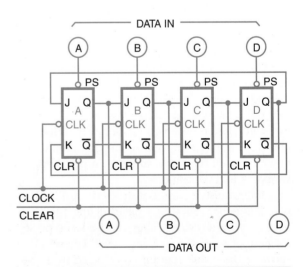

Figure 35–27 Loading a shift register using parallel input.

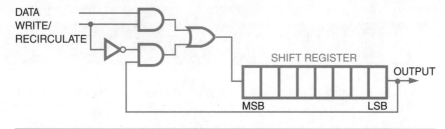

Figure 35–28 Shift register circuitry for maintaining and reading data.

plying the number by some power of 2. Shift registers are a simple and inexpensive means of performing multiplication and division of numbers.

Shift registers are often used for temporary storage. They are capable of storing one or more binary words. There are three requirements for this application of a shift register: First, it must be able to accept and store data. Second, it must be able to retrieve or read out the data on command. Third, when the data are read, they must not be lost. Figure 35–28 shows the external circuitry required to enable a shift register to read and maintain the data stored in it. The read/write line, when high, allows new data to be stored into the shift register. Once the data is stored, the read/write line goes low, enabling gate 2, which allows the data to recirculate while reading out the data.

35—3 Questions

1. What is the function of a shift register?
2. What is a significant feature of a shift register?
3. From what are shift registers constructed?
4. What is a common application of a shift register?
5. What arithmetic operations can a shift register perform, and how does it perform them?

SUMMARY

- A flip-flop is a bistable multivibrator whose output is either a high or a low.

- Types of flip-flops include:
  a. RS
  b. Clocked RS
  c. D
  d. JK
- Flip-flops are used in digital circuits such as counters.
- A latch is a temporary buffer memory.
- A counter is a logic circuit that can count a sequence of numbers or states.
- A single flip-flop produces a count sequence of 0 or 1.
- The maximum number of binary states a counter can have depends on the number of flip-flops contained in the counter.
- Counters can be either asynchronous or synchronous.
- Asynchronous counters are called ripple counters.
- Synchronous counters clock each stage at the same time.
- Shift registers are used to store data temporarily.
- Shift registers are constructed of flip-flops wired together.
- Shift registers can move data to the left or right.
- Shift registers are used for serial-to-parallel and parallel-to-serial data conversions.
- Shift registers can perform multiplication and division.

Chapter 35 Self-Test

1. Describe how an RS flip-flop changes states from a high on the Q output to a high on the $\overline{Q}$ output.

2. What is the major difference between the D flip-flop and the clocked RS flip-flop?

3. What components make up a counter, and how is it constructed?

4. Draw a schematic for a counter that will count to 10 and then repeat.

5. How does a shift register differ from a counter?

6. For what functions/applications can the shift register be used?

36

COMBINATIONAL LOGIC CIRCUITS

Objectives

After completing this chapter, the student will be able to:

- Describe the functions of encoders, decoders, multiplexers, adders, subtractors, and comparators.
- Identify the schematic symbols for encoders, decoders, multiplexers, adders, subtractors, and comparators.
- Identify applications for combinational logic circuits.
- Develop truth tables for the different combinational logic circuits.

Combinational logic circuits are circuits that combine the basic components AND gates, OR gates, and inverters to produce more sophisticated circuits. The output of a combinational logic circuit is a function of the states of the inputs, the types of gates used, and the interconnection of the gates. The most common of the combinational circuits are decoders, encoders, multiplexers, and arithmetic circuits.

36–1 Encoders

An **encoder** is a combinational logic circuit that accepts one or more inputs and generates a multibit binary output. *Encoding* is the process of converting any keyboard character or number as input to a coded output such as a binary or BCD form.

Figure 36–1 shows a **decimal-to-binary encoder.** Its function is to take a single digit (0 to 9) as input and to output a 4-bit code representation of the digit. This is referred to as a 10-line-to-4-line encoder. That is, if the digit 4 on the keyboard is typed in, this produces low or a 0 on line 4, which produces the 4-bit code 0100 as an output.

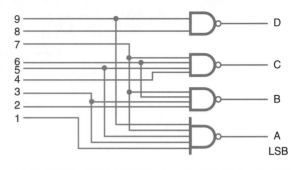

Figure 36–1 Decimal-to-binary encoder.

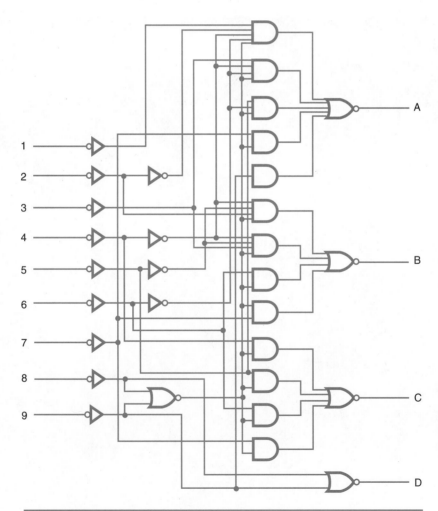

Figure 36–2 Decimal-to-binary priority encoder.

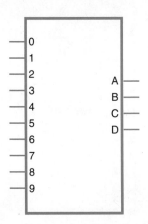

Figure 36–3 Logic symbol for a decimal-to-binary priority encoder.

Figure 36–2 shows a **decimal-to-binary priority encoder.** The priority function means that if two keys are pressed simultaneously the encoder produces a BCD output corresponding to the highest order decimal digit appearing on the input. For example, if both a 5 and a 2 are applied to the encoder, the BCD output is 1010 or the invert of decimal 5. This type of encoder is built into a single integrated circuit and consists of approximately thirty logic gates. Figure 36–3 shows the symbol for a priority encoder.

This type of encoder is used to translate the decimal input from a keyboard to an 8421 BCD code. The decimal-to-binary encoder and the decimal-to-binary priority encoder are found wherever there is keyboard input. This includes: calculators, computer keyboard inputs, electronic typewriters, and teletypewriters (TTY).

36–1 Questions

1. What is encoding?
2. What does an encoder accomplish?
3. What is the difference between a normal encoder and a priority encoder?
4. Draw the logic symbol for a decimal-to-binary priority encoder.
5. What are applications of decimal-to-binary encoders?

36–2 Decoders

A **decoder** is one of the most frequently used combinational logic circuits. It processes a complex binary code into a recognizable digit or character. For example, it might decode a BCD number into one of the ten possible decimal digits. The output of such a decoder is used to operate a decimal number readout or display. This type of decoder is called a 1-of-10 decoder or a 4-line-to-10-line decoder.

Figure 36–4 shows the ten NAND gates required for decoding a 4-bit BCD number to its approximate output (one decimal digit). When all the inputs to a NAND gate are 1, its output is 0. All other outputs from the NAND gates in the decoder are 1's. Rather than draw all the logic gates each time the circuit is used, the symbol shown in Figure 36–5 is used.

Two other types of decoder circuits are the 1-of-8 (octal, or base 8) decoder and the 1-of-16

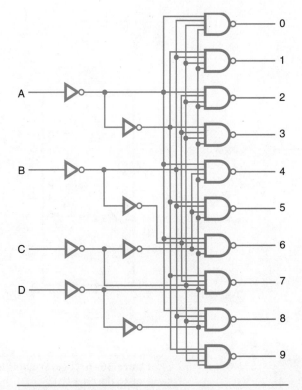

Figure 36–4 Binary-to-decimal decoder.

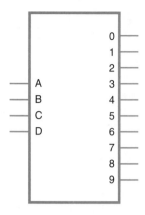

Figure 36–5 Logic symbol for a binary-to-decimal decoder.

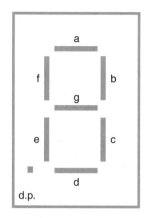

Figure 36–7 Seven-segment display configuration.

(hexadecimal, or base 16) decoder (Figure 36–6). The 1-of-8 decoder accepts a 3-bit input word and decodes it to one of eight possible outputs. The 1-of-16 decoder activates one of sixteen output lines by a 4-bit code word. It is also called a 4-line-to-16-line decoder.

A special type of decoder is the standard 8421 BCD-to-seven-segment decoder. It accepts a BCD input code and generates a special 7-bit output code to energize a seven-segment decimal readout display (Figure 36–7). The display consists of seven LED segments that are lit in different combinations

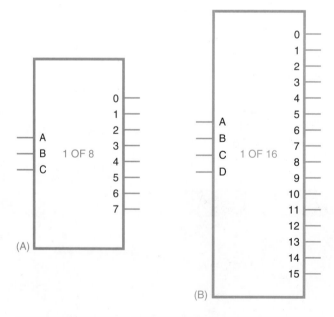

Figure 36–6 Logic symbols for a 1-of-8 decoder (A) and a 1-of-16 decoder (B).

Figure 36-8 Using the seven-segment display to form the ten decimal digits.

to produce each of the ten decimal digits, 0 through 9 (Figure 36-8). Besides seven-segment LED displays, there are incandescent and liquid crystal (LCD) displays.

Each of these displays operates on the same principle. A segment is activated by either a high or low voltage level. Figure 36-9 shows two types of LED displays: a common anode and a common cathode. In each case, the LED segment has to be forward biased for light to be emitted. For a common cathode, a high (1) lights up the segment, a low (0) does not.

Figure 36-10 shows the decoding logic circuit required to produce the output for a seven-segment display with a BCD input. Referring to Figure 36-7, notice that segment a is activated for digits 0, 2, 3, 5, 7, 8, and 9; segment b is activated for

digits 0, 1, 2, 3, 4, 7, 8, and 9; and so on. Boolean expressions can be formed to determine the logic circuitry needed to drive each segment of the display. The logic symbol for a BCD-to-seven-segment decoder is shown in Figure 36-11. This represents the circuit contained in an integrated circuit.

36-2 Questions

1. What is a decoder?
2. What are the uses of decoders?
3. Draw the logic symbol for a 1-of-10 decoder.
4. What is the purpose of a seven-segment decoder?
5. What codes can be used in decoders?

36-3 Multiplexers

A **multiplexer** is a circuit used to select and route any one of several input signals to a single output. An example of a nonelectronic circuit multiplexer is a single-pole, multiposition switch (Figure 36-12).

Multiposition switches are widely used in many electronic circuits. However, circuits that operate at high speed require the multiplexer to switch at high speed and to be automatically selected. A mechanical switch cannot perform this task satisfactorily. Therefore, multiplexers used to perform high-speed switching are constructed of electronic components.

Multiplexers handle two basic types of data: analog and digital. For analog applications, multiplexers are built of relays and transistor switches. For digital applications, multiplexers are built from standard logic gates.

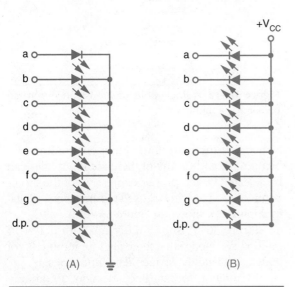

Figure 36-9 Differences between the two types of LED displays a common cathode (A), and a common anode (B).

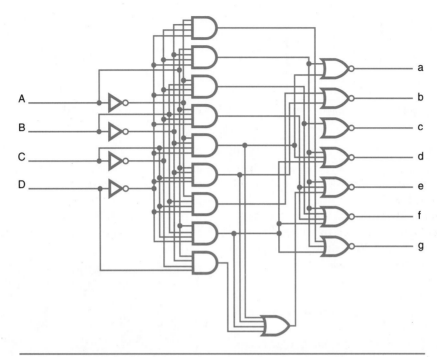

Figure 36–10 Binary-to-seven-segment display decoder.

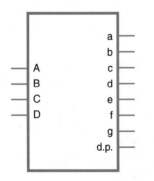

Figure 36–11 Logic symbol for a BCD-to-seven-segment decoder.

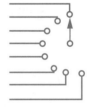

Figure 36–12 A single-pole, multiposition switch can be used as a multiplexer.

Digital multiplexers allow digital data from several individual sources to be routed through a common line for transmission to a common destination. A basic multiplexer has several input lines with a single output line. The input lines are activated by data selection input that identifies the line the data are to be received on. Figure 36–13 shows the logic circuit for an eight-input multiplexer.

Notice that there are three input-control lines, labeled A, B, and C. Any of the eight input lines can be selected by the proper expression of the input-control line. The symbol used to represent a digital multiplexer is shown in Figure 36–14.

Figure 36–15 shows the symbol for a 1-of-16 multiplexer. Notice that there are four input-control lines to activate the sixteen data input lines.

In addition to data line selection, a common application of a multiplexer is parallel-to-serial data conversion. A parallel binary word is applied to the input of the multiplexer. Then, by sequencing

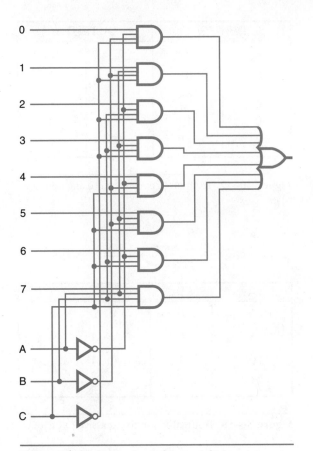

Figure 36-13 Logic circuit for an eight-input multiplexer.

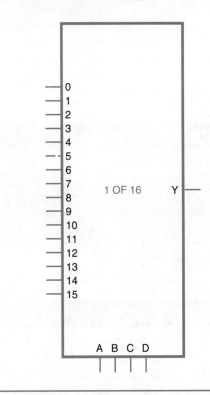

Figure 36-15 Logic symbol for a sixteen-input multiplexer.

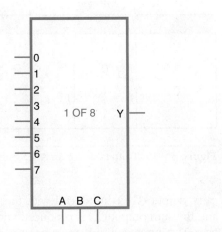

Figure 36-14 Logic symbol for an eight-input multiplexer.

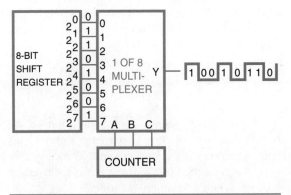

Figure 36-16 Using a multiplexer for parallel-to-serial conversion.

through the enabling codes, the output becomes a serial representation of the parallel input word.

Figure 36–16 shows a multiplexer set up for parallel-to-serial conversion. A 3-bit binary input word from a counter is used to select the desired input. The parallel input word is connected to each of the input lines of the multiplexer. As the counter is incremented, the input select code is sequenced through each of its states. The output of the multiplexer is equal to the parallel signal applied.

36–3 Questions

1. What is a multiplexer?
2. How are multiplexers used?
3. Draw a logic diagram for a multiplexer.
4. What type of data can a multiplexer handle?
5. How can a multiplexer be set up for serial-to-parallel conversion?

36–4 Arithmetic Circuits

Adder

An **adder** is the primary computation unit in a digital computer. Few routines are performed by a computer in which the adder is not used. Adders are designed to work in either serial or parallel circuits. Because the parallel adder is faster and used more often, it is covered in more detail here.

To understand how an adder works it is necessary to review the rules for adding:

$$
\begin{array}{cccc}
0 & 0 & 1 & 1 \\
+0 & +1 & +0 & +1 \\
\hline
0 & 1 & 1 & \text{Carry } 1 \quad 0 \\
\end{array}
$$

Figure 36–17 shows a truth table based on these rules. Note that the Greek letter sigma (Σ) is used to represent the sum column. The carry column is represented by C_0. These terms are used by industry when referring to an adder.

The sum column of the truth table is the same as the output column in the truth table of an exclusive OR gate (Figure 36–18). The carry column is the same as the output in a truth table for an AND gate (Figure 36–19).

Figure 36–20 shows an AND gate and an exclusive OR gate connected in parallel to provide the necessary logic function for single-bit addition. The

INPUTS		OUTPUTS	
A	B	Σ	C_0
0	0	0	0
1	0	1	0
0	1	1	0
1	1	0	1

Figure 36–17 Truth table constructed using addition rules.

A	B	Y
0	0	0
1	0	1
0	1	1
1	1	0

Figure 36–18 Truth table for an exclusive OR gate.

A	B	Y
0	0	0
1	0	0
0	1	0
1	1	1

Figure 36–19 Truth table for an AND gate.

carry output (C_0) is produced with the AND gate, and the sum output (Σ) is produced with the XOR gate. The inputs A and B are connected to both the AND gate and the XOR gate. The truth table for the circuit is the same as the truth table developed using

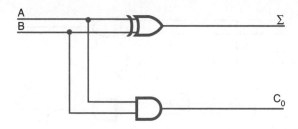

Figure 36–20 Half-adder circuit.

the binary addition rules (Figure 36–17). Because the circuit does not take into account any carries, it is referred to as a **half adder.** It can be used as the LSB adder for a binary addition problem.

An adder that takes into account the carry is called a **full adder.** A full adder accepts three inputs and generates a sum and carry output. Figure 36–21 shows the truth table for a full adder. The C_1 input represents the carry input. The C_0 output represents the carry output.

Figure 36–22 shows a full adder constructed of two half adders. The results of the first half adder are "ORed" with the second half adder to form the carry output. The carry output is a 1 if both inputs to the first XOR gate are 1's or if both inputs to the second XOR gate are 1's. Figure 36–23 shows the symbols used to represent a half adder and a full adder.

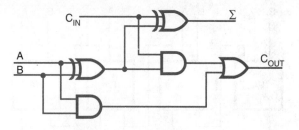

Figure 36–22 Logic circuit for a full adder using two half adders.

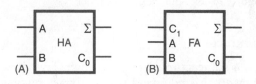

Figure 36–23 Logic symbols for half adder (A) and full adder (B).

A single full adder is capable of adding two single-bit numbers and an input carry. To add binary numbers with more than 1-bit, additional full adders must be used. Remember, when one binary number is added to another, each column that is added generates a sum and a carry of 1 or 0 to the next higher order column. To add two binary numbers, a full adder is required for each column. For example, to add a 2-bit number to another 2-bit number, two adders are required. Two 3-bit numbers require three adders, two 4-bit numbers require four adders, and so on. The carry generated by each adder is applied to the input of the next higher order adder. Because no carry input is required for the least significant position, a half adder is used.

Figure 36–24 shows a 4-bit parallel adder. The least significant input bits are represented by A_0 and B_0. The next higher order bits are represented by A_1 and B_1 and so on. The output sum bits are identified as Σ_0, Σ_1, Σ_2 and so on. Note that the carry output of each adder is connected to the carry input of the next higher order adder. The carry output of the final adder is the most significant bit of the answer.

INPUTS			OUTPUTS	
A	**B**	**C_1**	**Σ**	**C_0**
0	0	0	0	0
1	0	0	1	0
0	1	0	1	0
1	1	0	0	1
0	0	1	1	0
1	0	1	0	1
0	1	1	0	1
1	1	1	1	1

Figure 36–21 Truth table for a full adder.

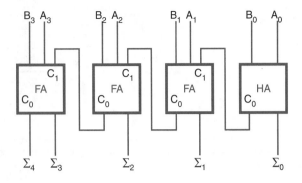

Figure 36–24 Four-bit parallel adder.

Subtractor

A **subtractor** allows subtraction of two binary numbers. To understand how a subtractor works it is necessary to review the rules for subtraction.

$$
\begin{array}{cccc}
0 & \text{Borrow 1} & 0 & 1 & 1 \\
-\,0 & & -\,1 & -\,0 & -\,1 \\
\hline
0 & & 1 & 1 & 0
\end{array}
$$

Figure 36–25 shows a truth table based on these rules. The letter D represents the difference column. The borrow column is represented by B_0.

Notice that the difference output (D) is 1 only when the input variables are not equal. Therefore, the difference can be expressed as the exclusive OR of the input variables. The output borrow is generated only when A is 0 and B is 1. Therefore, the borrow output is the complement of A "ANDed" with B.

INPUTS		OUTPUTS	
A	**B**	**D**	**B₀**
0	0	0	0
1	0	1	0
0	1	1	1
1	1	0	0

Figure 36–25 Truth table constructed using subtraction rules.

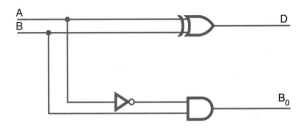

Figure 36–26 Logic circuit for a half subtractor.

Figure 36–26 shows a logic diagram of a **half subtractor.** It has two inputs and generates a difference and a borrow output. The difference is generated by an XOR gate and the borrow output is generated by an AND gate with $\overline{A}$ and B inputs. The $\overline{A}$ is achieved by using an inverter on the variable A input.

However, a half subtractor is identified as such because it does not have a borrow input. A **full subtractor** does. It has three inputs and generates a difference and a borrow output. A logic diagram and truth table for a full subtractor are shown in Figure 36–27. Figure 36–28 shows the symbols used to represent a half subtractor and full subtractor.

A full subtractor can handle only two 1-bit numbers. To subtract binary numbers with more than 1 bit, additional full subtractors must be used. Keep in mind that if a 1 is subtracted from a 0, a borrow must be made from the next higher order column. The output borrow of the lower order subtractor becomes the input borrow of the next higher order subtractor.

Figure 36–29 shows a block diagram of a 4-bit subtractor. A half subtractor is used in the least significant bit position because there is no input borrow.

Comparator

A **comparator** is used to compare the magnitudes of two binary numbers. It is a circuit used simply to determine if two numbers are equal. The output not only compares two binary numbers but also indicates whether one is larger or smaller than the other.

Figure 36–30 shows a truth table for a comparator. The only time an output is desired is when

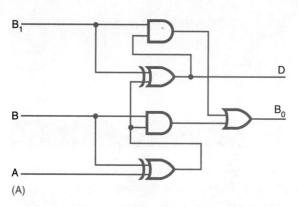

INPUT			OUTPUT	
A	B	B_1	D	B_0
0	0	0	0	0
1	0	0	1	0
0	1	0	1	1
1	1	0	0	0
0	0	1	1	1
1	0	1	0	0
0	1	1	0	1
1	1	1	1	1

(B)

Figure 36–27 Logic circuit (A) and truth table (B) for a full subtractor.

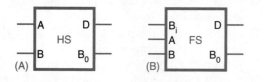

Figure 36–28 Logic symbols for half subtractors (A) and full subtractors (B).

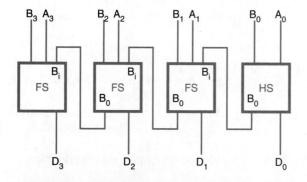

Figure 36–29 Four-bit subtractor.

INPUT		OUTPUT
A	B	Y
0	0	1
1	0	0
0	1	0
1	1	1

Figure 36–30 Truth table for a comparator.

more, additional XNOR gates are necessary. Figure 36–31 shows a logic diagram of a comparator for checking two 2-bit numbers. If the numbers are equal, a 1 is generated from the XNOR gate. The 1 is applied to the AND gate as a qualifying level. If both XNOR gates produce a 1 for the inputs to the AND gate, that identifies the numbers as equal and

both bits being compared are the same. The output column represents an exclusive OR with an inverter, also known as an exclusive NOR (XNOR) gate. An XNOR gate is essentially a comparator, because its output is a 1 only if the two inputs are the same. To compare numbers containing 2 bits or

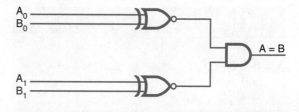

Figure 36–31 Comparing two 2-bit numbers.

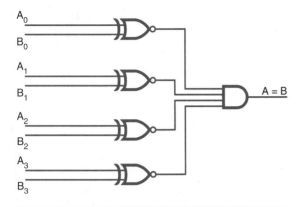

Figure 36–32 Comparing two 4-bit numbers.

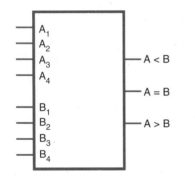

Figure 36–33 Diagram of a 4-bit comparator.

generates a 1 for the output of the AND gate. However, if the inputs to the XNOR gate are different, the XNOR gate generates a 0, which disqualifies the AND gate. The output of the AND gate is then a 0. Figure 36–32 shows a logic diagram of a comparator for checking two 4-bit numbers. Figure 36–33 is the diagram used to represent a 4-bit comparator.

36—4 Questions

1. What are the rules for binary addition?
2. What is the difference between a half adder and a full adder?
3. When is a half adder used?
4. What are the rules for binary subtraction?
5. Draw a block diagram of a 4-bit subtractor.
6. What is the function of a comparator?
7. Draw a logic diagram of a comparator.

SUMMARY

- An encoder accepts one or more inputs and generates a multibit binary output.
- A decimal-to-binary encoder takes a single digit (0 through 9) and produces a 4-bit output code that represents the digit.
- A priority encoder accepts the higher order key when two keys are pressed simultaneously.
- Decimal-to-binary encoders are used for keyboard encoding.
- A decoder processes a complex binary code into a digit or character that is easy to recognize.
- A BCD-to-seven-segment decoder is a special-purpose decoder to drive seven-segment displays.
- A multiplexer allows digital data from several sources to be routed through a common line for transmission to a common destination.
- Multiplexers can handle both analog and digital data.
- Multiplexers can be hooked up for parallel-to-serial conversion of data.
- The truth table for the adding rules of binary numbers is equivalent to the truth table for an AND gate and an XOR gate.
- A half adder does not take into account the carry.
- A full adder takes the carry into account.
- To add two 4-bit numbers requires three full adders and one half adder.
- The truth table for the subtracting rules of binary numbers is equivalent to the truth table for an AND gate with an inverter on one of the inputs and an XOR gate.
- A half subtractor does not have a borrow input.

■ A full subtractor has a borrow input.
■ A comparator is used to compare the magnitudes of two binary numbers.
■ A comparator generates an output only when the two bits being compared are the same.

■ A comparator can also determine whether one number is larger or smaller than the other.

Chapter 36 Self-Test

1. Why are encoders necessary in logic circuits?
2. What type of encoder is required for keyboard input?
3. Why are decoders important in logic circuits?
4. What are the applications for the different types of decoders?
5. Describe briefly how a digital multiplexer works.
6. For what applications can a digital multiplexer be used?
7. Draw a schematic using logic symbols for a half adder and full adder tied together for two-bit addition.
8. Explain how the adder drawn in question number 7 operates.

37

MICROCOMPUTER BASICS

Objectives

After completing this chapter, the student will be able to:

- Identify the basic blocks of a digital computer.
- Explain the function of each block of a digital computer.
- Describe what a program is and its relationship to both digital computers and microprocessors.
- Identify the basic registers in a microprocessor.
- Explain how a microprocessor operates.
- Identify the instruction groups associated with microprocessors.

The greatest application of digital circuits is in digital computers. A *digital computer* is a device that automatically processes data using digital techniques. Data are pieces of information. Processing refers to the variety of ways that data can be manipulated.

Digital computers are classified by size and computing power. The largest computers are called *mainframes*. These computers are expensive, having extensive memory and high-speed calculating capabilities. Smaller scale computers—the minicomputer and the microcomputer—are more widely used. Even though they represent a small percentage of the total computer dollars invested, smallscale computers represent the largest number of computers in use. The *microcomputer* is the smallest and least expensive of the digital computers that still retains all the features and characteristics of a computer.

Computers are also classified by function. The most common function is data processing. Industry, business, and government use computers to maintain records, perform accounting tasks, keep inventory, and provide a wide variety of other data processing functions.

Computers can be general purpose or special purpose. General-purpose computers are flexible and can be programmed for any task. Special-purpose, or dedicated, computers are designed to perform a single task.

37–1 Computer Basics

All **digital computers** consist of five basic blocks or sections: **control, arithmetic logic unit (ALU), memory, input,** and **output** (Figure 37–1). In some cases the input and output blocks are a single block identified as **input/output (I/O).** Because the control unit and the arithmetic logic unit are closely related and difficult to separate, they may be collectively referred to as the central processing unit (CPU) or microprocessing unit (MPU).

The *control unit* decodes each instruction that enters the computer. It then generates the necessary pulses to carry out the functions specified. If, for example, an instruction requires two numbers to be added together, the control unit sends pulses to the arithmetic logic unit (ALU) to perform the addition. If the instruction requires a word to be stored in memory, control sends the necessary pulses to memory to store the data.

Modern computers incorporate a means of incorporating several instructional commands into a single input instruction. This is accomplished by a program stored in memory. When the instruction is decoded by the control unit, it causes a sequence of instructions to be executed.

The control unit varies from computer to computer. Basically, the control unit consists of an address register, an instruction register, an instruction decoder, a program counter, a clock, and circuitry for generating the control pulses (Figure 37–2).

The instruction register stores the instruction word to be decoded. The word is decoded by the instruction decoder, which sends the appropriate logic signal to the control pulse generator. The control pulse generator produces a pulse when the appropriate clock signal is given. The output of the control pulse generator enables other circuitry in the computer to carry out the specific instruction.

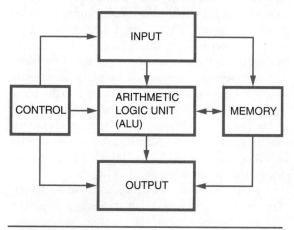

Figure 37–1 Basic blocks of a digital computer.

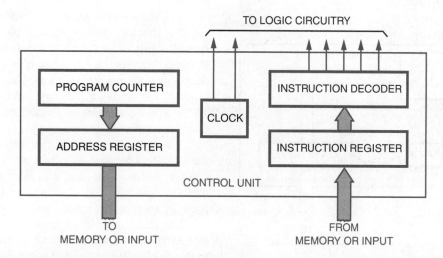

Figure 37–2 Control unit of a computer.

The program counter keeps track of the sequence of instructions to be executed. The instructions are stored in a program in memory. To begin the program, the starting address (specific memory location) of the program is placed in the program counter. The first instruction is read from memory, decoded, and performed. The program counter then automatically moves to the next instruction location. Each time an instruction is fetched and executed, the program counter advances one step until the program is completed.

Some instructions specify a jump or branch to another location in the program. The instruction register contains the address of the next instruction location and it is loaded into the address register.

The ALU performs math logic and decision-making operations. Most arithmetic logic units can do addition and subtraction. Multiplication and division are programmed in the control unit. The arithmetic logic unit can perform logic operations such as inversion, AND, OR, and exclusive OR. It can also make decisions by comparing numbers or test for specific quantities such as 0's, 1's or negative numbers.

Figure 37–3 shows an arithmetic logic unit. It consists of arithmetic logic circuitry and an accumulator register. All data to the accumulator and the ALU are sent via the data register. The accumulator register can be incremented (increased by one), decremented (decreased by one), shifted right (one position), or shifted left (one position). The accumulator is the same size as the memory word; the memory word is 8 bits wide, and the accumulator is also 8 bits wide in an 8-bit microprocessor.

The arithmetic logic circuitry is basically a binary adder. Both addition and subtraction can be done with the binary adder as well as logic operations. To add two binary numbers, one number is stored in the accumulator register and the other is stored in the data register. The sum of the two numbers is then placed in the accumulator register, replacing the original binary number.

Memory is the area where programs are stored. *Programs* contain the instructions that tell the computer what to do. A program is a sequential set of instructions to solve a particular problem.

A computer memory is simply a number of storage registers. Data can be loaded into the registers and then taken out, or "read out" to perform some operation without losing the register content. Each register or memory location is assigned a number called an address. The address is used to locate data in memory.

Figure 37–4 shows a typical memory layout. The memory registers retain the binary data. This memory, based on its ability to store (write) or retrieve (read) data is usually referred to as random-access read or write memory (RAM). Based on the ability of being able to read data only from the memory, it is referred to as Read Only Memory (ROM).

The memory address register allows access to specific memory locations by the memory address decoder. The size of the memory address register determines the maximum memory size for a computer. For example, a memory address register of 16 bits allows a maximum number of 2^{16} or 65,536 memory locations.

A word to be stored in memory is located in the data register and then placed in the desired memory location. To read data from memory, the memory location is determined, and data at the memory location are loaded into the shift register.

The input and output units of a computer allow it to receive and transmit information from and to the world outside the computer. An operator or peripheral equipment enters data into a computer

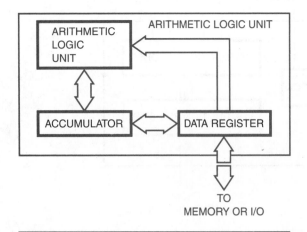

Figure 37–3 Arithmetic logic unit (ALU).

INPUT
DATA
BUS

MEMORY
CELL
SELECTION

OUTPUT
DATA
BUS

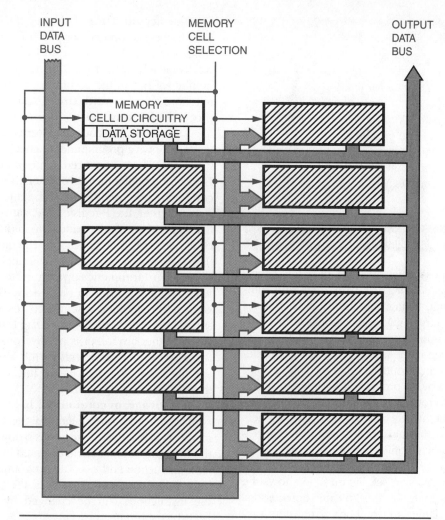

MEMORY
CELL ID CIRCUITRY
DATA STORAGE

Figure 37–4 Memory layout for a computer.

through the input unit. Data from the computer are passed to external peripheral equipment through the output unit.

The input and output units are under the control of the CPU. Special input/output (I/O) instructions are used to transfer data in and out of the computer.

Most digital computers can perform I/O operations at the request of an interrupt. An *interrupt* is a signal from an external device requesting service in the form of receiving or transmitting data. The interrupt results in the computer leaving the current program and jumping to another program.

When the interrupt request is accomplished, the computer returns to the original program.

37–1 Questions

1. Draw and label a block diagram of a digital computer.
2. What is the function of the following blocks in a digital computer?
   a. Control
   b. Arithmetic logic unit

c. Memory
d. Input
e. Output
3. What is the function of ROM in a computer?
4. What keeps track of the sequence of instructions to be executed?
5. What determines how much data can be stored in a computer?
6. Define a *program*.

37—2 Microprocessor Architecture

A **microprocessor** is the heart of a microcomputer. It contains four basic parts: registers, arithmetic logic unit, timing and control circuitry, and decoding circuitry. A microprocessor is designed so that an instruction, or program, can be fetched from memory, placed in the instruction register, and decoded. The timing, control, and decoding circuitry are all affected by the program. The program allows the operator to route data in or out of various registers or into the arithmetic logic unit. The registers and the arithmetic logic unit are used by the microprocessor for data or information manipulation.

Each microprocessor is different in its architecture and its instruction set. Figure 37–5 shows the basic parts of many of the 8-bit microprocessors. Because the names and number of registers vary from one microprocessor to the next, they are shown and identified separately.

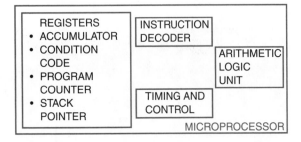

Figure 37–5 Parts of an 8-bit microprocessor.

The **accumulator** is the register most often used in the microprocessor. It is used to receive or store data from memory or an I/O device. It also works closely with the arithmetic logic unit. The number of bits in the accumulator determines the microprocessor's word size. In an 8-bit microprocessor the word size is 8 bits.

The **condition-code register** is an 8-bit register that allows a programmer to check the status of the microprocessor at a certain point in a program. Depending on the microprocessor, the name of the condition-code register may be the processor-status register, the P-register, the status register, or the flag register. An individual bit in the condition-code register is called a flag bit. The most common flags are carry, zero, and sign. The carry flag is used during an arithmetic operation to determine whether there is a carry or a borrow. The zero flag is used to determine whether the results of an instruction are all zeros. The sign flag is used to indicate whether a number is positive or negative. Of the 8 bits in the code register, the Motorola 6800 and the Zilog Z80 use 6 bits; the Intel 8080A uses 5; the MOS Technology 6502 uses 7.

The **program counter** is a 16-bit register that contains the address of the instruction fetched from memory. While the instruction is being carried out, the program counter is incremented by one for the next instruction address. The program counter can only be incremented. However, the sequence of the instructions can be changed by the use of branch or jump instructions.

The **stack pointer** is a 16-bit register that holds the memory location of data stored in the stack. The stack is discussed further later.

Most microprocessors have the same basic set of instructions with different machine codes and a few unique instructions. The basic instructions fall into nine categories:

1. Data movement
2. Arithmetic
3. Logic
4. Compare and test
5. Rotate and shift
6. Program control
7. Stack

8. Input/output
9. Miscellaneous

Data movement instructions move data from one location to another within the microprocessor and memory (Figure 37–6). Data are moved 8 bits at a time, in a parallel fashion (simultaneously), from the source to the destination specified. Microprocessor instructions use a symbolic notation that refers to how the data are moving. In the 6800 and 6502 microprocessor, the arrow moves from left to right. In the 8080A and the Z80, the arrow moves from right to left. In either case, the message is the same. The data move from the source to the destination.

Arithmetic instructions affect the arithmetic logic unit. The most powerful instructions are add, subtract, increment, and decrement. These instructions allow the microprocessor to compute and manipulate data. They differentiate a computer from a random logic circuit. The result of these instructions is placed in the accumulator.

Logic instructions are those instructions that contain one or more of the following Boolean operators: AND, OR, and exclusive OR. They are performed 8 bits at a time in the ALU, and the results are placed in the accumulator. Another logic operation is the complement instruction. This includes both 1's and 2's complement. Because complementing is done with additional circuitry, it is not included in all microprocessors. The 6502 has nei-

ther complement instruction. The 8080A has a 1's complement instruction. The 6800 and the Z80 have both 1's and 2's complement.

Complementing provides a method of representing signed numbers. Complementing numbers allow the ALU to perform subtract operations using an adder circuit. Therefore, the MPU can use the same circuits for addition and subtraction.

Compare instructions compare data in the accumulator with data from a memory location or another register. The result of the comparison is not stored in the accumulator, but a flag bit might change as a result of it. Comparison may be performed by masking or bit-testing. *Masking* is a process of subtracting two numbers and allowing only certain bits through. The mask is a predetermined set of bits that is used to determine if certain conditions exist within the MPU. There is a disadvantage with the masking procedure because it uses an AND instruction and therefore destroys the content of the accumulator. The bit-testing procedure, although it also uses an AND instruction, does not destroy the content of the accumulator. Not all microprocessors have a bit-testing instruction.

Rotate and shift instructions change the data in a register or memory by moving the data to the right or left one bit. Both instructions involve use of the carry bit. The difference between the instructions is that the rotate instruction saves the data, and the shift instruction destroys the data.

DESCRIPTION	MNEMONIC	NOTATION	SOURCE	DESTINATION
Load Accumulator	LDA	M → A	Memory	Accumulator
Load X-Register	LDX	M → X	Memory	X-Register
Store Accumulator	STA	A → M	Accumulator	Memory
Store X-Register	STX	X → M	X-Register	Memory
Transfer Accumulator to X-Register	TAX	A → X	Accumulator	X-Register
Transfer X-Register to Accumulator	TXA	X → A	X-Register	Accumulator

Figure 37–6 Data movement instructions.

Program-control instructions change the content of the program counter. These instructions allow the microprocessor to skip over memory locations to execute a different program or to repeat a portion of the same program. The instructions can be unconditional, where the content of the program counter changes, or conditional, where the state of a flag bit is first checked to determine whether the content of the program counter should change. If the condition of the flag bit is not met, the next instruction is executed.

Stack instructions allow the storage and retrieval of different microprocessor registers in the stack. The **stack** is a temporary memory location that is used for storing the contents of the program counter during a jump to a subroutine program. The difference between a stack and other forms of memory is the method in which the data is accessed or addressed. A push instruction stores the register content, and a pull instruction retrieves the register content. There is an advantage with the stack because data can be stored into it or read from it with single-byte instructions. All data transfers are between the top of the stack and the accumulator. That is, the accumulator communicates only with the top location of the stack.

In a 6800 and 6502 microprocessor, the content of the register is stored in the stack, and then the stack pointer is decremented by 1. This allows the stack pointer to point to the next memory location where data can be saved. The *stack pointer* is a 16-bit register that is used to define the memory location that acts as the top of the stack. When the pull instruction is used, the stack pointer is incremented by one, and the data are retrieved from the stack and placed in the appropriate register. In the 8080A, the top of the stack contains the pointer to the last memory location. The push instruction first decrements the stack pointer by 1 and then stores the register content in the stack.

Input/output instructions deal specifically with controlling I/O devices. The 8080A, 8085, and Z80 have I/O instructions. The 6800 and 6502 do not have specific I/O instructions. If a microprocessor uses an I/O instruction to deal with external devices, the technique is called *isolated I/O.*

Some instructions do not fall in any of the categories mentioned. These instructions are grouped together and are called **miscellaneous instructions.** Among these instructions are those used to enable or disable interrupt lines, clear or set flag bits, or allow the microprocessor to perform BCD arithmetic. Also included is the instruction to halt or break the program sequence.

37–2 Questions

1. What are the basic parts of a microprocessor?
2. What registers are located in the microprocessor?
3. What are the major categories of microprocessor instructions?

SUMMARY

■ Digital computers consist of a control section, an arithmetic logic unit, memory, and an input/output section.

■ The control section decodes the instructions and generates pulses to operate the computer.

■ The arithmetic logic unit performs math, logic, and decision-making operations.

■ Memory is an area where programs and data are stored while waiting for execution.

■ The input/output units allow data to be transmitted into and out of the computer.

■ The control section and arithmetic logic unit may be included in a single package called a microprocessor.

■ A program is a set of instructions arranged in a sequential pattern to solve a particular problem.

■ A microprocessor contains registers, arithmetic logic unit, timing and control circuitry, and decoding circuitry.

■ Instructions for a microprocessor fall into nine categories:
—Data movement
—Arithmetic
—Logic
—Compare and test
—Rotate and shift
—Program control
—Stack
—Input/output
—Miscellaneous

Chapter 37 Self-Test

1. Describe how a computer operates.
2. By what means does a computer that is interfaced to the real world perform data transmitted from an external device?
3. What is the difference between a microcomputer and a microprocessor?
4. What is the function of the microprocessor?

GLOSSARY

Accumulator The most used register in a microprocessor.

Active filters Inductorless filters using ICs.

Adder A logic circuit that performs addition of two binary numbers.

Alternating current (AC) Current that flows in one direction, then in the opposite direction.

Alternation The two halves of a cycle.

Ammeter A meter used to measure the amount of current flowing in a circuit.

Ampere Measure of current flow.

Ampere-hour A rating used to determine how well a battery delivers power.

Amplification Providing an output signal that is larger than the input.

Amplitude The maximum value of a sine wave or a harmonic in a complex waveform measured from zero.

Analog meter A meter that uses a graduated scale with a pointer.

AND Logic multiplication.

AND gate Performs the basic logic operation of multiplication.

Arithmetic instructions Allow the microprocessor to compute and manipulate data.

Arithmetic logic unit (ALU) Performs math logic and decision-making operations in the computer.

Armature The moving portion of a magnetic circuit.

Artificial magnet Magnet created by rubbing a piece of soft iron against another magnet.

Astable multivibrator A free-running multivibrator.

Asynchronous Not occurring at the same time.

Atom Smallest basic unit of matter.

Atomic number The number of protons in the nucleus of an atom.

Atomic weight The mass of an atom (the number of protons and neutrons).

Audio amplifier Amplifies AC signals in the audio range of 20 to 20,000 hertz.

Auto transformer Used to step the applied voltage up or down without isolation.

Barrier voltage Voltage created by the junction of P-type and N-type material.

Battery A combination of two or more cells.

Bias voltage The external voltage applied across the PN junction of a device.

Binary coded decimal A code of four binary digits to represent decimal numbers 0 through 9.

Binary number system A base 2 number system with two digits, 0 and 1.

Bipolar transistor A semiconductor device capable of amplifying voltage or power.

Bistable action Refers to locking onto one of two stable states.

Bistable multivibrator A multivibrator with two stable states.

Bit A term derived from the words *bi*nary dig*it*.

Block diagram A simplified outline of an electronic system in which circuits or parts are shown as rectangles.

Break over The point at which conduction starts to occur.

Bridge rectifier Operates both half cycles with full voltage out.

Capacitance The ability of a device to store energy in an electrostatic field.

Capacitive reactance Opposition a capacitor offers to applied AC voltage.

Capacitor A device that possesses capacitance.

Cell Method of producing electrical energy consisting of two dissimilar metals, copper and zinc, immersed in a salt, acid, or alkaline solution.

Channel The U-shaped region of a JFET.

Chemical cell A cell used to convert chemical energy into electrical energy.

Circuit breaker A device that performs the same function as a fuse but does not have to be replaced.

Clamping circuit Circuit used to clamp the top or bottom of a waveform to a DC level.

Class A amplifier Amplifier biased so that current flows throughout entire cycle.

Class AB amplifier Amplifier biased so that current flows for less than full but more than half cycle.

Class B amplifier Amplifier biased so that current flows for only one-half of input cycle.

Class C amplifier Amplifier biased so that current flows for less than half of input cycle.

Clipping circuit A circuit used to square off the peaks of an applied signal.

Clocked flip-flop A flip-flop that has a third input called the clock or trigger.

Closed circuit A complete path for current flow.

Closed loop mode An op-amp circuit that uses feedback.

Common-base amplifier Amplifier in which the base is common to both the input and output circuit.

Common-collector amplifier Amplifier in which the collector is common to both the input and output circuit.

Common-emitter amplifier Amplifier in which the emitter is common to both the input and output circuit.

Comparator A logic circuit used to compare the magnitude of two binary numbers.

Compare instructions Compare data in the accumulator with data from a memory location or another register.

Compound Combination of two or more elements.

Condition code register A register that keeps track of the status of the microprocessor.

Conductance The ability of a material to pass electrons.

Conductor A material that contains a large number of free electrons.

Continuity test A test for an open, short, or closed circuit using an ohmmeter.

Control unit Decodes each instruction as it enters the computer.

Coulomb Unit of electrical charge (represents 6.24×10^{18} electrons).

Counter A logic circuit that can count a sequence of numbers.

Coupling Joining of two amplifier circuits together.

Covalent bonding The process of sharing valence electrons.

Current Slow drift of electrons.

Cycle Two complete alternations of current with no reference to time.

D flip-flop A flip-flop that has only a single data input and a clock input.

DC amplifier *See* Direct-coupled amplifier.

Data movement instructions Move data from one location to another within the microprocessor and memory.

Decade counter A counter with a modulus of ten.

Decimal-to-binary encoder Converts a single decimal digit (0–9) to a 4-bit binary code.

Decimal-to-binary priority encoder Produces a binary output based on the highest decimal on the input.

Decoder Processes complex binary codes into recognizable characters.

Degenerative feedback A method of feeding back a portion of a signal to decrease amplification.

Degree of rotation The angle to which the armature has turned.

Depletion mode   A mode in which electrons are conducting until they are depleted by the gate bias voltage in a MOSFET.

Depletion MOSFET   A device that conducts when zero bias is applied to gate.

Depletion region   Region near the junction of N-type and P-type material.

DIAC   Acronym for *Di*ode *AC*, a bidirectional triggering diode.

Difference amplifier   Amplifier that subtracts one signal from another signal.

Difference of potential   The force that moves electrons in a conductor.

Differential amplifier   Amplifier with two separate inputs and one or two outputs.

Differentiator   A circuit used to produce a pip or peak waveform from a square wave.

Digital computer   A device that automatically processes data using digital techniques.

Digital meter   A meter that provides a readout in numerals.

Diode   A semiconductor device that allows current to flow only one way.

Direct-coupled amplifier   Provides amplification of low frequencies to DC voltage.

Direct current (DC)   Current that flows in only one direction.

Doping   The process of adding impurities to semiconductor material.

Dual in-line package (DIP)   Standard package for integrated circuits.

Duty cycle   Ratio of the pulse width to the period.

Effective value   Value of AC current that will produce same amount of heat as DC current.

Electromagnet   Magnet created by passing a current through a coil of wire.

Electromagnetic induction   Inducing a voltage from one coil into another coil.

Electromotive force (EMF)   Another name for difference of potential.

Electron   Negative charge in orbit around the nucleus of an atom.

Element   Basic building block of nature.

Emitter follower   Another name for a common emitter amplifier because the emitter voltage follows the base voltage.

Encoder   A logic circuit of one or more inputs that generates a binary output.

Enhancement mode   A mode in which electron flow is normally cut off until it is aided or enhanced by the bias voltage on the gate of a MOSFET.

Enhancement MOSFET   A device that only conducts when a bias is applied to the gate.

Exclusive NOR gate (XNOR)   Produces a high output only when both inputs are high.

Exclusive OR gate (XOR)   Produces a low output only when both inputs are high.

Fall time   Time it takes for a pulse to fall from 90% to 10% of maximum amplitude.

Farad   The basic unit of capacitance.

Faraday's law   The basic law of electromagnetism. The induced voltage in a conductor is directly proportional to the rate at which the conductor cuts the magnetic lines of force.

Filter   Converts pulsating DC voltage to a smooth DC voltage.

Fixed capacitor   A capacitor having a definite value that cannot be changed.

Flat response   Indicates that the gain of an amplifier varies only slightly within a stated frequency range.

Flip-flop   A bistable multivibrator whose output is either a high or a low. *See* Bistable multivibrator.

Forward bias   When the current flows in the forward direction.

Free-running   Does not require an external trigger to keep running.

Frequency   Number of cycles that occur in a specific period of time.

Frequency compensation   An internal network that does not require external components.

Frequency counter   Measures frequency by comparing against known frequency.

Frequency domain concept   All periodic waveforms are made of sine waves.

Full adder An adder circuit that takes into account a carry.

Full scale value The maximum value indicated by a meter.

Full subtractor A subtractor that takes into account the borrow input.

Full-wave rectifier Operates during both half cycles with half the output voltage.

Fundamental frequency Represents the repetition rate of the waveform.

Fuse A device designed to fail if an overload occurs.

Generator A device used to generate electricity by means of a magnetic field.

Germanium A gray-white element recovered from the ash of certain types of coal.

Ground The common point in electrical/electronic circuits.

Half adder An adder circuit that does not take into account any carries.

Half subtractor A subtractor that does not take into account any borrow input.

Half-wave rectifier Operates only during one-half of the input cycle.

Harmonics Multiples of the fundamental frequency.

Henry The unit of inductance.

Hertz Cycles per second, the unit of frequency.

High-pass filter Filter that passes high frequencies and attenuates low frequencies.

Hole The absence of an electron.

IF amplifier *See* Intermediate frequency amplifier.

Impedance Opposition to current flow in an AC circuit.

Inductance The property of a coil that prevents changes in current flow.

Induction The effect one body has on another body without physical contact.

Inductive reactance Opposition offered by an inductor in an AC circuit to current flow.

Inductor A device designed to have a specific inductance.

In phase When two signals change polarity at the same time.

Input/output (I/O) Allows the computer to accept and produce data.

Input/output instructions Control I/O devices.

Insulator A material that has few free electrons.

Integrated circuit (IC) A complete electronic circuit in a small package.

Integrator A circuit used to reshape a waveform.

Intermediate frequency amplifier A single-frequency amplifier.

Intrinsic material Pure material.

Ionization The process of gaining or losing electrons.

Isolation transformer Used to prevent electrical shocks; isolates the circuit being worked on from the main circuit.

JK flip-flop A flip-flop that incorporates all the features of other flip-flops.

Junction FET (JFET) Field effect transistor that can provide amplification.

Latch Lock onto.

Lead A connecting wire, such as a test lead, battery lead, and so forth.

Leading edge The front edge of a waveform.

LSB Least significant bit.

Light Electromagnetic radiation that is visible to the human eye.

Light-emitting diode (LED) Converts electrical energy directly into light energy.

Limiter circuit *See* Clipping circuit.

Load A device or circuit which work is performed on.

Loading down A load is placed on the output which affects the amount of output current.

Logic instructions Instructions that contain one or more of the Boolean operators (AND, OR, and exclusive OR).

Loudspeaker A device for converting audio frequency current into sound waves.

Low-pass filter Filter that passes low frequencies and attenuates high frequencies.

Magnet A piece of iron or steel that attracts other pieces of iron or steel.

Magnetic field The region around a magnet.

Magnetic induction   The effect a magnet has on an object without physical contact.

Magnetism   The property of a magnet.

Matter   Anything that occupies space.

Memory   Stores the program and data for operation by the computer.

Microampere   One-millionth of an ampere.

Microphone   An energy converter that changes sound energy into corresponding electrical energy.

Microprocessor   Contains ALU, timing, control, and decoding circuitry for the computer.

Milliampere   One-thousandth of an ampere.

Miscellaneous instructions   Instructions used to enable or disable interrupt lines, clear or set flag bits, or allow the microprocessor to perform BCD arithmetic.

Mixture   Physical combination of elements and compounds.

Modulus   The maximum number of states in a counting sequence.

Molecule   Smallest part of a compound that retains the properties of the compound.

Monostable multivibrator   A multivibrator with only one stable state.

MOSFET   *M*etal *o*xide *s*emiconductor *f*ield *e*ffect *t*ransistor.

MSB   *M*ost *s*ignificant *b*it.

Multimeter   A voltmeter, ammeter, and ohmmeter combined into a single meter.

Multiplexer   Selects and routes one of several inputs to a single output.

Multivibrator   A relaxation oscillator that has two temporary stable conditions.

Mutual inductance   An effect resulting when a magnetic field expands in the same direction as the current in the primary, aiding it and causing it to increase.

NAND gate   A combination of an inverter and an AND gate.

Negative feedback   Feeding back of output signal to oppose temperature change.

Negative ion   An atom that has gained one or more electrons.

Negative temperature coefficient   As temperature increases, resistance decreases.

Neutron   Electrically neutral particle in the nucleus of an atom.

Nonsinusoidal oscillators   Oscillators that do not produce a sine wave output.

NOR gate   A combination of an inverter and an OR gate.

NOT circuit   Performs the logic function of inversion.

N-type semiconductor material   Semiconductor material doped with a pentavalent material.

Nucleus   The center of the atom, which contains the mass of the atom.

Ohm   Unit of resistance.

Ohmmeter   A meter used to measure resistance.

Ohm's law   The relationship between current, voltage, and resistance.

One-shot multivibrator   *See* Monostable multivibrator.

Op-amp   *See* Operational amplifier.

Open circuit   Has infinite resistance because no current flows through it.

Open-loop mode   An op-amp circuit that does not use feedback.

Operational amplifier   A very high-gain DC amplifier.

Optical coupler   Used to isolate loads from their source.

OR   Logic addition.

OR gate   Performs the logic operation of addition.

Oscillator   A circuit that generates a repetitive AC signal.

Oscilloscope   Provides a visual display of what occurs in a circuit.

Overshoot   Occurs when the leading edge of the waveform exceeds the maximum value.

Over-voltage protection   Protects a load from voltage increases above a predetermined level.

Parallel cells   Cells with all positive and all negative terminals connected together.

Parallel circuit   Provides two or more paths for current flow.

Passive filters   Filters that use resistors, inductors, and capacitors.

Peak inverse voltage (PIV)   Maximum safe reverse voltage rating.

Peak-to-peak value Vertical distance from one peak to the other peak.

Peak value Absolute value of point on a waveform at the greatest amplitude.

Pentavalent Atoms with five valence electrons.

Period Time required to complete one cycle of a sine wave.

Periodic waveforms Waveforms that occur at regular intervals.

Permanent magnet Magnet that retains its magnetic properties.

Phase angle The exact phase shift between the input and the output.

Phase-shift network Shifts the phase of the output signal with respect to the input.

Photo cell *See* Photoconductive cell.

Photoconductive cell Has internal resistance that changes with light intensity.

Photodiode Used to control current flow by means of light energy.

Phototransistor Works like a photodiode but produces higher output current.

Photovoltaic cell A device used to convert light energy into electrical energy.

Piezoelectric effect A process that results when pressure is applied to a crystal.

PIN photodiode A photodiode with an intrinsic layer between the P and N regions.

Positive feedback Feeding back a part of the output signal that is in phase.

Positive ion An atom that has lost one or more electrons.

Potential The ability to do work.

Potentiometer A variable resistor used to control voltage.

Power The rate at which energy is dissipated by the resistance of a circuit.

Power amplifier An audio amplifier designed to drive a specific load.

Primary cell Chemical cell that cannot be recharged.

Program A list of computer instructions arranged sequentially to solve a problem.

Program-control instructions Change the content of the program counter.

Program counter Contains the instruction addresses for a program.

Proton Positively charged particle in the nucleus of an atom.

P-type semiconductor material Semiconductor material doped with a trivalent material.

Pulse width The duration the voltage is at its maximum or peak value until it drops to its minimum value.

Radio frequency amplifiers Amplify signals of 10,000 hertz to 30,000 megahertz.

Rectification The process of converting AC to DC.

Rectifier circuit Converts an AC voltage to a DC voltage.

Regenerative feedback A method of obtaining an increased output by feeding part of the output back to the input.

Relaxation oscillator Oscillator that stores energy during part of the oscillation cycle.

Relay An electromagnetic switch that opens and closes with an armature.

Resistance Opposition to electron flow in a circuit.

Resistivity The resistance a material offers to current flow.

Resistors Components manufactured to possess a specific value of resistance to the flow of current.

Reverse bias A device connected so that current does not flow across the PN junction.

RF amplifiers *See* Radio frequency amplifiers.

Rheostat A variable resistor used to control current.

Ringing Dampened oscillations occurring as the transient response of a resonant circuit to a shocked excitation.

Rise time Time it takes for a pulse to rise from 10% to 90% of maximum amplitude.

rms value Another term for effective value, rootmean-square.

Rotate and shift instructions Change the data in a register or memory by moving the data to the left or right one bit.

RS flip-flop A flip-flop with a set and reset input.

Schmitt trigger A modified bistable multivibrator that provides more regeneration.

Scientific notation The use of powers of ten to express large and small numbers.

Secondary cell Chemical cell that can be recharged.

Semiconductor A material that has four valence electrons.

Series-aiding cells Connecting positive terminal to negative terminal of cells.

Series circuit Provides a single path for current flow.

Series-opposing cells Connecting negative to negative terminal or positive to positive terminal of a cell.

Series-parallel cells Used to increase current and voltage outputs above that of a single cell.

Series-parallel circuit A combination of the series circuit and the parallel circuit.

Series voltage regulator Connected in series with the load.

Shell Orbit about the nucleus.

Shift register Constructed of flip-flops and used to store data temporarily.

Short circuit A circuit made inactive by a low resistance path across the circuit. Has zero ohms resistance.

Shunt Any part connected in parallel with some other part.

Siemen The unit of conductance.

Silicon Mineral recovered from silicon dioxide, found extensively in the earth's crust.

Silicon control rectifier (SCR) A thyristor that controls current flow in only one direction.

Sinusoidal oscillator Oscillator that produces sine wave outputs.

Solar cell *See* Photovoltaic cell.

Solenoid A cylindrical coil with a movable plunger in the center.

Stack An area of memory used for temporary storage.

Stack instructions Allow the storage and retrieval of different microprocessor registers in the stack.

Stack pointer A register that holds memory locations of data stored in the stack.

Step-down transformer Transformer that produces a secondary voltage less than the primary one.

Step-up transformer Transformer that produces a secondary voltage greater than the primary one.

Substrate Forms the base for a semiconductor device.

Subtractor A logic circuit that performs subtraction of two binary numbers.

Summing amplifier Used when mixing audio signals together.

Switch A device for directing or controlling the current flow in a circuit.

Synchronous Occurring at the same time.

Tank circuit Formed by connecting an inductor and a capacitor in parallel.

Tap A connection point made in the body of the secondary of a transformer.

Temporary magnets Magnets that retain only a small portion of their magnetic properties.

Thermocouple A device used to convert heat into electrical energy.

Thyristors A broad range of semiconductor components used for electronically controlled switches.

Time constant The time relationship in L/R and RC circuits.

Time domain concept All waveforms can be changed from one shape to another using circuits.

Tolerance An indication of the amount a resistor may vary and still be acceptable.

Trailing edge The back edge of a waveform.

Transformer A device for transferring energy from one circuit to another.

Transient A temporary component of current existing in a circuit during adjustment to a load change, voltage source difference, or line impulse.

Transistor *See* Bipolar transistor.

TRIAC Acronym for *tri*ode *AC* semiconductor, provides full control of AC.

Trivalent Atoms with three valence electrons.

Turns ratio Determined by dividing number of turns in secondary by number of turns in primary.

Up-down counter A counter that can count in either direction.

Undershoot Occurs when the trailing edge exceeds its normal minimum value.

Valence Indication of an atom's ability to gain or lose electrons.

Valence shell The outer electron shell of an atom.

Van de Graaf generator A device used to generate electricity by means of friction.

Variable capacitor A capacitor whose value can be changed either by varying the space between plates (trimmer capacitor) or by varying the amount of meshing between two sets of plates (tuning capacitor).

Vector A graphic representation by an arrow of a quantity having magnitude and direction.

Vector addition The adding of vectors using observed rules.

Veitch diagram A chart used to simplify complex Boolean expressions.

Video amplifier Wideband amplifiers used to amplify video signals to 6 megahertz.

Volt Unit for measuring voltage.

Volt-ampere (VA) A rating used for transformers, similar to a power rating.

Voltage Another name for difference of potential.

Voltage amplifier An audio amplifier used to produce a high gain.

Voltage doubler Produces a DC output voltage twice the peak value of the input.

Voltage drop Occurs when current flows in a circuit.

Voltage multiplier A circuit capable of producing higher DC voltages without a transformer.

Voltage regulator Produces a constant output voltage regardless of change of load.

Voltage rise The voltage applied to a circuit.

Voltage tripler Produces an output voltage three times the peak value of the input.

Voltmeter Device used to measure the voltage between two points in a circuit.

Watt Unit of power.

Zener diode Designed to operate at voltages that exceed breakdown voltage.

Zener region Region above breakdown, where zener diodes operate.

APPENDICES

TERM	ABBREVIATION	TERM	ABBREVIATION
Alpha	α	Kilo	k
Alternating Current	ac, AC	Mega	M
Ampere	A, amp	Micro	μ
Beta	β	Milli	m
Capacitance	C	Nano	n
Capacitive Reactance	X_c	Ohm	Ω
Charge	Q	Peak-to-Peak	p-p
Conductance	G	Period	T
Coulomb	C	Pico	p
Current	I, i	Power	P
Cycles per Second	Hz	Reactance	X
Degrees Celsius	°C	Resistance	R, r
Degrees Fahrenheit	°F	Resonant Frequency	f_r
Direct Current	dc, DC	Root Mean Square	rms
Electromotive Force	emf, E, e	Second	s
Farad	F	Siemens	S
Frequency	f	Time	t
Giga	G	Var	var
Henry	H	Volt	V, v
Hertz	Hz	Volt-Ampere	VA
Impedance	Z	Wavelength	λ
Inductance	L	Watt	W
Inductive Reactance	X_L		

Appendix 1 Electronics abbreviations.

PERIODIC TABLE OF THE ELEMENTS

Table of Selected Radioactive Isotopes

SARGENT-WELCH SCIENTIFIC COMPANY
7300 NORTH LINDER AVENUE, SKOKIE, ILLINOIS 60077

Side 1

Appendix 2   Periodic Table of the Elements.

GREEK LETTER		GREEK NAME
A	α	alpha
B	β	beta
Γ	γ	gamma
Δ	δ	delta
E	ε	epsilon
Z	ζ	zeta
H	η	eta
Θ	θ	theta
I	ι	iota
K	κ	kappa
Λ	λ	lambda
M	μ	mu
N	ν	nu
Ξ	ξ	xi
O	o	omicron
Π	π	pi
P	ρ	rho
Σ	σ	sigma
T	τ	tau
Y	υ	upsilon
Φ	φ	phi
X	χ	chi
Ψ	ψ	psi
Ω	ω	omega

Appendix 3 The Greek alphabet.

PREFIX	SYMBOL	VALUE
Mega-	M	10^6
Kilo-	k	10^3
Milli-	m	10^{-3}
Micro-	μ	10^{-6}
Pico-	p	10^{-12}

Appendix 4 Commonly used prefixes.

			TOLERANCE	
Black	0	Gray	0.05%
Brown	1	Violet	0.10%
Red	2	Blue	0.25%
Orange	3	Green	0.5%
Yellow	4	Brown	1%
Green	5	Red	2%
Blue	6	Gold	5%
Violet	7	Silver	10%
Gray	8	None	20%
White	9		

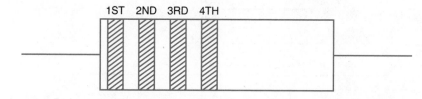

COLOR	1ST BAND 1ST DIGIT	2ND BAND 2ND DIGIT	3RD BAND NUMBER OF ZEROS	4TH BAND TOLERANCE
Black	0	0	—	—
Brown	1	1	0	1%
Red	2	2	00	2%
Orange	3	3	000	—
Yellow	4	4	0000	—
Green	5	5	00000	0.5%
Blue	6	6	000000	0.25%
Violet	7	7		0.10%
Gray	8	8		0.05%
White	9	9		—
Gold	—	—	x.1	5%
Silver	—	—	x.01	10%
None	—	—	—	20%

Appendix 5 Resistor color codes.

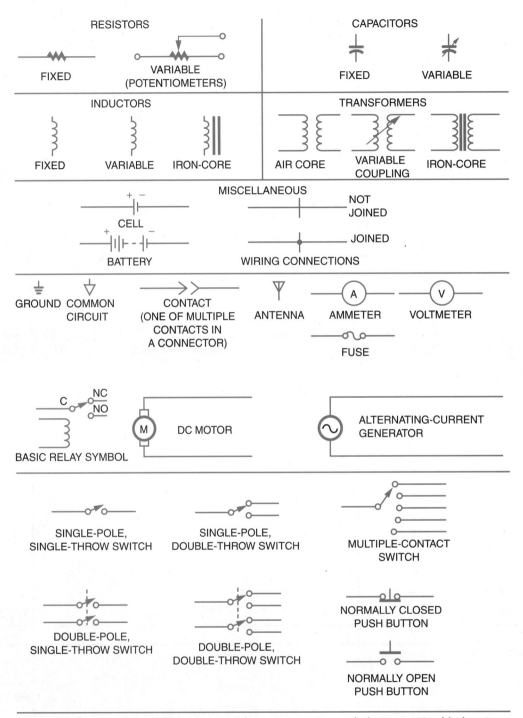

Appendix 6 Electronics symbols. *(Adapted from Perozzo,* Practical Electronics Troubleshooting, © *1985 by Delmar Publishers Inc. Used with permission.)*

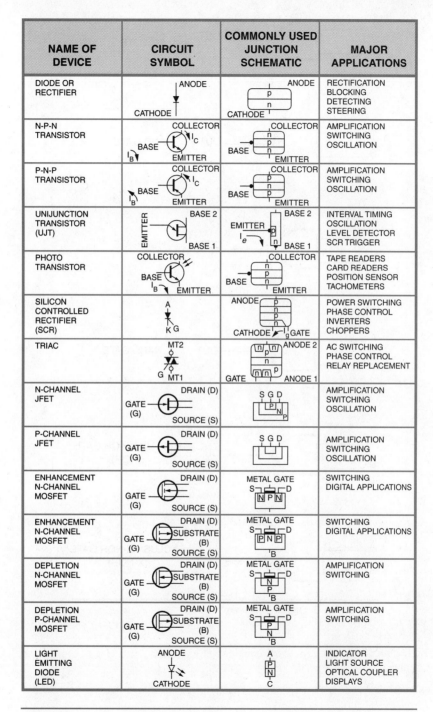

NAME OF DEVICE	CIRCUIT SYMBOL	COMMONLY USED JUNCTION SCHEMATIC	MAJOR APPLICATIONS
DIODE OR RECTIFIER	ANODE / CATHODE	ANODE p n CATHODE	RECTIFICATION BLOCKING DETECTING STEERING
N-P-N TRANSISTOR	COLLECTOR I_C BASE I_B EMITTER	COLLECTOR n p n BASE EMITTER	AMPLIFICATION SWITCHING OSCILLATION
P-N-P TRANSISTOR	COLLECTOR I_C BASE I_B EMITTER	COLLECTOR p n p BASE EMITTER	AMPLIFICATION SWITCHING OSCILLATION
UNIJUNCTION TRANSISTOR (UJT)	EMITTER BASE 2 BASE 1	BASE 2 EMITTER I_e p n BASE 1	INTERVAL TIMING OSCILLATION LEVEL DETECTOR SCR TRIGGER
PHOTO TRANSISTOR	COLLECTOR BASE I_B EMITTER	COLLECTOR n p n BASE EMITTER	TAPE READERS CARD READERS POSITION SENSOR TACHOMETERS
SILICON CONTROLLED RECTIFIER (SCR)	A K G	ANODE p n p n CATHODE I_g GATE	POWER SWITCHING PHASE CONTROL INVERTERS CHOPPERS
TRIAC	MT2 G MT1	ANODE 2 n p n n GATE n p n ANODE 1	AC SWITCHING PHASE CONTROL RELAY REPLACEMENT
N-CHANNEL JFET	DRAIN (D) GATE (G) SOURCE (S)	S G D P N P	AMPLIFICATION SWITCHING OSCILLATION
P-CHANNEL JFET	DRAIN (D) GATE (G) SOURCE (S)	S G D	AMPLIFICATION SWITCHING OSCILLATION
ENHANCEMENT N-CHANNEL MOSFET	DRAIN (D) GATE (G) SOURCE (S)	METAL GATE S D N P N	SWITCHING DIGITAL APPLICATIONS
ENHANCEMENT N-CHANNEL MOSFET	DRAIN (D) GATE (G) SUBSTRATE (B) SOURCE (S)	METAL GATE S D P N P B	SWITCHING DIGITAL APPLICATIONS
DEPLETION N-CHANNEL MOSFET	DRAIN (D) GATE (G) SUBSTRATE (B) SOURCE (S)	METAL GATE S D N P B	AMPLIFICATION SWITCHING
DEPLETION P-CHANNEL MOSFET	DRAIN (D) GATE (G) SUBSTRATE (B) SOURCE (S)	METAL GATE S D P N B	AMPLIFICATION SWITCHING
LIGHT EMITTING DIODE (LED)	ANODE CATHODE	A P N C	INDICATOR LIGHT SOURCE OPTICAL COUPLER DISPLAYS

Appendix 7 Semiconductor schematic symbols. (*Reprinted with permission Microwave Products Department, General Electric Company, Owenboro, Kentucky.*)

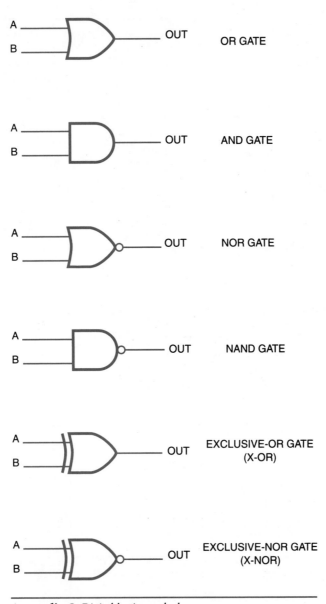

A ——
B —— OUT   OR GATE

A ——
B —— OUT   AND GATE

A ——
B —— OUT   NOR GATE

A ——
B —— OUT   NAND GATE

A ——
B —— OUT   EXCLUSIVE-OR GATE
               (X-OR)

A ——
B —— OUT   EXCLUSIVE-NOR GATE
               (X-NOR)

Appendix 8 Digital logic symbols.

SELF-TEST ANSWERS

CHAPTER 1   FUNDAMENTALS OF ELECTRICITY

1. The number of free electrons that are available.

2. How many electrons are in the valence shell; less than four-conductor, four-semiconductor, more than four-insulators.

3. To understand how electricity flows or does not flow through various materials.

4. The flow of electrons is current, the force that moves the electrons is voltage, and the opposition to the flow of electrons is resistance.

5. Resistance is measured in ohms and the amount of resistance that allows one ampere of current to flow when one volt is applied is one ohm.

CHAPTER 2   CURRENT

1. *Given:*   *Solution:*

   $I = ?$

   $Q = 7$ coulombs $I = \dfrac{Q}{t}$

   $t = 5$ seconds

   $I = \dfrac{7}{5}$

   $I = 1.4$ amperes

2. Electrons flow from the negative terminal of the potential through the conductor, moving from atom to atom, to the positive terminal of the potential.

3. a. $235 = 2.35 \times 10^2$
    b. $0.002376 = 2.376 \times 10^{-3}$
    c. $56323.786 = 5.6323786 \times 10^4$

4. a. Milli means to divide by 1000 or to multiply by 0.001.
    b. Micro means to divide by 1,000,000 or to multiply by 0.000001.

CHAPTER 3   VOLTAGE

1. The actual work accomplished in a circuit (the movement of electrons) is the result of the difference of potential (voltage).

2. Electricity can be produced by friction, magnetism, chemicals, light, heat, and pressure.

3. Secondary cells are rated in ampere-hours.

4.

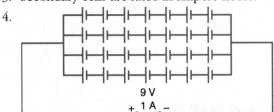

9 V
+ 1 A −

5. *Given:*   *Solution:*

$E_T = 9$ V   Draw the circuit:

L₁ = 3 V rating

L₂ = 3 V rating

L₃ = 6 V rating

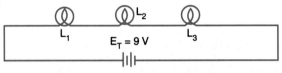

1/2 the voltage would be dropped across L_1 and L_2, and the other half of the voltage would be dropped across L_3.

Therefore:

$L_1 + L_2$ = 6-V drop L_3 would drop 4.5 V

L_3 = 6-V drop L_2 would drop 2.25 V

$9 \times \frac{1}{2} = \frac{9}{2} = 4.5$ V L_1 would drop 2.25 V

Total voltage 9.00 V

CHAPTER 4   RESISTANCE

1. The resistance of a material depends on the size, shape, and temperature of the material. It is determined by measuring a 1-foot length of wire made of the material that is 1 mil in diameter and at a temperature of 20 degrees celcius.

2. *Given:*   *Solution:*

   Resistance = 2200 ohms $2200 \times 0.10 = 220$ ohms

   Tolerance = 10% $2200 - 220 = 1980$ ohms

                                $2200 + 220 = 2420$ ohms

   Tolerance range is:

   1980 ohms to 2420 ohms

3. a. Green, Blue, Red, Gold
   b. Brown, Green, Green, Silver
   c. Red, Violet, Gold, Gold
   d. Brown, Black, Brown, None
   e. Yellow, Violet, Yellow, Silver

4.

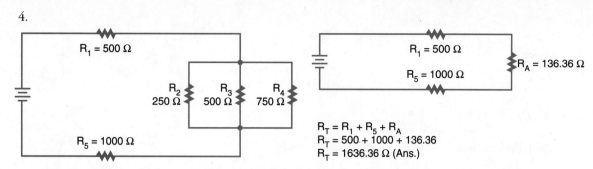

$R_T = R_1 + R_5 + R_A$
$R_T = 500 + 1000 + 136.36$
$R_T = 1636.36 \, \Omega$ (Ans.)

5. The current flows from the negative side of the voltage source through the series components, dividing among the branches of the parallel components, recombining to flow through any more series or parallel components and then returns to the positive side of the voltage source.

CHAPTER 5 OHM'S LAW

1. *Given:*
   I = ?
   E = 9 V
   R = 4500Ω

   Solution:
   $I = \dfrac{E}{R}$

   $I = \dfrac{9}{4500}$

   I = 0.002 A or 2 mA

2. *Given:*
   I = 250 mA = 0.250 A
   E = ?
   R = 470 Ω

   Solution:
   $I = \dfrac{E}{R}$

   $0.250 = \dfrac{E}{470}$

   $\dfrac{0.250}{1} \times \dfrac{E}{470}$

   (1)(E) = (0.250)(470)

   E = 117.5 V

3. *Given:*
   I = 10 A
   E = 240 V
   R = ?

   Solution:
   $I = \dfrac{E}{R}$

   $10 = \dfrac{240}{R}$

   $\dfrac{10}{1} \times \dfrac{240}{R}$

   (1)(240) = (10)(R)

   $\dfrac{240}{10} = \dfrac{\cancel{10} R}{\cancel{10}}$

   $\dfrac{240}{10} = 1R$

   24 Ω = R

4. a.

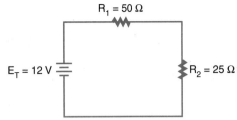

First, find the total resistance of the circuit (series).

$$R_T = R_1 + R_2$$
$$R_T = 50 + 25$$
$$R_T = 75 \ \Omega$$

Second, redraw the circuit using the total equivalent resistance.

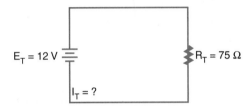

Third, find the total current of the circuit.

Given: *Solution:*

$I_T = ?$

$E_T = 12 \ V$ $I_T = \dfrac{E_T}{R_T}$

$R_T = 75 \ \Omega$

$$I_T = \dfrac{12}{75}$$

$$I_T = 0.16 \text{ A or } 160 \text{ mA}$$

b.

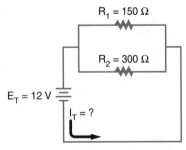

First, find the total resistance of the circuit (parallel).

$$\frac{1}{R_T} = \frac{1}{R_1} + \frac{1}{R_2}$$

$$\frac{1}{R_T} = \frac{1}{150} + \frac{1}{300}$$

$$\frac{1}{R_T} = \frac{2}{300} + \frac{1}{300}$$

$$\frac{1}{R_T} = \frac{3}{300}$$

$$(3)(R_T) = (1)(300)$$

$$\frac{(3)(R_T)}{3} = \frac{300}{3}$$

$$1\,R_T = \frac{300}{3}$$

$$R_T = 100\ \Omega$$

Second, redraw the circuit with equivalent resistance.

Third, find the total resistance of the circuit.

Given: *Solution:*

$I_T = ?$ $I_T = \dfrac{E_T}{R_T}$

$E_T = 12\ V$

$R_T = 100\ \Omega$ $I_T = \dfrac{12}{100}$

$I_T = 0.12\ A$ or 120 mA

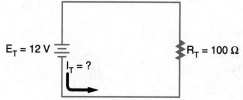

c.

First, find the equivalent resistance for the parallel portion of the circuit.

$$\frac{1}{R_A} = \frac{1}{R_1} + \frac{1}{R_2}$$

$$\frac{1}{R_A} = \frac{1}{75} + \frac{1}{75}$$

$$\frac{1}{R_A} = \frac{2}{75}$$

$$(2)(R_A) = (1)(75)$$

$$\frac{(2)(R_A)}{2} = \frac{(1)(75)}{2}$$

$$1R_A = \frac{75}{2}$$

$$R_A = 37.5\ \Omega$$

Second, redraw the circuit with equivalent resistance.

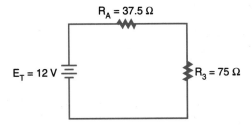

Third, find the total resistance of the circuit.

$$R_T = R_A + R_3$$
$$R_T = 37.5 + 75$$
$$R_T = 112.5 \ \Omega$$

Now, find the total current of the circuit.

Given:

$I_T = ?$

$E_T = 12 \ V$

$R_T = 112.5 \ \Omega$

Solution:

$$I_T = \frac{E_T}{R_T}$$

$$I_T = \frac{12}{112.5}$$

$$I_T = 0.107 \ A \ or \ 107 \ mA$$

CHAPTER 6   ELECTRICAL MEASUREMENTS—METERS

1. Digital
2. Analog
3. a. 23 volts

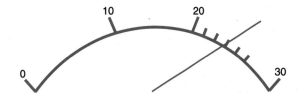

   b. 220 milliamperes

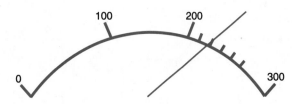

c. 2700 ohms

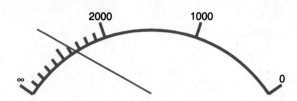

4. One meter can be used to measure voltage, current, and resistance.

CHAPTER 7   POWER

1. *Given:*
 P = ?
  I = 40 mA = 0.04 A
 E = 30 V

 Solution:
 P = IE
 P = (0.04)(30)
 P = 1.2 W

2. *Given:*
 P = 1 W
  I = 10 mA = 0.01 A
 E = ?

 Solution:
 P = IE
 1 = (0.01)(E)

 $$\frac{1}{0.01} = \frac{(0.01)(E)}{(0.01)}$$

 $$\frac{1}{0.01} = 1E$$

 100 V = E

3. *Given:*
 P = 12.3 W
 I = ?
 E = 30 V

 Solution:
 P = IE
 12.3 = (I)(30)

 $$\frac{12.3}{30} = \frac{(I)(30)}{30}$$

 $$\frac{12.3}{30} = 1I$$

 0.41 A = I
  I = 0.41 A or 410 mA

4. a.

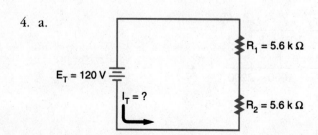

First, find the total resistance of the circuit (series).

$$R_T = R_1 + R_2$$
$$R_T = 5600 + 5600$$
$$R_T = 11,200 \ \Omega$$

Second, redraw the circuit using total resistance.

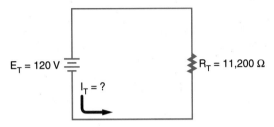

Third, find total circuit current.

Given:

$I_T = ?$

$E_T = 120 \ V$

$R_T = 11,200 \ \Omega$

Solution:

$$I_T = \frac{E_T}{R_T}$$

$$I_T = \frac{120}{11,200}$$

$$I_T = 0.0107 \ A \ \text{or} \ 10.7 \ mA$$

Now, find total circuit power.

$$P_T = I_T E_T$$
$$P_T = (0.0107)(120)$$
$$P_T = 1.3 \ W$$

b.

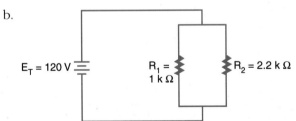

First, find the total resistance of the circuit (parallel).

$$\frac{1}{R_T} = \frac{1}{R_1} + \frac{1}{R_2}$$

$$\frac{1}{R_T} = \frac{1}{1000} + \frac{1}{2200}$$

$$\frac{1}{R_T} = 0.001 + 0.000455$$

$$\frac{1}{R_T} = 0.001455$$

$$\frac{1}{R_T} = \frac{0.001455}{1}$$

$$(0.001455)(R_T) = (1)(1)$$

$$R_T = \frac{1}{0.001455}$$

$$R_T = 687.29 \ \Omega$$

Second, redraw the circuit using total resistance.

$E_T = 120 \ V$ $R_T = 687.29 \ \Omega$

Third, find the total circuit current.

Given: *Solution:*

$I_T = ?$

$E_T = 120 \ V$ $I_T = \dfrac{E_T}{R_T}$

$R_T = 687.29 \ \Omega$ $I_T = \dfrac{120}{687.29}$

$\quad\quad\quad\quad\quad\quad I_T = 0.175 \ A \ or \ 175 \ mA$

Now, find total circuit power.

Given: *Solution:*

$P_T = ?$ $P_T = I_T E_T$

$I_T = 0.175 \ A$ $P_T = (0.175)(120)$

$E_T = 120 \ V$ $P_T = 21 \ W$

c.

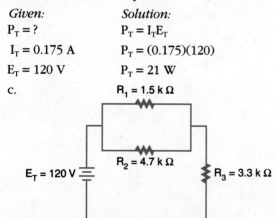

$R_1 = 1.5 \ k\Omega$

$R_2 = 4.7 \ k\Omega$

$E_T = 120 \ V$ $R_3 = 3.3 \ k\Omega$

First, find the equivalent resistance for the parallel portion of the circuit.

$$\frac{1}{R_A} = \frac{1}{1500} + \frac{1}{4700}$$

$$\frac{1}{R_A} = 0.000667 + 0.000213$$

$$\frac{1}{R_A} = 0.000880$$

$$\frac{1}{R_A} = \frac{0.000880}{1}$$

$$(0.000880)(R_A) = (1)(1)$$

$$R_A = \frac{1}{0.000880}$$

$$R_A = 1{,}136.36\ \Omega$$

Second, redraw the circuit using the equivalent resistance.

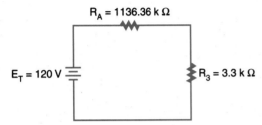

$R_A = 1136.36\ k\ \Omega$

$E_T = 120\ V$

$R_3 = 3.3\ k\ \Omega$

Third, find the total resistance of the circuit.

$$R_T = R_A + R_3$$

$$R_T = 1136.36 + 3300$$

$$R_T = 4436.36\ \Omega$$

Fourth, find the total current for the circuit.

Given:

$I_T = ?$

$E_T = 120\ V$

$R_T = 4436.36\ \Omega$

Solution:

$$I_T = \frac{E_T}{R_T}$$

$$I_T = \frac{120}{4436.36}$$

$$I_T = 0.027\ A\ or\ 27\ mA$$

Fifth, find the total power for the circuit.

Given:

$P_T = ?$

$I_T = 0.027\ A$

$E_T = 120\ V$

Solution:

$P_T = I_T E_T$

$P_T = (0.027)(120)$

$P_T = 3.24\ W$ (with rounding)

3.25 W (without rounding)

CHAPTER 8   DC CIRCUITS

1. a.

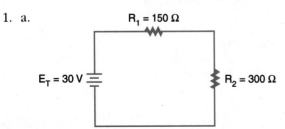

$R_1 = 150\ \Omega$

$E_T = 30\ V$

$R_2 = 300\ \Omega$

Find total circuit resistance.

$$R_T = R_1 + R_2$$
$$R_T = 150 + 300$$
$$R_T = 450 \ \Omega$$

Redraw the equivalent circuit.

$E_T = 30 \ V$ $R_T = 450 \ \Omega$

Find total circuit current.

$$I_T = \frac{E_T}{R_T}$$
$$I_T = \frac{30}{450}$$
$$I_T = 0.0667 \ A \ or \ 66.7 \ mA$$

Find the voltage drop across each resistor.

$I_T = I_1 + I_2$ (the current flow in a series circuit is the same throughout the circuit)

$$I_{R_1} = \frac{E_{R_1}}{R_1}$$

$$0.0667 = \frac{E_{R_1}}{150}$$

$$\frac{0.0667}{1} = \frac{E_{R_1}}{150}$$

$$(1)(E_{R_1}) = (0.0667)(150)$$

$$E_{R_1} = (0.0667)(150)$$

$$E_{R_1} = 10V$$

$$I_{R_2} = \frac{E_{R_2}}{R_2}$$

$$0.0667 = \frac{E_{R_2}}{300}$$

$$\frac{0.0667}{1} = \frac{E_{R_2}}{300}$$

$$(1)(E_{R_2}) = (0.0667)(300)$$

$$E_{R_2} = (0.0667)(300)$$

$$E_{R_2} = 20V$$

Find the power for each resistor.

$$P_{R_1} = I_{R_1}E_{R_1} \qquad P_{R_2} = I_{R_2}E_{R_2}$$
$$P_{R_1} = (0.0667)(10) \qquad P_{R_2} = (0.0667)(20)$$
$$P_{R_1} = 0.667 \ W \qquad P_{R_2} = 1.334 \ W$$

Find the total power of the circuit.

$$P_T = I_T E_T \qquad P_T = P_{R_1} + P_{R_2}$$
$$P_T = (0.0667)(30) \qquad or \ P_T = 0.667 + 1.334$$
$$P_T = 2.001 \ W \qquad P_T = 2.001 \ W$$

b.

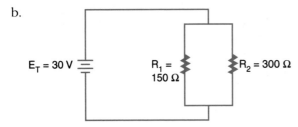

Find the total circuit resistance.

$$\frac{1}{R_T} = \frac{1}{R_1} + \frac{1}{R_2}$$

$$\frac{1}{R_T} = \frac{1}{150} + \frac{1}{300}$$

$$\frac{1}{R_T} = \frac{2}{300} + \frac{1}{300}$$

$$\frac{1}{R_T} = \frac{3}{300}$$

$$(3)(R_T) = (1)(300)$$

$$\frac{\cancel{(3)}(R_T)}{\cancel{3}} = \frac{(1)(300)}{3}$$

$$R_T = \frac{300}{3}$$

$$R_T = 100 \ \Omega$$

Redraw the equivalent circuit.

Find the total circuit current.

$$I_T = \frac{E_T}{R_T}$$

$$I_T = \frac{30}{100}$$

$$I_T = 0.3 \ A \text{ or } 300 \ mA$$

Find the current through each branch of the parallel circuit. The voltage is the same across each branch of the parallel circuit.

$$E_T = E_1 = E_2$$

$$I_{R_1} = \frac{E_{R_1}}{R_1} \qquad\qquad I_{R_2} = \frac{E_{R_2}}{R_2}$$

$$I_{R_1} = \frac{30}{150} \qquad\qquad I_{R_2} = \frac{30}{300}$$

$$I_{R_1} = 0.2 \text{ A} \qquad\qquad I_{R_2} = 0.1 \text{ A}$$

Find the power for each resistor.

$$P_{R_1} = I_{R_1}E_{R_1} \qquad\qquad P_{R_2} = I_{R_2}E_{R_2}$$

$$P_{R_1} = (0.2)(30) \qquad\qquad P_{R_2} = (0.1)(30)$$

$$P_{R_1} = 6 \text{ W} \qquad\qquad P_{R_2} = 3 \text{ W}$$

Find the total power of the circuit.

$$P_T = I_T E_T$$

$$P_T = (0.3)(30)$$

$$P_T = 9W$$

c.

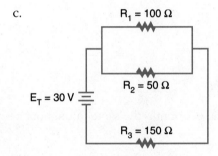

Find the equivalent resistance for the parallel portion of the circuit.

$$\frac{1}{R_A} = \frac{1}{R_1} + \frac{1}{R_2}$$

$$\frac{1}{R_A} = \frac{1}{100} + \frac{1}{50}$$

$$\frac{1}{R_A} = \frac{1}{100} + \frac{2}{100}$$

$$\frac{1}{R_A} = \frac{3}{100}$$

$$(3)(R_A) = (1)(100)$$

$$\frac{(3)(R_A)}{3} = \frac{(1)(100)}{3}$$

$$R_A = \frac{100}{3}$$

$$R_A = 33.3 \text{ } \Omega$$

Redraw the circuit.

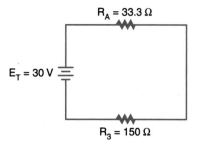

Now find the total circuit resistance.

$$R_T = R_A + R_3$$
$$R_T = 33.3 + 150$$
$$R_T = 183.3 \ \Omega$$

Find the current flow (I_T) for the equivalent circuit.

$$I_T = \frac{E_T}{R_T}$$
$$I_T = \frac{30}{183.3}$$
$$I_T = 0.164 \text{ A or } 164 \text{ mA}$$

Find the voltage drop across resistors in equivalent circuit. (Current is the same throughout a series circuit.)

$$I_T = I_{R_A} = I_{R_3}$$

$$I_{R_A} = \frac{E_{R_A}}{R_A}$$

$$0.164 = \frac{E_{R_A}}{33.3}$$

$$(1)(E_{R_A}) = (0.164)(33.3)$$

$$E_{R_A} = 5.46 \text{ V}$$

$$I_{R_3} = \frac{E_{R_3}}{R_3}$$

$$0.164 = \frac{E_{R_3}}{150}$$

$$(1)(E_{R_3}) = (0.164)(150)$$

$$E_{R_3} = 24.6 \text{ V}$$

Find the current across each of the resistors in the parallel portion of the circuit.

$$I_{R_1} = \frac{E_{R_1}}{R_1}$$

$$I_{R_1} = \frac{5.46}{100}$$

$$I_{R_1} = 0.056 \text{ A}$$

$$I_{R_2} = \frac{E_{R_2}}{R_2}$$

$$I_{R_2} = \frac{5.46}{50}$$

$$I_{R_2} = 0.109 \text{ A}$$

Find the power across each component and total power.

$$P_T = I_T E_T$$
$$P_T = (0.164)(30)$$
$$P_T = 4.92 \text{ W}$$

$$P_{R_1} = I_{R_1} E_{R_1}$$
$$P_{R_1} = (0.056)(5.46)$$
$$P_{R_1} = 0.298 \text{ W}$$

$$P_{R_2} = I_{R_2} E_{R_2}$$
$$P_{R_2} = (0.109)(5.46)$$
$$P_{R_2} = 0.595 \text{ W}$$

$$P_{R_3} = I_{R_3} E_{R_3}$$
$$P_{R_3} = (0.164)(24.6)$$
$$P_{R_3} = 4.034 \text{ W}$$

CHAPTER 9 MAGNETISM

1. The domain theory of magnetism can be verified by jarring the domains into a random arrangement by heating or hitting with a hammer. The magnet will lose its magnetism.

2. The strength of an electromagnet can be increased by increasing the number of turns of wire, increasing the current flow, and by inserting a ferromagnetic core in the center of the coil.

3. When the loop is rotated from position A to position B, a voltage is induced when the motion is at right angles to the magnetic field. As the loop is rotated to position C, the induced voltage decreases to zero volts. As the loop continues to position D, a voltage is again induced, but the commutator reverses the output polarity so it is the same as was first output by the DC generator. The output pulsates in one direction varying twice during each revolution between zero and maximum.

CHAPTER 10 INDUCTANCE

1. The magnetic field around an inductor can be increased by using an iron core.

2.

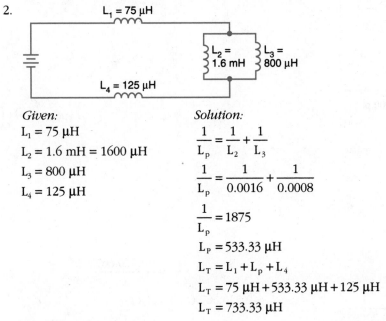

Given:

$L_1 = 75\ \mu H$

$L_2 = 1.6\ mH = 1600\ \mu H$

$L_3 = 800\ \mu H$

$L_4 = 125\ \mu H$

Solution:

$$\frac{1}{L_p} = \frac{1}{L_2} + \frac{1}{L_3}$$

$$\frac{1}{L_p} = \frac{1}{0.0016} + \frac{1}{0.0008}$$

$$\frac{1}{L_p} = 1875$$

$$L_p = 533.33\ \mu H$$

$$L_T = L_1 + L_p + L_4$$

$$L_T = 75\ \mu H + 533.33\ \mu H + 125\ \mu H$$

$$L_T = 733.33\ \mu H$$

3. First, draw the circuit:

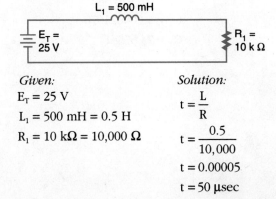

Given:

$E_T = 25\ V$

$L_1 = 500\ mH = 0.5\ H$

$R_1 = 10\ k\Omega = 10,000\ \Omega$

Solution:

$$t = \frac{L}{R}$$

$$t = \frac{0.5}{10,000}$$

$$t = 0.00005$$

$$t = 50\ \mu sec$$

100 μsec = 2 time constants, energized 86.5%

25 × 86.5% = 21.63 V

This voltage represents E_R on rise.

$$E_L = E_T - E_R$$
$$E_L = 25 - 21.63$$
$$E_L = 3.37 \text{ V}$$

CHAPTER 11   CAPACITANCE

1. The charge is stored on the plates of the capacitor.

2. First, draw the circuit:

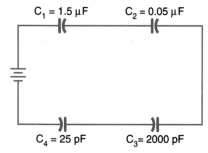

Given:

$C_1 = 1.5 \ \mu F$

$C_2 = 0.05 \ \mu F$

$C_3 = 2000 \text{ pF} = 0.002 \ \mu F$

$C_4 = 25 \text{ pF} = 0.000025 \ \mu F$

Solution:

$$\frac{1}{C_T} = \frac{1}{C_1} + \frac{1}{C_2} + \frac{1}{C_3} + \frac{1}{C_4}$$

$$\frac{1}{C_T} = \frac{1}{1.5} + \frac{1}{0.05} + \frac{1}{0.002} + \frac{1}{0.000025}$$

$$\frac{1}{C_T} = 0.667 + 20 + 500 + 40000$$

$$\frac{1}{C_T} = 40,520.667$$

$$\frac{1}{C_T} = \frac{40,520.667}{1}$$

$$(40,520.667)(C_T) = (1)(1)$$

$$C_T = \frac{1}{40,520.667}$$

$$C_T = 0.000024678$$

$$C_T = 24.678 \text{pF}$$ μF

3. First draw the circuit.

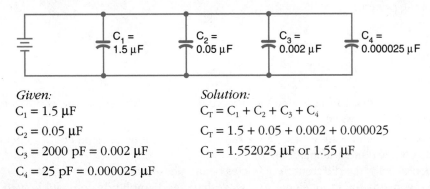

Given:

$C_1 = 1.5\ \mu F$

$C_2 = 0.05\ \mu F$

$C_3 = 2000\ pF = 0.002\ \mu F$

$C_4 = 25\ pF = 0.000025\ \mu F$

Solution:

$C_T = C_1 + C_2 + C_3 + C_4$

$C_T = 1.5 + 0.05 + 0.002 + 0.000025$

$C_T = 1.552025\ \mu F$ or $1.55\ \mu F$

CHAPTER 12   ALTERNATING CURRENT

1. A conductor must be placed in a magnetic field in order for magnetic induction to occur.

2. To apply the left-hand rule, the thumb is pointed in the direction of the conductor movement, the index finger (extended at right angles to the thumb) indicates the direction of the magnetic lines of flux from north to south, and the middle finger (extended at a right angle to the index finger) indicates the direction of current flow in the conductor. The left-hand rule is used to determine the direction of current flow in a conductor that is being passed through a magnetic field.

3. The peak-to-peak value is the vertical distance between the two peaks of a waveform.

4. The effective value of alternating current is the amount that will produce the same degree of heat in a given resistance as an equal amount of direct current.

5. a. Square wave:

   b. Triangular wave:

   c. Sawtooth wave:

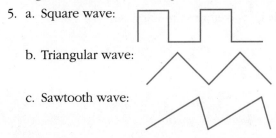

6. Nonsinusoidal waveforms can be considered as being constructed of many sine waves having different frequencies (harmonics) and amplitudes.

CHAPTER 13   AC MEASUREMENTS

1. A DC meter movement can be used to measure AC by using rectifiers to convert the AC signal to a DC current.

2. The clamp-on ammeter uses a split-core transformer. The core can be opened and placed around the conductor. A voltage is induced into the core which is also cut by a coil. The induced voltage creates a current flow which is rectified and sent to a meter movement.

3. An oscilloscope can provide the following information about an electronic circuit: the frequency of a signal, the duration of a signal, the phase relationship between signal waveforms, the shape of a signal's waveform, and the amplitude of a signal.

4. Initially set the oscilloscope controls as follows: Intensity, focus, astigmatism, and position controls (set to the center of their range)

Triggering: INT +
Level: Auto
Time/CM: 1 msec
Volts/CM: 0.02
Power: On

Connect the oscilloscope probe to the test jack of the voltage calibrator. Adjust the controls for a sharp, stable image of a square wave.

5. A frequency counter consists of a time base, an input-signal conditioner, a gate-control circuit, a main gate, a decade counter, and a display.

Time base: compensates for the different frequencies being measured.

Signal conditioner: converts the input signal to a waveshape and amplitude compatible with the circuitry in the counter.

Gate-control circuitry: acts as the synchronization center of the counter. It opens and closes the main gate and provides a signal to latch the count at the end of the counting period and resets the circuitry for the next count.

Main gate: passes the conditioned input signal to the counter circuit.

Decade counter: keeps a running tally of all the pulses that pass through the main gate.

Display: provides a visual readout of the frequency being measured.

6. The integrated circuit has been the primary force for moving the frequency counter from the laboratory to the work bench. It has reduced the physical size of the counter.

CHAPTER 14   RESISTIVE AC CIRCUITS

1. In a pure resistive AC circuit the current and voltage waveforms are in phase.

2. *Given:*

$I_T = 25$ mA $= 0.025$ A

$E_T = ?$

$R_T = 4.7$ k$\Omega = 4700$ Ω

Solution:

$$I_T = \frac{E_T}{R_T}$$

$$0.025 = \frac{E_T}{4700}$$

$$\frac{0.025}{1} \diagup\!\!\!\!\!\times \frac{E_T}{4700}$$

$$(1)(E_T) = (0.025)(4700)$$

$$E_T = 117.5 \text{ V}$$

3.

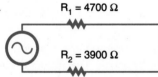

$R_1 = 4700$ Ω

$R_2 = 3900$ Ω

Given:

$I_T = ?$

$E_T = 12$ V

$R_T = ?$

$R_1 = 4.7$ k$\Omega = 4700$ Ω

$R_2 = 3.9$ k$\Omega = 3900$ Ω

$E_1 = ?$

$E_2 = ?$

Solution:

$R_T = R_1 + R_2$

$R_T = 4700 + 3900$

$R_T = 8600$

$I_T = I_1 = I_2$

$$I_T = \frac{E_T}{R_T}$$

$$I_T = \frac{12}{8600}$$

$$I_T = 0.0014 \text{ A}$$

or 1.4 mA

$$I_1 = \frac{E_1}{R_1}$$

$$\frac{0.0014}{1} \diagdown \frac{E_1}{4700}$$

$$I_2 = \frac{E_2}{R_2}$$

$$(1)(E_2) = (0.0014)(3900)$$

$$0.0014 = \frac{E_1}{4700}$$

$$(1)(E_1) = (0.0014)(4700)$$

$$0.0014 = \frac{E_2}{3900}$$

$$E_2 = 5.46 \text{ V}$$

$$E_1 = 6.58 \text{ V}$$

$$\frac{0.0014}{1} \diagdown \frac{E_2}{3900}$$

4. *Given:*

$E_T = 120$ V

$R_1 = 2.2\text{k}\Omega = 2200\ \Omega$

$R_2 = 5.6\ \text{k}\Omega = 5600\ \Omega$

$I_1 = ?$

$I_2 = ?$

Solution:

$$E_T = E_1 = E_2$$

$$I_1 = \frac{E_1}{R_1}$$

$$I_1 = \frac{120}{2200}$$

$$I_1 = 0.055 \text{ A or } 55 \text{ mA}$$

$$I_2 = \frac{E_2}{R_2}$$

$$I_2 = \frac{120}{5600}$$

$$I_2 = 0.021 \text{ A or } 21 \text{ mA}$$

5. The rate at which energy is delivered to a circuit or energy (heat) is dissipated determines the power consumption in an AC circuit, the same as it is in a DC circuit.

6. *Given:*

$I_T = ?$

$E_T = 120$ V

$R_T = 1200\ \Omega$

$P_T = ?$

Solution:

$$I_T = \frac{E_T}{R_T}$$

$$I_T = \frac{120}{1200}$$

$$I_T = 0.1 \text{ A or } 100 \text{ mA}$$

$$P_T = I_T E_T$$

$$P_T = (0.1)(120)$$

$$P_T = 12 \text{ W}$$

CHAPTER 15   CAPACITIVE AC CIRCUITS

1. In a capacitive AC circuit, the current leads the applied voltage.

2. *Given:*

$X_c = ?$

$\pi = 3.14$

$f = 60$ Hz

$C = 1000\ \mu\text{F} = 0.001$ F

Solution:

$$X_C = \frac{1}{2\pi fC}$$

$$X_C = \frac{1}{(2)(3.14)(60)(0.001)}$$

$$X_C = \frac{1}{0.3768}$$

$$X_C = 2.65\ \Omega$$

3. *Given:*

$I_T = ?$

$E_T = 12$

$X_C = 2.65\ \Omega$

Solution:

$$I_T = \frac{E_T}{X_C}$$

$$I_T = \frac{12}{2.65}$$

$$I_T = 4.53 \text{ A}$$

4. Capacitive AC circuits can be used for: filtering, coupling, and phase shifting.

5. Capacitive coupling circuits allow AC components of a signal to pass through a coupling network, while at the same time blocking the DC components of the signal.

CHAPTER 16   INDUCTIVE AC CIRCUITS

1. In an inductive circuit, the current lags the applied voltage.

2. The inductive reactance of an inductive circuit is affected by the inductance of the inductor and the frequency of the applied voltage.

3. *Given:*   *Solution:*
   $X_L = ?$ $X_L = 2\pi fL$
   $\pi = 3.14$ $X_L = (2)(3.14)(60)(0.1)$
   $f = 60$ Hz $X_L = 37.68\ \Omega$
   $L = 100mH = 0.1$ H

4. *Given:*   *Solution:*
   $I_T = ?$
   $E_T = 24$ V $I_T = \dfrac{E_T}{X_L}$
   $X_L = 37.68\ \Omega$
                          $I_T = \dfrac{24}{37.68}$

   $I_T = 0.64$ A or 640 mA

5. Applications for inductors in circuits include filtering and phase shifting.

6. The frequency above or below the frequencies passed or attenuated in an inductive circuit is called the cut-off frequency.

CHAPTER 17   RESONANCE CIRCUITS

1.

$R_1 = 56\ \Omega$

$E_T = 120$ V
$f = 60$ Hz

$L_1 = 750$ mH

$C_1 = 10\ \mu F$

Find capacitive reactance.

$$X_C = \frac{1}{2\pi fC}$$

$$X_C = \frac{1}{(6.28)(60)(0.000010)}$$

$$X_C = 265.39\ \Omega$$

Find inductive reactance.

$$X_L = 2\pi fL$$
$$X_L = (6.28)(60)(0.750)$$
$$X_L = 282.60 \ \Omega$$

Now, solve for X.

$$X = X_L - X_c$$
$$X = 282.6 - 265.39$$
$$X = 17.2 \ \Omega \ \text{(inductive)}$$

Using X, solve for Z.

$$Z^2 = X^2 + R^2$$
$$Z^2 = (17.21)^2 + (56)^2$$
$$Z^2 = 296.18 + 3136$$
$$Z^2 = 3432.18$$
$$Z = \sqrt{3432.18}$$
$$Z = 58.58 \ \Omega$$

Solve for total current.

$$I_T = \frac{E_T}{Z}$$
$$I_T = \frac{120}{58.58}$$
$$I_T = 2.05 \ A$$

2.

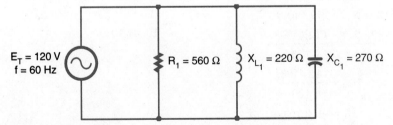

$E_T = 120$ V
$f = 60$ Hz
$R_1 = 560 \ \Omega$ $X_{L_1} = 220 \ \Omega$ $X_{C_1} = 270 \ \Omega$

Find individual branch current.

$$I_R = \frac{E_R}{R} \qquad\qquad I_{X_L} = \frac{E_{X_L}}{X_L} \qquad\qquad I_{X_C} = \frac{E_{X_C}}{X_C}$$

$$I_R = \frac{120}{560} \qquad\qquad I_{X_L} = \frac{120}{220} \qquad\qquad I_{X_C} = \frac{120}{270}$$

$$I_T = 0.214 \ A \qquad\qquad I_{X_L} = 0.545 \ A \qquad\qquad I_{X_C} = 0.444 \ A$$

Find I_X and I_Z using I_R, I_{X_L} and I_{X_C}.

$$I_X = I_{X_L} - I_{XC}$$
$$I_X = 0.545 - 0.444$$
$$I_X = 0.101 \text{ A (inductive)}$$

$$I_Z^2 = (I_R)^2 + (I_X)^2$$
$$I_Z^2 = (0.214)^2 + (0.101)^2$$
$$I_Z^2 = 0.0459 + 0.0102$$
$$I_Z^2 = 0.0561$$
$$I_Z = \sqrt{0.561}$$
$$I_Z = 0.237 \text{A}$$

CHAPTER 18   TRANSFORMERS

1. When two electrically isolated coils are placed next to each other, and an AC voltage is applied across one coil, the changing magnetic field induces a voltage into the second coil.

2. Transformers are rated in volt-amperes rather than in watts because of the different types of loads that can be placed on the secondary winding. A pure capacitive load will cause an excessive current to flow and a power rating would have little meaning.

3. If a transformer is connected without a load, there is no secondary current flow. The primary windings act like an inductor in an AC circuit. When a load is connected across the secondary winding, a current is induced into the secondary. The current in the secondary establishes its own magnetic field which cuts the primary, inducing a voltage back into the primary. This induced field expands in the same direction as the current in the primary, aiding it and causing it to increase.

4. *Given:*

$N_P = 400$ turns

$N_S = ?$

$E_P = 120$ V

$E_S = 12$ V

Solution:

$$\frac{E_S}{E_P} = \frac{N_S}{N_P}$$

$$\frac{12}{120} \diagup \frac{N_S}{400}$$

$$(120)(N_S) = (12)(400)$$

$$\frac{(120)(N_S)}{120} = \frac{(12)(400)}{120}$$

$$1N_S = \frac{(12)(400)}{120}$$

$$N_S = 40 \text{ turns}$$

$$\text{turn ratio} = \frac{N_S}{N_P}$$

$$= \frac{40}{400}$$

$$= \frac{1}{10} \text{ or } 10:1$$

5. *Given:*

$N_P = ?$

$N_S = ?$

$Z_P = 16$

$Z_S = 4$

Solution:

$$\frac{Z_P}{Z_S} = \left(\frac{N_P}{N_S}\right)^2$$

$$\frac{16}{4} = \left(\frac{N_P}{N_S}\right)^2$$

$$\sqrt{4} = \frac{N_P}{N_S}$$

$$\frac{2}{1} = \frac{N_P}{N_S}$$

The turns ratio is 2:1.

6. Transformers are important for transmitting electrical power because of power loss. The amount of power loss is related to the amount of resistance of the power lines and current. The easiest way to reduce power losses is to keep the current low by stepping up the voltage with transformers.

7. An isolation transformer prevents connecting to ground either side of the power line for equipment being worked on.

CHAPTER 19   SEMICONDUCTOR FUNDAMENTALS

1. Silicon has more resistance to heat than germanium, making it preferable.

2. Covalent bonding is the process of atoms sharing electrons. When semiconductor atoms share electrons, their valence shell becomes full with eight electrons, thereby obtaining stability.

3. In pure semiconductor materials, the valence electrons are held tightly to the parent atom at low temperatures and do not support current flow. As the temperature increases, the valence electrons become agitated and break the covalent bond, allowing the electrons to drift randomly from one atom to the next. As the temperature continues to increase, the material begins to behave like a conductor. At only extremely high temperatures will silicon conduct current as ordinary conductors do.

4. To convert a block of pure silicon to N-type material, the silicon is doped with atoms having five valence electrons, called *pentavalent materials,* such as arsenic and antimony.

5. When a voltage is applied to N-type material, the free electrons contributed by the donor atoms flow toward the positive terminal. Additional electrons break away from their covalent bonds and also flow toward the positive terminal.

CHAPTER 20   PN JUNCTION DIODES

1. A PN junction diode allows current to flow in only one direction.

2. A diode will conduct when it is forward biased. That is, the positive terminal of the voltage source is connected to the P-type material, and the negative terminal of the voltage source is connected to the N-type material.

3.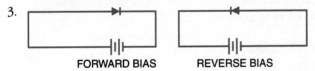

   FORWARD BIAS   REVERSE BIAS

CHAPTER 21   ZENER DIODES

1. In a zener diode voltage regulator, the zener diode is connected in series with a resistor with the output taken across the zener diode. The zener diode opposes an increase in input voltage, because when the current increases the resistance drops. The change in input voltage appears across the series resistor.

2. A power supply, current-limiting resistor, an ammeter and a voltmeter are required for testing a zener diode. The output of the power supply is connected across the limiting resistor in series with the zener diode and ammeter. The voltmeter is connected across the zener diode. The output voltage is slowly increased until the specified current is flowing through the zener diode. The current is then varied on either side of the specified zener current. If the voltage remains constant, the zener diode is operating properly.

CHAPTER 22 BIPOLAR TRANSISTORS

1. The emitter-based junction is forward biased, and the collector-based junction is reverse biased.

2. When testing a transistor with an ohmmeter, a good transistor will show a low resistance when forward biased and a high resistance when reverse biased across each junction.

3. A voltmeter, not an ohmmeter, is used to determine whether a transistor is silicon or germanium by measuring the voltage drop across the junction. The leads would be difficult to determine because it would be hard to say which end is the emitter or collector. However, the base would be determined as low resistance when forward biased to either emitter or collecter and high resistance when reverse biased. A PNP or NPN transistor could be determined.

4. The collector voltage determines whether the device is an NPN or PNP transistor. If the wrong type of transistor is substituted, a failure in the device will result.

5. Testing a transistor with a transistor tester reveals more information about a transistor than when testing with an ohmmeter.

CHAPTER 23 FIELD EFFECT TRANSISTORS (FETs)

1. The pinch-off voltage is the voltage required to pinch off the drain current for a JFET.

2. The pinch-off voltage is given by the manufacturer for a gate-source voltage of zero.

3. Depletion-mode MOSFETs conduct when zero bias is applied to the gate. They are considered to be normally on devices.

4. Enhancement-mode MOSFETs are normally off and only conduct when a suitable bias voltage is applied to the gate.

5. Safety precautions for a MOSFET include:
   ■ Keep the leads shorted together prior to installation.
   ■ Use a metallic wrist band to ground the hand being used.
   ■ Use a grounded tip soldering iron.
   ■ Always ensure the power is off prior to installation of a MOSFET.

CHAPTER 24 THYRISTORS

1. The PN junction diode has one junction and two leads (anode and cathode); an SCR has three junctions and three leads (anode, cathode, and gate).

2. The anode supply voltage will keep the SCR turned on even after the gate voltage is removed. This allows a current to flow continuously from cathode to anode.

3. The load resistor is in series with the SCR to limit the cathode-to-anode current.

4. An SCR can be tested with an ohmmeter or a commercial transistor tester. To test the SCR with an ohmmeter, connect the positive lead to the cathode and the negative lead to the anode. A high resistance reading in excess of 1 megohm should be read. Reverse the leads so the positive lead is on the anode and the negative lead is on the cathode. Again, a high resistance reading in excess of 1 megohm should be read. Short the gate to the anode, and the resistance reading should drop to less than 1000 ohms. Remove the short, and the low resistance reading should remain. Remove the leads and repeat the test.

5. A DIAC is used as a triggering device the TRIACs. It prevents the TRIAC from turning on until a certain gate voltage is reached.

CHAPTER 25   INTEGRATED CIRCUITS

1. A hybrid integrated circuit contains monolithic, thin-film, and discrete components.
2. A chip is the semiconductor material that comprises the integrated circuit and is about one-eighth of an inch square.
3. Resistors and capacitors in an integrated circuit are formed by methods other than monolithic because of the accuracy required. It is not possible to adjust the values as well as with thin and thick film techniques.

CHAPTER 26   OPTOELECTRIC DEVICES

1. The photodiode has the fastest response time to light changes of any photosensitive devices.
2. The phototransistor lends itself to a wider range of applications because it is able to produce a higher gain. It does not, however, respond to light changes as fast as the photodiode.
3. The more current that flows through an LED, the brighter the light emitted. However, a series resistor must be used with LEDs to limit the current flow or else damage to the LED will result.

CHAPTER 27   POWER SUPPLIES

1. Concerns when selecting a transformer for a power supply include primary power rating, frequency of operation, secondary voltage and current rating, and power handling capabilities.
2. Transformers are used to isolate the power supply from the AC voltage source. They can also be used to step-up or step down voltage.
3. The rectifier in a power supply converts the incoming AC voltage to a DC voltage.
4. A disadvantage of the full-wave rectifier is that it requires a center-tapped transformer. An advantage is that it only requires two diodes. An advantage of the bridge rectifier is that it does not require a transformer; however, it does require four diodes. Both rectifier circuits are more efficient and easier to filter than the half-wave rectifier.
5. A filter capacitor charges when current is flowing, then discharges when current stops flowing, keeping a constant current flow on the output.
6. Capacitors are selected for filtering to give a long RC time constant. The slower discharge gives a higher output voltage.
7. A series regulator compensates for higher voltages on the input by increasing the series resistance, thereby dropping more voltage across the series resistance so the output voltage remains the same. It also senses lower voltage on the input, decreasing the series resistance dropping less voltage resulting in the output remaining the same.
8. The voltage and load current requirements must be known when selecting an IC voltage regulator.
9. Voltage multipliers allow the voltage of a circuit to be stepped up without the use of a step-up transformer.
10. The full-wave voltage doubler is easier to filter than the half-wave voltage doubler. In addition, the capacitors in the full-wave voltage doubler are subjected to only the peak value of the input signal.
11. An over-voltage protection circuit called a *crowbar* is used to protect the load from failure of the power supply.
12. Overcurrent protection devices include fuses and circuit breakers.

CHAPTER 28 AMPLIFIER BASICS

1. A transistor provides amplification by using an input signal to control current flow in the transistor to control the voltage through a load.

2. A common emitter circuit provides both voltage and current gain and a high power gain. Neither of the other two circuit configurations provides this combination.

3. Temperature changes affect the gain of a transistor. Degenerative or negative feedback compensates for this condition.

4. Class A amplifiers are biased so that the output flows throughout the entire cycle. Class B amplifiers are biased so that the output flows for only half of the input cycle. Class AB amplifiers are biased so that the output flows for more than half but less than the full input cycle. Class C amplifiers are biased so the output flows for less than half of the input cycle.

5. When connecting two transistor amplifiers together, the bias voltage from one amplifier must be prevented from affecting the operation of the second amplifier.

6. If capacitors or inductors are used for coupling, the reactance of the device will be affected by the frequency being transmitted.

CHAPTER 29 AMPLIFIER APPLICATIONS

1. DC or direct-coupled amplifiers are used to amplify frequencies from DC (0 hertz) to many thousands of hertz.

2. Temperature stability with DC amplifiers is achieved by using a differential amplifier.

3. Audio voltage amplifiers provide a high voltage gain, whereas audio power amplifiers provide high power gains to a load.

4. The complementary push-pull amplifier requires matched NPN and PNP transistors. The quasicomplementary amplifier does not require matched transistors.

5. A video amplifier has a wider frequency range than an audio amplifier.

6. A factor that limits the output of a video amplifier is the shunt capacitance of the circuit.

7. An RF amplifier amplifies frequencies from 10,000 hertz to 30,000 megahertz.

8. An IF amplifier is a single-frequency amplifier used to increase a signal to a usable level.

9. An op-amp consists of an input stage (differential amplifier), high-gain voltage amplifier, and an output amplifier. It is a high-gain DC amplifier, capable of output gains of 20,000 to 1,000,000 times the input signal.

10. Op-amps are amplifiers used for comparing, inverting, and noninverting a signal, and summing. They are also used as an active filter and a difference amplifier.

CHAPTER 30 OSCILLATORS

1. The parts of an oscillator include the frequency-determining circuit called the tank circuit, an amplifier to increase the output signal from the tank circuit, and a feedback circuit to deliver part of the output signal back to the tank circuit to maintain oscillation.

2. The tank circuit can sustain oscillation by feeding back part of the output signal in the proper phase to replace the energy losses caused by the resistance of the components in the tank circuit.

3. The major types of sinusoidal oscillators are the Hartley oscillator, the Colpitts oscillator, and the Clapp oscillator.

4. Crystals have a natural frequency of vibration and are ideal for oscillator circuits. The crystal frequency is used to control the tank circuit frequency.

5. Nonsinusoidal oscillators do not produce sine-wave outputs. Typically, all nonsinusoidal oscillators are some form of a relaxation oscillator.

6. Blocking oscillators, multivibrators, RC networks, and integrated circuits are all used in nonsinusoidal oscillators.

CHAPTER 31   WAVESHAPING CIRCUITS

1. The frequency-domain concept states that all periodic waveforms are made up of sine waves. A periodic waveform can be made by superimposing a number of sine waves having different amplitudes, phases, and frequencies.

2. Overshoot, undershoot, and ringing occur in waveshaping because of imperfect circuits.

3. A differentiator is used to produce a pip or peaked waveform for timing and synchronizing circuits. An integrator is used for waveshaping.

4. The DC reference level of a signal can be changed by using a clamped circuit to clamp the waveform to a DC voltage.

5. A monostable circuit has only one stable state, and produces one output pulse for each input pulse. A bistable circuit has two stable stages, and requires two input pulses to complete a cycle.

6. A flip-flop can produce a square or rectangular waveform for gating and timing signals or for switching applications.

CHAPTER 32   BINARY NUMBER SYSTEM

1.
Digital	Binary	Digital	Binary	Digital	Binary	Digital	Binary
0	00000	7	00111	14	01110	21	10101
1	00001	8	01000	15	01111	22	10110
2	00010	9	01001	16	10000	23	10111
3	00011	10	01010	17	10001	24	11000
4	00100	11	01011	18	10010	25	11001
5	00101	12	01100	19	10011	26	11010
6	00110	13	01101	20	10100	27	11011

2. Seven binary bits are required to represent the decimal number 100 (1100100).

3. To convert a decimal number to a binary number, progressively divide the decimal number by 2, writing down the remainder after each division. The remainders, taken in reverse order, form the binary number.

4. a. 100101.001011 = 37.171875
   b. 111101110.11101110 = 494.9296875
   c. 10000001.00000101 = 129.0195312

5. Convert each decimal digit to a binary digit (0–9) using a BCD binary code.

6. a. 0100 0001 0000 0110 = 4106
   b. 1001 0010 0100 0011 = 9243
   c. 0101 0110 0111 1000 = 5678

CHAPTER 33   BASIC LOGIC GATES

1.
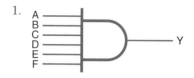

2.

A	B	C	D	Y
0	0	0	0	0
1	0	0	0	0
0	1	0	0	0
1	1	0	0	0
0	0	1	0	0
1	0	1	0	0
0	1	1	0	0
1	1	1	0	0
0	0	0	1	0
1	0	0	1	0
0	1	0	1	0
1	1	0	1	0
0	0	1	1	0
1	0	1	1	0
0	1	1	1	0
1	1	1	1	1

3.
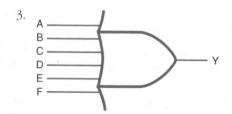

4.

A	B	C	D	Y
0	0	0	0	0
1	0	0	0	1
0	1	0	0	1
1	1	0	0	1
0	0	1	0	1
1	0	1	0	1
0	1	1	0	1
1	1	1	0	1
0	0	0	1	1
1	0	0	1	1
0	1	0	1	1
1	1	0	1	1
0	0	1	1	1
1	0	1	1	1
0	1	1	1	1
1	1	1	1	1

5. The NOT circuit is used to perform inversion or complementation.

6. The circle or bubble is placed at the input for inversion of the input signal, and placed at the output for output inversion.

7.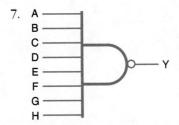

8.

A	B	C	D	Y
0	0	0	0	1
1	0	0	0	1
0	1	0	0	1
1	1	0	0	1
0	0	1	0	1
1	0	1	0	1
0	1	1	0	1
1	1	1	0	1
0	0	0	1	1
1	0	0	1	1
0	1	0	1	1
1	1	0	1	1
0	0	1	1	1
1	0	1	1	1
0	1	1	1	1
1	1	1	1	0

9.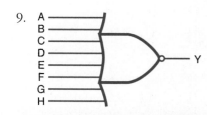

10.

A	B	C	D	Y
0	0	0	0	1
1	0	0	0	0
0	1	0	0	0
1	1	0	0	0
0	0	1	0	0
1	0	1	0	0
0	1	1	0	0
1	1	1	0	0
0	0	0	1	0
1	0	0	1	0
0	1	0	1	0
1	1	0	1	0
0	0	1	1	0
1	0	1	1	0
0	1	1	1	0
1	1	1	1	0

11. An XOR gate generates an output only when the inputs are different. If the inputs are both 0's or 1's, the output is a zero.

12. An XNOR gate has a maximum of two inputs.

CHAPTER 34 SIMPLIFYING LOGIC CIRCUITS

1. Use the Veitch diagram as follows:
   a. Draw the diagram based on the number of variables.
   b. Plot the logic functions by placing an X in each square representing a term.
   c. Obtain the simplified logic function by looping adjacent groups of X's in groups of eight, four, or two. Continue to loop until all X's are included in a loop.
   d. "OR" the loops with one term per loop.
   e. Write the simplified expression.

2.

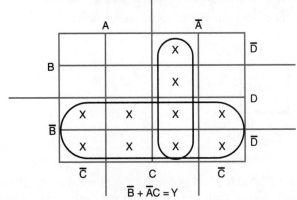

$$\overline{B} + \overline{A}C = Y$$

CHAPTER 35 SEQUENTIAL LOGIC CIRCUITS

1. To change the output of an RS flip-flop requires a high or 1 to be placed on the R input. This changes the state of the flip-flop to a 0 on the Q output and a 1 on the $\overline{Q}$ output.

2. The major difference between the D flip-flop and the clocked RS flip-flop is that the D flip-flop has a single data input and a clock input.

3. A counter is constructed of flip-flops connected in either an asynchronous or synchronous count mode. In the asynchronous mode, the $\overline{Q}$ output of the first stage is connected to the clock input of the next stage. In the synchronous mode, all the clock inputs of each of the stages are connected in parallel.

4.

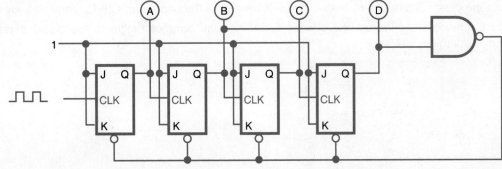

5. A shift register is designed to store data temporarily. Data can be loaded into the shift register either serially or in parallel.

6. Shift registers can be used to store data, for serial-to-parallel and parallel-to-serial data conversion, and to perform such arithmetic functions as division and multiplication.

CHAPTER 36 COMBINATIONAL LOGIC CIRCUITS

1. Encoders allow decoding of keyboard inputs into binary outputs.

2. A decimal-to-binary priority encoder is required for keyboard inputs.

3. Decoders allow the processing of complex binary codes into a recognizable digit or character.

4. Types of decoders include 1-of-10 decoders, 1-of-8 decoders, 1-of-16 decoders, and BCD-to-seven-segment decoders.

5. Multiplexers are used to select and route one of several input signals to a single output.

6. Multiplexers can be used for data line selection, and parallel-to-serial conversions.

7.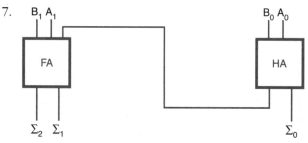

8. The half-adder accepts the two binary digits to be added and generates a sum and a carry. The carry is fed to the next stage and added to the two binary digits, generating a sum and a carry. The answer is the result of the carry and the two sum outputs.

CHAPTER 37 MICROCOMPUTER BASICS

1. A computer consists of a control unit, an arithmetic logic unit (ALU), memory, and an input/output unit (I/O). The control unit decodes the instructions and generates the necessary pulses to carry out the specified function. The arithmetic logic unit performs all the math logic and decision-making operations. Memory is where the programs and data are stored. The input/output unit allows data to be entered into and removed from the computer.

2. An *interrupt* signal from an external device lets the computer know that it would like data or wants to send data.

3. A microprocessor is part of a microcomputer. It consists of the control unit and arithmetic logic unit.

4. The microprocessor performs the control functions, and handles the math logic and decision-making for a microcomputer.

INDEX